Elements of Algebraic Coding Theory

CHAPMAN & HALL MATHEMATICS SERIES

OTHER TITLES IN THE SERIES INCLUDE

Dynamical Systems
Differential equations, maps and chaotic behaviour
D. K. Arrowsmith and C. M. Place

Network Optimization
V. K. Balakrishnan

Algebraic Numbers and Algebraic Functions
P. M. Cohn

Elements of Linear Algebra
P. M. Cohn

Control and Optimization
B. D. Craven

Sets, Functions and Logic
A foundation course in mathematics
Second edition
K. Devlin

Functions of Two Variables
S. Dineen

The Dynamic Cosmos
M. S. Madsen

Full information on the complete range of Chapman & Hall mathematics books is available from the publishers.

Elements of Algebraic Coding Theory

L. R. Vermani

Professor of Mathematics
Kurukshetra University
Kurukshetra, India

CHAPMAN & HALL
London · Weinheim · New York · Tokyo · Melbourne · Madras

Published by Chapman & Hall, 2–6 Boundary Row, London SE1 8HN, UK

Chapman & Hall, 2–6 Boundary Row, London SE1 8HN, UK

Chapman & Hall GmbH, Pappelallee 3, 69469 Weinheim, Germany

Chapman & Hall USA, 115 Fifth Avenue, New York, NY 10003, USA

Chapman & Hall Japan, ITP-Japan, Kyowa Building, 3F, 2-2-1 Hirakawacho, Chiyoda-ku, Tokyo 102, Japan

Chapman & Hall Australia, 102 Dodds Street, South Melbourne, Victoria 3205, Australia

Chapman & Hall India, R. Seshadri, 32 Second Main Road, CIT East, Madras 600035, India

First edition 1996

Typeset in 10/12 pt Times by Thomson Press (India) Ltd, New Delhi, India
Printed in Great Britain by St Edmundsbury Press Ltd, Bury St Edmunds, Suffolk

ISBN 0 412 57380 6

A catalogue record for this book is available from the British Library

Library of Congress Catalog Card Number: 96-84046

∞ Printed on permanent acid-free text paper, manufactured in accordance with ANSI/NISO Z39.48-1992 and ANSI/NISO Z39.48-1984 (Permanence of Paper).

Contents

Preface vii

1 Group codes **1**
1.1 Elementary properties 1
1.2 Matrix encoding techniques 8
1.3 Generator and parity check matrices 14

2 Polynomial codes **24**
2.1 Definition of vector space and polynomial ring 24
2.2 Polynomial codes 26
2.3 Generator and parity check matrices – general case 34

3 Hamming codes **39**
3.1 Binary representation of numbers 39
3.2 Hamming codes 41

4 Finite fields and BCH codes **47**
4.1 Finite fields 47
4.2 Some examples of primitive polynomials 62
4.3 Bose–Chaudhuri–Hocquenghem codes 65

5 Linear codes **81**
5.1 Generator and parity check matrices 81
5.2 Dual code of a linear code 87
5.3 Weight distribution of the dual code of a binary linear code 97
5.4 New codes obtained from given codes 102

6 Cyclic codes **107**
6.1 Cyclic codes 107
6.2 Check polynomial 111

6.3 BCH and Hamming codes as cyclic codes 114
6.4 Non-binary Hamming codes 119
6.5 Idempotents 129
6.6 Some solved examples and an invariance property 131
6.7 Cyclic codes and group algebras 135
6.8 Self dual binary cyclic codes 137

7 Factorization of polynomials **140**
7.1 Factors of $X^n - 1$ 140
7.2 Factorization through cyclotomic cosets 143
7.3 Berlekamp's algorithm for factorization of polynomials 149
7.4 Berlekamp's algorithm – a special case 157

8 Quadratic residue codes **172**
8.1 Introduction 172
8.2 Some examples of quadratic residue codes 176
8.3 Extended quadratic residue codes and distance properties 180
8.4 Idempotents of quadratic residue codes 194
8.5 Some examples 202

9 Maximum distance separable codes **208**
9.1 Necessary and sufficient conditions for MDS codes 208
9.2 The weight distribution of MDS codes 215
9.3 An existence problem 218
9.4 Reed–Solomon codes 220

10 Automorphism group of a code **223**
10.1 Automorphism group of a binary code 223
10.2 Automorphism group of a non-binary code 229
10.3 Automorphism group – its relation with minimum distance 234

11 Hadamard matrices and Hadamard codes **242**
11.1 Hadamard matrices 242
11.2 Hadamard codes 248

Bibliography **251**

Index **253**

Preface

Coding theory came into existence in connection with some engineering problems in the late 1940s (1948–50 to be precise). The subject developed by using sophisticated mathematical techniques including algebraic. The aspect of the subject using algebraic techniques came to be known as Algebraic Coding Theory. The subject is concerned with devising 'efficient' encoding and decoding procedures. There are by now about half a dozen books written on the subject besides a couple of books on Applied Modern Algebra containing some aspects of the subject. The present book is mainly based on a course of lectures given at Kurukshetra University to mathematics students during the last few years. For giving this course of lectures, the books by MacWilliams and Sloane (1978), Van Lint (1971), Birkhoff and Bartee (1970), Dornhoff and Hohn (1978) were used extensively. The object of the present book is to present only the fundamentals of the subject keeping a first-year student in view. However, an effort is made to give a rigorous treatment with full details (even though sometimes trivial and except for some results from Algebra which are accepted without proofs) and the material covered may be regarded as a first course on the subject.

We start with the definition of a block code and of distance between words of equal length. Using the maximum likelihood decoding procedure, we obtain necessary and sufficient conditions for a code to (i) detect, (ii) correct any set of k or fewer errors. Two very important and useful algebraic methods of defining codes (encoding procedures) are through matrix and polynomial multiplication. The codes obtained are called respectively matrix codes and polynomial codes. These two types of codes are studied in Chapters 1 and 2. Generator and parity check matrices are also discussed here.

Hamming codes are single error correcting codes which are studied using a constructive approach in Chapter 3. For defining Hamming codes, we need the binary representation of numbers which is discussed in the first section of this chapter.

One of the most important classes of codes invented so far is that of Bose–Chaudhuri–Hocquenghem (BCH) codes. These are polynomial codes and are discussed in Chapter 4. For defining BCH codes, we need quite a few results from finite fields. Construction of finite fields is of paramount importance for these codes and is discussed at length although some results needed from rings are assumed. Some BCH codes of smaller lengths are constructed.

Linear codes are subspaces of finite dimensional vector spaces over a finite field and are discussed in Chapter 5. The concept of dual code is introduced and MacWilliams's identity relating the weight enumerator of the dual of a binary linear code with that of the code is given.

Cyclic codes can be identified as ideals in a certain quotient ring of a polynomial ring and are discussed in Chapter 6. Among other results, it is proved that BCH codes and Hamming codes are cyclic codes. Non-binary Hamming codes are defined and it is proved that Hamming codes (binary as well as non-binary) are perfect codes. A couple of examples of binary cyclic self dual codes are given. Study of cyclic codes raises the problem of factorization of the polynomial $X^m - 1$ as a product of irreducible polynomials and is discussed in Chapter 7. Berlekamp's Algorithm (1968) regarding factorization of any polynomial over a finite field is also discussed. A number of examples illustrating the algorithm are given—in particular factorization of the binary polynomial $X^{61} - 1$ is obtained.

In Chapter 8, we study quadratic residue (QR) codes. Binary Golay code $\mathcal{G}_{23}$ and ternary Golay code $\mathcal{G}_{11}$ occur as examples of quadratic residue codes. Certain minimum distance properties of QR codes and the relationship between extended QR codes and duals of QR codes are obtained. Idempotents of binary and ternary QR codes are explicitly given.

Maximum distance separable (MDS) codes are discussed in Chapter 9 giving among others a necessary and sufficient condition for a linear code to be MDS. The problem of existence of largest possible n for which there is an $[n, k, d]$ MDS code over GF(q) for a given value of k and q is also considered.

Automorphism group of a code is useful in giving information about the minimum distance of the code. Automorphism group of a code is defined and some simple properties of these are obtained in Chapter 10. It is proved that in a binary cyclic code which is invariant under a certain group of permutations, the weights of all the code words cannot be divisible by 4.

All the codes studied in Chapters 1–10 are group/linear codes. To avoid the impression that perhaps all codes are 'linear', we introduce Hadamard matrices and then define Hadamard codes which are non-linear.

To my wife Raj
and
daughters Vandana & Shalini

1

Group codes

1.1 ELEMENTARY PROPERTIES

Definition 1.1 – groups
A non-empty set G with a binary composition is called a **group** if the following hold.

(i) The composition in G is associative, i.e. $(ab)c = a(bc) \forall a, b, c \in G$.
(ii) There exists an element $e \in G$ such that $ea = ae = a \forall a \in G$.
(iii) For every $a \in G$, there exists an element $b \in G$ such that $ab = ba = e$.

It is fairly easy to prove that element $e \in G$ satisfying condition (ii) above is uniquely determined and, then, is called the **identity** of G. Also for $a \in G$, element $b \in G$ satisfying $ab = ba = e$ is uniquely determined and is called the **inverse** of a, denoted by a^{-1}.

Definition 1.2 – Abelian groups
A group G is called **Abelian** if $ab = ba \forall a, b \in G$.

Definition 1.3 – rings
A non-empty set R with two binary compositions, say addition and multiplication, defined on it is called a **ring** if:

(i) R is an Abelian group w.r.t. the additive composition;
(ii) multiplication in R is associative, i.e. $(ab)c = a(bc) \forall a, b, c \in R$; and
(iii) the two distributive laws hold, i.e. $\forall a, b, c \in R$, $a(b + c) = ab + ac$ and $(a + b)c = ac + bc$.

A ring R which also has the property

$$ab = ba \forall a, b \in R$$

is called a **commutative ring**. If R is a ring having an element $1 \in R$ such that $1a = a = a1$ for every $a \in R$, then R is called a **ring with identity**.

Definition 1.4 – fields

A set F having *at least* two elements with two compositions, say addition and multiplication defined on it, is called a **field** if:

(i) F, w.r.t. the additive and multiplicative composition, is a commutative ring with identity; and
(ii) every non-zero element of F is invertible w.r.t. multiplication.

It is then immediate that F^*, the set of all non-zero elements of F, is an Abelian group w.r.t. multiplication. Observe that $\mathbb{Q}$, the set of all rational numbers, w.r.t. the addition and multiplication of the number system, is a field, and one having an infinite number of elements. However, there do exist fields having only a finite number of elements; these are called finite or **Galois fields**.

If p is a prime integer, consider the set $F_p = \{0, 1, 2, \ldots, p-1\}$ of p elements in which addition $\oplus$ and multiplication $\bigcirc$ are defined modulo p. Explicitly, for a, $b \in F_p$, $a \oplus b = c$ and $a \bigcirc b = d$ where c, d are respectively the least non-negative remainders when the sum $a + b$ and product ab of the integers a, b are divided by p. When there is no danger of confusion, we also write $a + b$ for $a \oplus b$ and ab (or $a \times b$) for $a \bigcirc b$. For $p = 2$, we denote the field F_2 by $\mathbb{B}$ ('b' for binary). Thus $\mathbb{B} = \{0, 1\}$ with addition and multiplication defined by $0 + 0 = 0$, $1 + 0 = 0 + 1 = 1, 1 + 1 = 0, 0 \times 0 = 1 \times 0 = 0 \times 1 = 0, 1 \times 1 = 1$. Throughout this chapter, we are concerned with the field $\mathbb{B}$ of two elements. (There also exist other finite fields and we will study these later.)

Let $\mathbb{B}^n$, where n is a positive integer, denote the set of all ordered n-tuples or sequences of length n with entries belonging to the field $\mathbb{B}$. Define sum of two sequences of length n component-wise, i.e. if $a = a_1 \text{---} a_n$, $b = b_1 \text{---} b_n$, then $a + b = c_1 \text{---} c_n$, where $c_i \in \mathbb{B}$ are defined by $c_i = a_i + b_i$, $1 \le i \le n$. Observe that with this composition $\mathbb{B}^n$ becomes an Abelian group. The zero sequence (the sequence of length n with every component or entry zero) is the identity of $\mathbb{B}^n$ and every element of $\mathbb{B}^n$ is its own inverse.

Definition 1.5 – code words

A **binary block** (m, n)**-code** consists of an **encoding function** E: $\mathbb{B}^m \to \mathbb{B}^n$ and a **decoding function** D: $\mathbb{B}^n \to \mathbb{B}^m$. The elements of $\operatorname{Im} E$ (image of E) are called **code words**.

One of the earliest examples (although outdated) of codes that we might come across in our daily life is **Morse code**. To send a message telegraphically, the message to be conveyed is first written as English text. In the Morse code, each character of the English language is identified by a sequence of dots and dashes. Using this coding system, the message is converted into a sequence of dots and dashes and there results a code word which is then transmitted through the machine to another (say in Delhi). The operator working on the machine in Delhi receives a sequence of dots and dashes which is then translated, using the inverse process, back into English text. Some of the dots and dashes transmitted may have been wrongly received in Delhi due to disturbance in the channel. The operator in Delhi will then have to depend on his vocabulary in English to decipher the message as accurately as possible.

Another example in daily life is provided by telephone. When we speak into the microphone, the speech is converted into an electric wave. To ensure that this electric wave does not become very weak when it travels through the wire over a long distance or due to resistance of the wire, this wave is superimposed by a (modulating) electric wave (the net result becoming a code word). On the receiving end, the added wave is filtered out (i.e. the code word is decoded) and the rest of the electric wave which represents the original speech is then converted into sound wave and received by the receiver.

In coding theory, we are concerned with devising methods of encoding and decoding so that the errors which occur due to disturbance in channel, if not altogether eliminated, are minimized. One of the assumptions about the channel is that it does not increase or decrease the length of the sequence that passes through it, although it may garble a few dots and dashes (or 0s and 1s). The second assumption is that the probability of a 0 being garbled (or not) is the same as that of 1 being garbled (or not), i.e. it is a binary symmetric channel.

Since we are mostly concerned with binary codes, we suppress the word binary and, unless stated otherwise by a code, we shall mean a binary code throughout this chapter.

Also the decoding function D is not, in general, defined from $\mathbb{B}^n \to \mathbb{B}^m$ but is defined from $\mathbb{B}^n \to C$, where C is the set of all code words. The process/problem of coding theory may be pictorially depicted as:

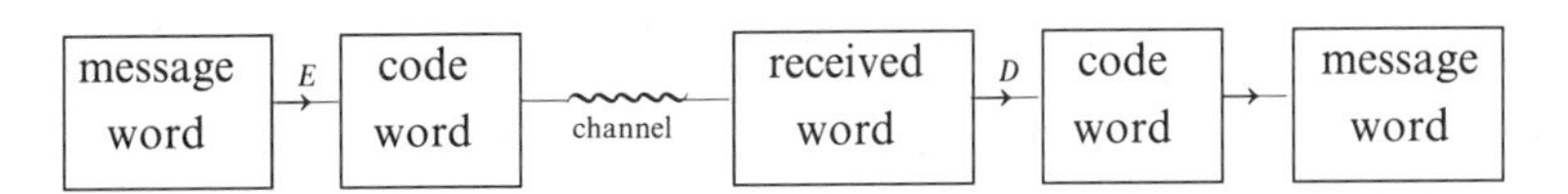

Definition 1.6 – distance function

If $a, b \in \mathbb{B}^n$, we define the **distance** $d(a, b)$ between a and b by

$$d(a,b) = \sum_{i=1}^{n} x_i \qquad \begin{cases} x_i = 0 & \text{if } a_i = b_i \\ x_i = 1 & \text{if } a_i \neq b_i \end{cases}$$

where $a = a_1 a_2 \text{---} a_n$ and $b = b_1 b_2 \text{---} b_n$.

For example:

(i) if $a = 10011011$ and $b = 11001101$, then $a_1 = b_1$, $a_2 \neq b_2$, $a_3 = b_3$, $a_4 \neq b_4$, $a_5 = b_5$, $a_6 \neq b_6$, $a_7 \neq b_7$, $a_8 = b_8$ and so $d(a, b) = 4$

(ii) if $a = 111001$ and $b = 101010$, then $d(a, b) = 3$

Observe that $d(a, b) = d(b, a) \forall a, b \in \mathbb{B}^n$.

Definition 1.7 – weight function
If $a \in \mathbb{B}^n$, we define the **weight** $\mathrm{wt}(a)$ of a as the number of non-zero components of the sequence a.

For example:

(i) if $a = 10011011$, then $\mathrm{wt}(a) = 5$
(ii) if $a = 11001101$, then $\mathrm{wt}(a) = 5$
(iii) if $a = 111001$, then $\mathrm{wt}(a) = 4$
(iv) if $a = 101010$, then $\mathrm{wt}(a) = 3$

Lemma 1.1
If $a, b \in \mathbb{B}^n$, then $d(a, b) = \mathrm{wt}(a + b)$.

Proof
Let $a = a_1 a_2 \cdots a_n$ and $b = b_1 b_2 \cdots b_n$. For any i, $1 \le i \le n$, $a_i + b_i = 1$ iff $a_i \ne b_i$. Hence the pair (a_i, b_i) contributes 1 to $\mathrm{wt}(a + b)$ iff it contributes 1 to $d(a, b)$. Therefore, $d(a, b) = \mathrm{wt}(a + b)$.

Corollary
If $a, b, c \in \mathbb{B}^n$, then $d(a + c, b + c) = d(a, b)$.

Lemma 1.2
If $a, b, c \in \mathbb{B}^n$, then $d(a, b) \le d(a, c) + d(b, c)$.

Proof
Let $a = a_1 a_2 \cdots a_n$, $b = b_1 b_2 \cdots b_n$ and $c = c_1 c_2 \cdots c_n$. For any i, $1 \le i \le n$, define

$$d(a_i, b_i) = \begin{cases} 1 & \text{if } a_i \ne b_i \\ 0 & \text{if } a_i = b_i \end{cases}$$

Similarly define $d(a_i, c_i)$ and $d(b_i, c_i)$. Then

$$d(a, b) = \sum_{i=1}^{n} d(a_i, b_i)$$

If $a_i = b_i$, then $d(a_i, b_i) \le d(a_i, c_i) + d(b_i, c_i)$ trivially. Suppose that $a_i \ne b_i$. Then $d(a_i, b_i) = 1$. If $a_i = c_i$, then necessarily $b_i \ne c_i$ while if $b_i = c_i$, then $a_i \ne c_i$. Thus, in either case

$$d(a_i, b_i) \le d(a_i, c_i) + d(b_i, c_i)$$

and therefore

$$\begin{aligned} d(a, b) &= \sum_{i=1}^{n} d(a_i, b_i) \\ &\le \sum_{i=1}^{n} d(a_i, c_i) + \sum_{i=1}^{n} d(b_i, c_i) \\ &= d(a, c) + d(b, c) \end{aligned}$$

Definition 1.8 – nearest-neighbour decoding principle

The **nearest-neighbour decoding principle** states that if a word $r \in \mathbb{B}^n$ is received and it happens to be a code word, we put $D(r) = r$. If r is not a code word, we find the distance of r from all the code words and among these distances, we find the least. Suppose that d is the least distance. Then there exists a code word a (say) such that $d(a, r) = d$. If a is the only code word with $d(a, r) = d$, we put $D(r) = a$. If there is more than one code word with distance from r equal to d, we say that there is a **decoding failure**.

In the case of binary symmetric channel (as in our case) in which all code words are equally likely to be transmitted and the single-symbol error probability is less than $\frac{1}{2}$, the principle is called the **maximum likelihood decoding principle**.

Definition 1.9 – detected and undetected errors

We say that an error vector (or word) e is **detected** by a code if $a + e$ is not a code word for any code word a. If, for some code word a, $a + e$ is again a code word, we say that the error vector (word or pattern) e goes **undetected**.

When a word of length n is transmitted and k of these entries are incorrectly received, we say that k transmission errors have occurred. By taking 1 at positions corresponding to k errors and zero everywhere else, we create a word (error word/vector) e of weight k. Also, the k non-zero positions of an error vector of weight k give a set of k errors. This gives a one-to-one correspondence between error words of weight k and the sets of k errors. We say that a set of k errors is detected if the corresponding error word of weight k is detected.

Theorem 1.1

For a code to detect all sets of k or fewer errors, it is necessary and sufficient that the minimum distance between any two code words be $k + 1$ or more.

Proof

Let C be the set of all code words (of length n) of the given code. Suppose that $\forall b, b' \in C, d(b, b') \geq k + 1$. Let $b \in C$ be transmitted and suppose that the channel introduces an error word $e = e_1 e_2 \cdots e_n$ with

$$\mathrm{wt}(e) = \left(\sum_{i=1}^{n} e_i \right) \leq k$$

Then the received word is $b + e$ and

$$d(b + e, b) = \mathrm{wt}(b + e + b) = \mathrm{wt}(e + 2b) = \mathrm{wt}(e) \leq k$$

This shows that $b + e$ is not a code word. As such, we know that some transmission error has occurred and this proves that the error vector e is detected.

Conversely, suppose that the code is able to detect all sets of k or fewer errors. This means that whatever word e with $\mathrm{wt}(e) \leq k$ and code word b be given, $b+e$ is not a code word. Suppose that b, $b' \in C$ and that $d(b,b') \leq k$. Let $e = b + b'$. Then $\mathrm{wt}(e) \leq k$. Also $b+e = b+b+b' = b'$ – a code word. This proves that the error vector e goes undetected. This contradiction proves that

$$d(b,b') \geq k+1 \,\forall\, b, b' \in C$$

Definition 1.10 – corrected errors

We say that an error word e is **corrected** by a code if the decoding function D of the code is such that $D(b+e) = b$ for every code word b. We also say that a set of k errors is **corrected** if the corresponding error word of weight k is corrected.

Theorem 1.2

For a code (D, E) to correct all sets of k or fewer errors, it is necessary that the minimum distance between code words be at least $2k+1$ (it being given that the nearest neighbour decoding principle holds).

Proof

Suppose that $b = b_1 \text{---} b_n$, $a = a_1 \text{---} a_n$ are two code words with $d(a,b) \leq 2k$. Then $\mathrm{wt}(a+b) = l \leq 2k$. We can then find words e, e' such that $\mathrm{wt}(e) \leq k$, $\mathrm{wt}(e') \leq k$ and $a+b = e+e'$. Then $a+e = b+e'$.

Case (i): $l = 2t+1$

Then we can suppose that $\mathrm{wt}(e) = t$ and $\mathrm{wt}(e') = t+1 \leq k$. Therefore,

$$d(a+e, a) = \mathrm{wt}(a+e+a) = \mathrm{wt}(e) < \mathrm{wt}(e') = d(b+e', b)$$

Suppose that b is the code word transmitted and the channel adds to it the error vector e'. Then

$$d(b+e', a) = d(a+e, a) < d(b+e', b)$$

and by the nearest neighbour decoding principle, $b+e'$ will be decoded into a or some other code word but never b. Thus, the error vector e' with $\mathrm{wt}(e') \leq k$ is not corrected. Hence, the distance between any two code words must be at least $2k+1$.

Case (ii): $l = 2t$

We can then suppose that $\mathrm{wt}(e) = \mathrm{wt}(e') = t$. Now, $d(a+e, a) = \mathrm{wt}(e) = t = w(e')$ $= d(b+e', b)$ so that the received word $a+e$ is equidistant from two code words a and b and either the received word $b+e'$ is decoded into a word different from b or the code is unable to decide about the transmitted word – again a contradiction.

Example

Consider the (2, 3) parity check code. Then the encoding function E is

$$00 \to 000, 01 \to 011, 10 \to 101, 11 \to 110$$

Thus C, the set of all code words, is

$$\{000, 011, 101, 110\}$$

There are only three possible error vectors of weight 1

$$001, 010, 100$$

and any one of these added to any code word does not yield a code word. Thus every single error is detected by this code.

Suppose that 011 is transmitted and the channel adds to it the error vector 100. Then the received word is 111. But the same word will be received if 101 were transmitted and the channel had added the error vector 010. The received word is in fact equidistant from three code words and, so, this error is not corrected.

Theorem 1.3

A binary code with minimum distance $2k+1$ is capable of correcting any pattern of k or fewer errors.

Proof

Let C be a code of length n with minimum distance $2k+1$ and e be an error pattern of weight at most k. Let b be a code word that is transmitted so that the received word is $r = b + e$. If b^* is any code word $b^* \neq b$, then

$$\begin{aligned} d(r, b^*) &= d(b+e, b^*) \\ &= \mathrm{wt}(b+e+b^*) \\ &= \mathrm{wt}(b+b^*+e) \end{aligned}$$

Now

$$\mathrm{wt}(b+b^*) = d(b, b^*) \geq 2k+1$$

and $\mathrm{wt}(e) \leq k$. Therefore

$$\mathrm{wt}(b+b^*+e) \geq k+1$$

i.e.

$$d(r, b^*) \geq k+1$$

Also

$$d(r, b) = \mathrm{wt}(r+b) = \mathrm{wt}(b+b+e) = \mathrm{wt}(e) \leq k$$

Hence b is the nearest code word to r, the received word and by the maximum likelihood decoding principle $D(r) = b$. Thus, the error vector e with $\text{wt}(e) \leq k$ is corrected.

Definition 1.11 – (m, 3m) triple repetition code
The code of length $3m$ in which the encoding function $E: \mathbb{B}^m \to \mathbb{B}^{3m}$ is defined by

$$E(a) = E(a_1 a_2 \cdots a_m)$$
$$= a_1 a_2 \cdots a_m a_1 a_2 \cdots a_m a_1 a_2 \cdots a_m$$

where $a = a_1 a_2 \cdots a_m \in \mathbb{B}^m$, is called a **triple repetition code**.

If $a = a_1 a_2 \cdots a_m$, $b = b_1 b_2 \cdots b_m$ are distinct words of length m, then $d(a, b) \geq 1$ and, therefore,

$$d(E(a), E(b)) \geq 3$$

Thus, the triple repetition code is capable of detecting any two errors and correcting any single error.

1.2 MATRIX ENCODING TECHNIQUES

One systematic algebraic technique for encoding binary words is by **matrix multiplication**.

Recall that if $\mathbf{A} = (a_{ij})$ is an $m \times n$ matrix and $\mathbf{B} = (b_{rs})$ is an $n \times k$ matrix, then the product $\mathbf{AB}$ of $\mathbf{A}$ and $\mathbf{B}$ is an $m \times k$ matrix (c_{ij}) where

$$c_{ij} = \sum_{r=1}^{n} a_{ir} b_{rj} \quad 1 \leq i \leq m, 1 \leq j \leq k$$

Also recall that if $\mathbf{A}$, $\mathbf{B}$ are two groups then a map $f: \mathbf{A} \to \mathbf{B}$ satisfying the property

$$f(xy) = f(x) f(y) \forall x, y \in \mathbf{A}$$

is called a **homomorphism**. A homomorphism $f: \mathbf{A} \to \mathbf{B}$ is called:

(i) a **monomorphism** if the map f is one–one; and
(ii) an **isomorphism** if the map f is both one–one and onto.

An $m \times n$ matrix, with $m < n$ over $\mathbb{B}$ is called an **encoding matrix** (or **generator matrix**) if the first m columns of it form the identity matrix $\mathbf{I}_m$. Given a generator matrix $\mathbf{G}$, we define an encoding function $E: \mathbb{B}^m \to \mathbb{B}^n$ by

$$E(x) = x\mathbf{G} \quad x \in \mathbb{B}^m$$

Since the first m columns of $\mathbf{G}$ form an identity matrix, the initial part of $x\mathbf{G}$ is x itself. Thus, the matrix encoding method gives distinct code words corresponding to different message words and the map E is one–one. Again, both

$\mathbb{B}^m$ and $\mathbb{B}^n$ are additive Abelian groups and for $x, y \in \mathbb{B}^m$

$$E(x + y) = (x + y)\mathbf{G} = x\mathbf{G} + y\mathbf{G} = E(x) + E(y)$$

Thus E is a homomorphism and we have the following proposition.

Proposition 1.1
For any $m \times n$ generator matrix $\mathbf{G}$, the encoding function $E: \mathbb{B}^m \to \mathbb{B}^n$ given by $E(x) = x\mathbf{G}$, $x \in \mathbb{B}^m$, is a monomorphism.

A code given by a generating matrix is called a **matrix code**.

Definition 1.12
When the code words in a block code form an additive group, the code is called a **group code**.

Corollary
A matrix code is a group code.

To consider other examples of group codes we first need another definition.

Definition 1.13
An $(m, m + 1)$ parity check code is the code given by the encoding function $E: \mathbb{B}^m \to \mathbb{B}^{m+1}$ defined by

$$E(a_1 a_2 \cdots a_m) = a_1 a_2 \cdots a_m a_{m+1}$$

where

$$a_{m+1} = \begin{cases} 1 & \text{if } \operatorname{wt}(a) = \operatorname{wt}(a_1 a_2 \cdots a_m) \text{ is odd} \\ 0 & \text{if } \operatorname{wt}(a) \text{ is even} \end{cases}$$

Lemma 1.3
$(m, m + 1)$ parity check code is a group code.

Proof
Let $a = a_1 a_2 \cdots a_m$, $a' = a'_1 a'_2 \cdots a'_m$ be two message words and $b = b_1 \cdots b_m b_{m+1}$, $b' = b'_1 \cdots b'_m b'_{m+1}$ be the corresponding code words in the parity check scheme. Now, $b + b' = c_1 \cdots c_m c_{m+1}$, where $c_i = b_i + b'_i$, $1 \le i \le m + 1$.

$$c_1 + c_2 + \cdots + c_m = (b_1 + \cdots + b_m) + (b'_1 + \cdots + b'_m)$$

which is odd iff one of $(b_1 + \cdots + b_m)$ and $(b'_1 + \cdots + b'_m)$ is odd and the other is even. But, in this case, either $b_{m+1} = 1$ and $b'_{m+1} = 0$ or $b_{m+1} = 0$ and $b'_{m+1} = 1$. Hence, in this case $c_{m+1} = 1$.

Again $c_1 + \cdots + c_m$ is even iff either both of $b_1 + \cdots + b_m$ and $b'_1 + \cdots + b'_m$ are odd or both are even. If this is so, then $b_{m+1} + b'_{m+1} = 0$, i.e. $c_{m+1} = 0$.

This proves that c is a code word under the parity check scheme. This gives a composition in the set of all code words. The 0 word is the identity and every word is its own inverse. Hence, the set of all code words forms a group.

Lemma 1.4
The triple repetition $(m, 3m)$ code is a group code.

Proof
If

$$b = a_1 \text{---} a_m a_1 \text{---} a_m a_1 \text{---} a_m$$

and

$$b' = a'_1 \text{---} a'_m a'_1 \text{---} a'_m a'_1 \text{---} a'_m$$

are two code words, then

$$b + b' = c_1 \text{---} c_m c_1 \text{---} c_m c_1 \text{---} c_m$$

where $c_i = b_i + b'_1$, which is again a block of length m repeated three times. Hence $b + b'$ is a code word. 0 is the identity and every code word is its own inverse. Hence, the code is a group code.

Exercise 1.1

1. Prove that an $(m, m + 1)$ parity check code is a matrix code. What is the generator matrix of this code?
2. Give an example of a group code which is not a matrix code.

Proposition 1.2
For a group code, the minimum distance equals the minimum of the weights of the non-zero code words.

Proof
Let d be the minimum distance of the group code. Then there exist code words b, b' such that $d = d(b, b') = \text{wt}(b + b')$. But the code being a group code, $b + b'$ is a code word. Let t be the minimum of the weights of non-zero code words. Then, by the above,

$$d \geq t$$

t being the minimum among weights of non-zero code words, there exists a non-zero code word b'' such that

$$t = \text{wt}(b'') = d(b'', 0) \geq d$$

Hence $d = t$.

Remark
In group codes, the error patterns that pass undetected are precisely those which correspond to non-zero code words.

Proof

If $e = b^*$ is a code word, then for any code word b, $b + e$ is again a code word and the error pattern e goes undetected.

Conversely, suppose that an error pattern e goes undetected. Then there exists a code word b such that $b + e = b^*$ (say) is again a code word. This shows that $e = b + b^*$ – again a code word.

Example

Consider the 3×6 generating matrix

$$\mathbf{G} = \begin{pmatrix} 1 & 0 & 0 & 1 & 1 & 0 \\ 0 & 1 & 0 & 0 & 1 & 1 \\ 0 & 0 & 1 & 1 & 1 & 1 \end{pmatrix}$$

All the code words of the code generated by this matrix are:

$$\begin{array}{llll} 000 \to 000000 & & 001 \to 001111 & \\ 100 & 100110 & 110 & 110101 \\ 010 & 010011 & 101 & 101001 \\ 011 & 011100 & 111 & 111010 \end{array}$$

This is a group code with 4 code words of weight 3, 3 code words of weight 4 and only the zero code word of weight 0. Thus, the minimum distance of the code is 3. Hence, this is a code which is capable of correcting any single error and detecting any error of weight 2.

Next, we give a decoding procedure for group codes with the help of which the probability of an error passing undetected is minimized. This decoding procedure uses decomposition of a finite group into cosets which we describe briefly.

Recall that a non-empty subset N of a group M is called a **subgroup** of M if:

(i) the composition in M induces a composition in N, i.e. wherever $a, b \in N$, then $ab \in N$; and
(ii) N is a group w.r.t. the induced composition.

For example, if $\mathbf{G}$ is an $m \times n$ matrix over $\mathbb{B}$ then $\{a\mathbf{G} \mid a \in \mathbb{B}^m\}$ is a subgroup of $\mathbb{B}^n$. The order of the subgroup is at most 2^m whereas if $\mathbf{G}$ is a generator matrix, then the order of the subgroup is precisely 2^m (Proposition 1.1).

Let $n > 1$ and C be a subgroup of $\mathbb{B}^n$. For $a \in \mathbb{B}^n$, $a + C = \{a + c : c \in C\}$ is a subset (and not in general a subgroup) of $\mathbb{B}^n$ called a **coset** of C in $\mathbb{B}^n$. If $b \in a + C$, then $b = a + c$ for some $c \in C$. Therefore, for any $c' \in C$,

$$b + c' = a + (c + c') \in a + C$$

Thus $b + C \subseteq a + C$. Again $b = a + c$ implies $a = b + c$ and, as above, it follows that $a + C \subseteq b + C$. Hence $a + C = b + C$. On the other hand, if $a + C = b + C$, then $b = b + 0 \in b + C = a + C$. We thus have

$$b \in \mathbb{B}^n \text{ is in } a + C \text{ iff } a + C = b + C \qquad (1.1)$$

Now, consider two cosets $a + C$ and $b + C$ of C in $\mathbb{B}^n$. If $(a + C) \cap (b + C) \neq \emptyset$, there exists an $x \in a + C$ and $x \in b + C$. It follows from (1.1) above that $a + C = x + C$ and $b + C = x + C$ and therefore, $a + C = b + C$. So

$$\text{two cosets of } C \text{ in } \mathbb{B}^n \text{ are either disjoint or identical} \tag{1.2}$$

Observe that the number of elements in any coset $a + C$ of C in $\mathbb{B}^n$ is equal to the order of the subgroup C (which equals the number of elements in C). Every element of $\mathbb{B}^n$ is in some coset (in fact in a unique coset in view of (1.2) above) of C in $\mathbb{B}^n$, e.g. if $a \in \mathbb{B}^n$, then $a \in a + C$. Also, the group $\mathbb{B}^n$ being finite, the number of distinct cosets of C in $\mathbb{B}^n$ is finite. If $a^1 + C, \ldots, a^k + C$ are all the distinct cosets of C in $\mathbb{B}^n$, then we have

$$\mathbb{B}^n = \bigcup_{i=1}^{k} (a^i + C) \qquad \text{and} \qquad (a^i + C) \cap (a^j + C) = \emptyset \text{ for } i \neq j \tag{1.3}$$

Consider an (m, n) group code and let C be the set of all code words of this code. Then order of C is 2^m. $\mathbb{B}^n$ is the set of all words of length n which as seen earlier is a group and C is a subgroup of it. We can then write $\mathbb{B}^n$ as a disjoint union of cosets of C in $\mathbb{B}^n$. In each coset of C in $\mathbb{B}^n$, we choose a word b^i of least weight and call it a **coset leader**. Observe that

$$\mathrm{wt}(b^i) \leq \mathrm{wt}(b^i + c^i) \forall c^i \in C$$

Any element c of $\mathbb{B}^n$ can be uniquely written as $c = b^i + c^j$ for some $c^j \in C$. We define a decoding function D by putting

$$D(c) = c^j$$

For any code word, $c^k \neq c^j$, we have

$$\begin{aligned} d(c, c^k) &= d(b^i + c^j, c^k) = \mathrm{wt}(b^i + c^j + c^k) \\ &\geq \mathrm{wt}(b^i) \\ &= d(b^i + c^j, c^j) = d(c, c^j) \end{aligned}$$

Thus, there is no code word lying within the circle with centre at c and radius equal to $d(c, c^j)$.

This decoding procedure or process is known as **decoding by coset leaders.**

Theorem 1.4

In group codes, decoding by coset leaders corrects precisely those error patterns which are coset leaders.

Proof

Suppose that an error pattern e is corrected by this method of decoding. Let c^i be a code word transmitted so that the received word is $b = c^i + e$. Then $b = b^k + c^r$ for some code word c^r and coset leader b^k. By the decoding process,

$D(b) = c^r$ and since the error is corrected, we must also have $D(b) = c^i$. Hence, $c^r = c^i$. Thus, $b^k + c^i = c^i + e$ or $e = b^k$ – a coset leader.

Conversely, suppose that $e = b^k$ is a coset leader. Then, for any code word, c^i, the received word is $c^i + e = b^k + c^i$ and $D(b^k + c^i) = c^i$. Hence the error pattern is corrected.

Example

Consider first the (3, 4)-parity check code $\mathscr{C}$:

$$\begin{array}{ll} 000 \longrightarrow 0000 & 011 \longrightarrow 0110 \\ 001 \longrightarrow 0011 & 101 \longrightarrow 1010 \\ 010 \longrightarrow 0101 & 110 \longrightarrow 1100 \\ 100 \longrightarrow 1001 & 111 \longrightarrow 1111 \end{array}$$

Coset decomposition of $\mathscr{C}$ in $\mathbb{B}^4$ is:

Coset leader	The coset							
0000	0000	0011	0101	1001	0110	1010	1100	1111
0001	0001	0010	0100	1000	0111	1011	1101	1110

Observe that we could have taken 0010 or 0100 as coset leader for the coset $0001 + \mathscr{C}$. Having chosen 0001 as coset leader, if 1011 is the word received, then decoding by coset leaders decodes this word into $1011 + 0001 = 1010$ which is the word at the head of the column in which the received word 1011 lies.

Next, consider the (2, 6) triple repetition code $\mathscr{C}$:

$$00 \to 000000,\ 01 \to 010101,\ 10 \to 101010,\ 11 \to 111111$$

Here $\mathscr{C}$ is a subgroup of $\mathbb{B}^6$ and, using Lagrange's theorem, we may represent the elements of $\mathbb{B}^6$ in tabular form as follows:

Coset leader	The coset			
000000	000000	010101	101010	111111
000001	000001	010100	101011	111110
000010	000010	010111	101000	111101
000100	000100	010001	101110	111011
001000	001000	011101	100010	110111
010000	010000	000101	111010	101111
100000	100000	110101	001010	011111
000011	000011	010110	101001	111100
001001	001001	011100	100011	110110
100001	100001	110100	001011	011110
000110	000110	010011	101100	111001
010010	010010	000111	111000	101101
001100	001100	011001	100110	110011
100100	100100	110001	001110	011011
011000	011000	001101	110010	100111
110000	110000	100101	011010	001111

In this case also, we find that if the received word $r = b + e$, then decoding by coset leaders decodes this word into b, the code word which lies at the head of the column in which r lies.

The above observed principle holds in general. Thus, if all the words of $\mathbb{B}^n$ are written in a tabular form each row being a coset $b^i + C$ of C in $\mathbb{B}^n$ with b^i a coset leader and the first row representing the words of C, to decode a received word r we locate it in the table. The word r is decoded into the code word which appears at the head of the column in which r occurs.

1.3 GENERATOR AND PARITY CHECK MATRICES

Let us consider, once again, the matrix code given by the generator matrix

$$\mathbf{G} = \begin{pmatrix} 1 & 0 & 0 & 1 & 1 & 0 \\ 0 & 1 & 0 & 0 & 1 & 1 \\ 0 & 0 & 1 & 1 & 1 & 1 \end{pmatrix}$$

If $a_1a_2a_3a_4a_5a_6$ is the code word in this code corresponding to the message word $a_1a_2a_3$, then

$$(a_1 \quad a_2 \quad a_3 \quad a_4 \quad a_5 \quad a_6) = (a_1 \quad a_2 \quad a_3)\mathbf{G}$$

and thus

$$a_4 = a_1 + a_3$$
$$a_5 = a_1 + a_2 + a_3$$
$$a_6 = a_2 + a_3$$

These equations may be rewritten as

$$a_1 + a_3 + a_4 = 0$$
$$a_1 + a_2 + a_3 + a_5 = 0$$
$$a_2 + a_3 + a_6 = 0$$

These are called **parity check equations**. In matrix form, these equations may be written as

$$\begin{pmatrix} 1 & 0 & 1 & 1 & 0 & 0 \\ 1 & 1 & 1 & 0 & 1 & 0 \\ 0 & 1 & 1 & 0 & 0 & 1 \end{pmatrix} \begin{pmatrix} a_1 \\ a_2 \\ a_3 \\ a_4 \\ a_5 \\ a_6 \end{pmatrix} = 0$$

The matrix

$$\mathbf{H} = \begin{pmatrix} 1 & 0 & 1 & 1 & 0 & 0 \\ 1 & 1 & 1 & 0 & 1 & 0 \\ 0 & 1 & 1 & 0 & 0 & 1 \end{pmatrix}$$

is called the **parity check matrix** of the code. Let

$$\mathbf{A} = \begin{pmatrix} 1 & 1 & 0 \\ 0 & 1 & 1 \\ 1 & 1 & 1 \end{pmatrix}$$

Then $\mathbf{G} = (\mathbf{I}_3 \quad \mathbf{A})$, where $\mathbf{I}_3$ is the identity matrix of order 3. Also

$$\mathbf{A}' = \mathbf{A}^t = \begin{pmatrix} 1 & 0 & 1 \\ 1 & 1 & 1 \\ 0 & 1 & 1 \end{pmatrix}$$

and $\mathbf{H} = (\mathbf{A}' \quad \mathbf{I}_3)$.

We shall later prove this relation between the generator matrix and the corresponding parity check matrix in the general case. The matrix $\mathbf{H}$ has the property that for any code word a, $\mathbf{Ha} = 0$. (Note that $\mathbf{a}$ is the vector formed by taking the elements of the code word a.)

We observe that the $(m, m+1)$ parity check code we considered earlier has $a_1 a_2 \text{---} a_{m+1}$ as a code word provided

$$a_{m+1} = \begin{cases} 0 & \text{if } a_1 + a_2 + \cdots + a_m \text{ is even} \\ 1 & \text{if } a_1 + a_2 + \cdots + a_m \text{ is odd} \end{cases}$$

Then $a_1 + a_2 + \cdots + a_m + a_{m+1} = 0$.

Observe that this code is a matrix code given by the generator matrix

$$\mathbf{G} = \begin{pmatrix} 1 & 0 & \cdots & 0 & 1 \\ 0 & 1 & \cdots & 0 & 1 \\ \vdots & \vdots & \cdots & \vdots & \vdots \\ 0 & 0 & \cdots & 1 & 1 \end{pmatrix}$$

The parity check matrix of this code is the $1 \times (m+1)$ matrix $\mathbf{H} = (1 \quad 1 \quad \cdots \quad 1)$.

Next, we consider the (3, 6) matrix code given by the generator matrix

$$\mathbf{G} = \begin{pmatrix} 1 & 0 & 0 & 1 & 1 & 1 \\ 0 & 1 & 0 & 0 & 1 & 1 \\ 0 & 0 & 1 & 1 & 0 & 1 \end{pmatrix}$$

For any code word $a_1 a_2 \text{---} a_6$ in this code we have

$$(a_1 \quad a_2 \quad a_3 \quad a_4 \quad a_5 \quad a_6) = (a_1 \quad a_2 \quad a_3)\mathbf{G}$$

and so

$$a_4 = a_1 + a_3$$
$$a_5 = a_1 + a_2$$
$$a_6 = a_1 + a_2 + a_3$$

or

$$a_1 + a_3 + a_4 = 0$$
$$a_1 + a_2 + a_5 = 0$$
$$a_1 + a_2 + a_3 + a_6 = 0$$

In matrix notation, these parity check equations may be written as

$$\mathbf{H}\begin{pmatrix} a_1 \\ a_2 \\ a_3 \\ a_4 \\ a_5 \\ a_6 \end{pmatrix} = 0$$

where

$$\mathbf{H} = \begin{pmatrix} 1 & 0 & 1 & 1 & 0 & 0 \\ 1 & 1 & 0 & 0 & 1 & 0 \\ 1 & 1 & 1 & 0 & 0 & 1 \end{pmatrix}$$

Again, observe that if

$$\mathbf{A} = \begin{pmatrix} 1 & 1 & 1 \\ 0 & 1 & 1 \\ 1 & 0 & 1 \end{pmatrix}$$

then

$$\mathbf{A}^t = \begin{pmatrix} 1 & 0 & 1 \\ 1 & 1 & 0 \\ 1 & 1 & 1 \end{pmatrix}$$

and

$$\mathbf{G} = (\mathbf{I}_3 \quad \mathbf{A}) \qquad \text{while} \qquad \mathbf{H} = (\mathbf{A}^t \quad \mathbf{I}_3)$$

We now define a parity check matrix in general.

Definition 1.14 – parity check matrix
If $m < n$, then any $(n - m) \times n$ matrix $\mathbf{H}$, whose last $n - m$ columns form the identity matrix $\mathbf{I}_{n-m}$ is called a **parity check matrix**.

Let $\mathbf{H}$ be an $(n - m) \times n$ parity check matrix and $a = a_1 \cdots a_m$ be a sequence of length m. Let $b = b_1 \cdots b_n$ be a word of length n with $b_i = a_i$, $1 \le i \le m$ and $\mathbf{Hb}^t = 0$. Suppose that $\mathbf{H} = (\mathbf{A} \quad \mathbf{I}_{n-m})$. Then $\mathbf{Hb}^t = 0$ implies

$$(\mathbf{A} \quad \mathbf{I}_{n-m}) \begin{pmatrix} \mathbf{a}^t \\ \bar{\mathbf{b}}^t \end{pmatrix} = 0$$

where $\bar{\mathbf{b}}$ is the vector formed from the sequence $\bar{b} = b_{m+1} \cdots b_n$ and so $\mathbf{Aa}^t + \mathbf{I}_{n-m}\bar{\mathbf{b}}^t = 0$, i.e. $\mathbf{Aa}^t + \bar{\mathbf{b}}^t = 0$. Hence, $\bar{\mathbf{b}}^t = \mathbf{Aa}^t$ and $b_{m+1} \cdots b_n$ are uniquely determined by a. This proves that for every $a \in \mathbb{B}^m$ there is a uniquely determined word $b \in \mathbb{B}^n$ with $a_i = b_i$, $1 \le i \le m$ and $\mathbf{Hb}^t = 0$. We can thus define an encoding function $E: \mathbb{B}^m \to \mathbb{B}^n$ as follows: for $a \in \mathbb{B}^m$, define $E(a) = b$, where b is the uniquely determined element of $\mathbb{B}^n$ with $a_i = b_i$, $1 \le i \le m$ and $\mathbf{Hb}^t = 0$. We define the **syndrome** of a word $r \in \mathbb{B}^n$ by $\mathbf{s} = \mathbf{Hr}^t$.

Observe that the syndrome of a code word is zero. Using syndrome, we see how the parity check matrix $\mathbf{H}$ associated with a generator matrix $\mathbf{G}$ helps in correcting errors that occur in transmission. The **syndrome** (or the **parity check) decoding procedure** is defined as follows:

Let $r = r_1 \cdots r_m r_{m+1} \cdots r_n$ be the word received and $\mathbf{s} = \mathbf{Hr}^t$ be its syndrome.

(i) If $\mathbf{s} = 0$, we assume that r is the code word sent and the original message word is $r_1 \cdots r_m$.

(ii) If $\mathbf{s}$ matches the ith column of $\mathbf{H}$, we assume that an error in transmission has occurred in the ith position and take

$$c = r_1 \cdots r_{i-1}(r_i + 1)r_{i+1} \cdots r_n$$

as the code word transmitted. The original message is the sequence formed by the initial m entries of c.

(iii) If $\mathbf{s}$ is neither 0 nor a column of $\mathbf{H}$ then at least two errors occurred in transmission.

Theorem 1.5

An $(n - m) \times m$ parity check matrix $\mathbf{H}$ will decode all single errors correctly iff the columns of $\mathbf{H}$ are non-zero and distinct.

Proof

Suppose that the ith column of $\mathbf{H}$ is zero. Let e be the word of weight 1 with 1 in the ith position and 0 everywhere else. Then for any code word b,

$$\mathbf{H}(\mathbf{b} + \mathbf{e})^t = \mathbf{Hb}^t + \mathbf{He}^t = 0 + 0 = 0$$

and, so, by our decoding procedure $D(b + e) = b + e$ and the error vector $\mathbf{e}$ goes undetected.

Next suppose that the ith and jth columns of $\mathbf{H}$ are identical. Let e^i (respectively e^j) be the word of length n with 1 in the ith (respectively jth) position and 0 everywhere else. For any code word b, we have

$$\begin{aligned}\mathbf{H}(\mathbf{b}+\mathbf{e}^i)^t &= \mathbf{H}\mathbf{b}^t + \mathbf{H}(\mathbf{e}^i)^t \\ &= \mathbf{H}(\mathbf{e}^i)^t \\ &= i\text{th column of } \mathbf{H} \\ &= j\text{th column of } \mathbf{H} \\ &= \mathbf{H}(\mathbf{b}+\mathbf{e}^j)^t\end{aligned}$$

Thus we are unable to decide if the error occurred in the ith position or the jth.

Conversely, suppose that the columns of $\mathbf{H}$ are non-zero and distinct. For any error vector $\mathbf{e}$ of weight 1 with 1 in the ith position and any code word b,

$$\mathbf{H}(\mathbf{b}+\mathbf{e})^t = \mathbf{H}(\mathbf{b}^t+\mathbf{e}^t) = \mathbf{H}\mathbf{b}^t + \mathbf{H}\mathbf{e}^t = 0 + \mathbf{H}\mathbf{e}^t = i\text{th column of } \mathbf{H}$$

and by our decoding procedure $D(b+e) = b$. Hence every single error is corrected.

Theorem 1.6

(i) If $\mathbf{G} = (\mathbf{I}_m \quad \mathbf{A})$ is an $m \times n$ generator matrix of a code, then

$$\mathbf{H} = (\mathbf{A}^t \quad \mathbf{I}_{n-m})$$

is the unique parity check matrix for the same code.

(ii) If $\mathbf{H} = (\mathbf{B} \quad \mathbf{I}_{n-m})$ is an $(n-m) \times n$ parity check matrix, then

$$\mathbf{G} = (\mathbf{I}_m \quad \mathbf{B}^t)$$

is the unique generator matrix for the same code.

Proof

Let $a \in \mathbb{B}^m$ and b be the code word corresponding to this message word in the code given by the generator matrix $\mathbf{G}$. Then $\mathbf{b} = \mathbf{aG}$. Suppose that

$$b = b_1 \text{---} b_m b_{m+1} \text{---} b_n \qquad \text{and} \qquad a = a_1 \text{---} a_m$$

As the first m columns of $\mathbf{G}$ form the identity matrix, the relation $\mathbf{b} = \mathbf{aG}$ shows that $a_i = b_i$, $1 \le i \le m$. We write $\bar{b}$ for the word $b_{m+1} \text{---} b_n$ so that $\mathbf{b} = (\mathbf{a} \quad \bar{\mathbf{b}})$. Now

$$\begin{aligned}\mathbf{H}\mathbf{b}^t &= (\mathbf{A}^t \quad \mathbf{I}_{n-m})(\mathbf{aG})^t \\ &= (\mathbf{A}^t \quad \mathbf{I}_{n-m})\mathbf{G}^t\mathbf{a}^t \\ &= (\mathbf{A}^t \quad \mathbf{I}_{n-m})(\mathbf{I}_m \quad \mathbf{A})^t\mathbf{a}^t \\ &= (\mathbf{A}^t \quad \mathbf{I}_{n-m})\begin{pmatrix}\mathbf{I}_m \\ \mathbf{A}^t\end{pmatrix}\mathbf{a}^t \\ &= (\mathbf{A}^t \quad \mathbf{I}_m + \mathbf{I}_{n-m}\mathbf{A}^t)\mathbf{a}^t \\ &= (\mathbf{A}^t + \mathbf{A}^t)\mathbf{a}^t = 0 \times \mathbf{a}^t = 0\end{aligned}$$

Hence, b is the code word corresponding to the message word a in the code given by the parity check matrix $\mathbf{H}$.

Conversely, suppose that $c = c_1 \cdots c_n$ is the code word corresponding to the message word $a = a_1 \cdots a_m$ in the code determined by the parity check matrix $\mathbf{H} = (\mathbf{A}^t \quad \mathbf{I}_{n-m})$. Then $c_i = a_i$, $1 \le i \le m$ and

$$\mathbf{H}\mathbf{c}^t = 0 \qquad \text{or} \qquad \mathbf{H}\begin{pmatrix} \mathbf{a} \\ \bar{\mathbf{c}}^t \end{pmatrix} = 0$$

where $\bar{c} = c_{m+1} \cdots c_n$, or

$$(\mathbf{A}^t \quad \mathbf{I}_{n-m})\begin{pmatrix} \mathbf{a} \\ \bar{\mathbf{c}}^t \end{pmatrix} = 0 \qquad \text{or} \qquad \mathbf{A}^t\mathbf{a}^t + \mathbf{I}_{n-m}\bar{\mathbf{c}}^t = 0$$

This implies that $\bar{\mathbf{c}} = \mathbf{aA}$. Therefore,

$$\mathbf{c} = (\mathbf{a} \quad \bar{\mathbf{c}}) = (\mathbf{aI}_m \quad \mathbf{aA}) = \mathbf{a}(\mathbf{I}_m \quad \mathbf{A}) = \mathbf{aG}$$

showing thereby that c is the code word corresponding to the message word a in the code defined by the generator matrix $\mathbf{G}$.

This proves that codes defined by the parity check matrix $\mathbf{H}$ and the generator matrix $\mathbf{G}$ are identical.

Suppose that to the generator matrix $\mathbf{G} = (\mathbf{I}_m \quad \mathbf{A})$ corresponds another parity check matrix $\mathbf{H}_1 = (\mathbf{B} \quad \mathbf{I}_{n-m})$. Let e^i be the message word with 1 in the ith position and 0 everywhere else. The corresponding code word is $\mathbf{e}^i\mathbf{G}$, i.e. the ith row of $\mathbf{G}$. We may write $\mathbf{e}^i\mathbf{G} = (\mathbf{e}^i \quad \tilde{\mathbf{e}}^i)$ where $\tilde{\mathbf{e}}^i$ is the ith row of $\mathbf{A}$. Since $\mathbf{H}_1$ is a parity check matrix of the code defined by $\mathbf{G}$,

$$\mathbf{H}_1(\mathbf{e}^i \quad \tilde{\mathbf{e}}^i)^t = 0$$

or

$$(\mathbf{B} \quad \mathbf{I}_{n-m})\begin{pmatrix} (\mathbf{e}^i)^t \\ (\tilde{\mathbf{e}}^i)^t \end{pmatrix} = 0$$

or

$$\mathbf{B}(\mathbf{e}^i)^t + (\tilde{\mathbf{e}}^i)^t = 0$$

so that $(\tilde{\mathbf{e}}^i)^t = \mathbf{B}(\mathbf{e}^i)^t$ matches the ith column of $\mathbf{B}$ and $\tilde{\mathbf{e}}^i$ matches the ith row of $\mathbf{B}^t$. Hence the ith row of $\mathbf{A}$ equals the ith column of $\mathbf{B}$. Since this is true for every i, $1 \le i \le m$, we have $\mathbf{B} = \mathbf{A}^t$ and so, $\mathbf{H}_1 = \mathbf{H}$. Hence, corresponding to a given generator matrix $\mathbf{G}$, there corresponds a uniquely determined parity check matrix $\mathbf{H} = (\mathbf{A}^t \quad \mathbf{I}_{n-m})$.

If we had started with a given parity check matrix $\mathbf{H}$, the above argument shows that the corresponding generator matrix is also uniquely determined.

Exercise 1.2

1. Proceeding by first principles (i.e. without using the above theorem) obtain the parity check matrices of the matrix codes given by the following

generator matrices:

(a)

$$\begin{pmatrix} 1 & 0 & 0 & 1 & 1 & 0 \\ 0 & 1 & 0 & 0 & 1 & 1 \\ 0 & 0 & 1 & 1 & 1 & 1 \end{pmatrix}$$

(b)

$$\begin{pmatrix} 1 & 0 & 0 & 1 & 1 & 1 \\ 0 & 1 & 0 & 0 & 1 & 1 \\ 0 & 0 & 1 & 1 & 1 & 0 \end{pmatrix}$$

(c)

$$\begin{pmatrix} 1 & 0 & 0 & 0 & 1 & 1 \\ 0 & 1 & 0 & 1 & 1 & 1 \\ 0 & 0 & 1 & 1 & 1 & 0 \end{pmatrix}$$

(d)

$$\begin{pmatrix} 1 & 0 & 0 & 0 & 1 & 1 & 0 \\ 0 & 1 & 0 & 0 & 0 & 1 & 1 \\ 0 & 0 & 1 & 0 & 1 & 0 & 1 \\ 0 & 0 & 0 & 1 & 1 & 1 & 1 \end{pmatrix}$$

(e)

$$\begin{pmatrix} 1 & 0 & 0 & 0 & 1 & 1 & 1 \\ 0 & 1 & 0 & 0 & 0 & 1 & 1 \\ 0 & 0 & 1 & 0 & 1 & 0 & 1 \\ 0 & 0 & 0 & 1 & 1 & 1 & 1 \end{pmatrix}$$

2. Determine the number of errors that are (a) detected, and (b) corrected by the matrix codes defined by the generator matrices given in question 1 above.
3. Define a code $E: \mathbb{B}^3 \to \mathbb{B}^6$ with the parity check matrix

(a)

$$\mathbf{H} = \begin{pmatrix} 1 & 1 & 0 & 1 & 0 & 0 \\ 1 & 0 & 0 & 0 & 1 & 0 \\ 1 & 1 & 1 & 0 & 0 & 1 \end{pmatrix}$$

(b)

$$\mathbf{H} = \begin{pmatrix} 1 & 1 & 0 & 1 & 0 & 0 \\ 1 & 1 & 1 & 0 & 1 & 0 \\ 1 & 0 & 0 & 0 & 0 & 1 \end{pmatrix}$$

(c)

$$\mathbf{H} = \begin{pmatrix} 1 & 1 & 0 & 1 & 0 & 0 \\ 1 & 1 & 1 & 0 & 1 & 0 \\ 1 & 0 & 1 & 0 & 0 & 1 \end{pmatrix}$$

Determine all the code words of the three codes. Does each of these codes correct all single errors?

Definition 1.15 – dual codes

Let C be the (m, n) code with generator matrix $\mathbf{G} = (\mathbf{I}_m \quad \mathbf{A})$. Then $(n-m, n)$ matrix code defined by the parity check matrix $\mathbf{H} = (\mathbf{A} \quad \mathbf{I}_m)$ is called the **dual code** of C and it is denoted by $C^{\perp}$.

For example, if C is the code defined by the matrix

$$\mathbf{G} = \begin{pmatrix} 1 & 0 & 1 & 1 & 0 \\ 0 & 1 & 1 & 0 & 1 \end{pmatrix}$$

all the code words of this code are

0 0 0 0 0   1 0 1 1 0   0 1 1 0 1   1 1 0 1 1

The minimum distance of the code is 3 and so it is a double error detecting, single error correcting code. Writing $\mathbf{G} = (\mathbf{I} \quad \mathbf{A})$ we find that $\mathbf{H} = (\mathbf{A} \quad \mathbf{I})$. The code words of the code with $\mathbf{H}$ as the parity check matrix are the same as the code words of the code with generator matrix

$$\mathbf{G}_1 = (\mathbf{I}_3 \quad \mathbf{A}^t) = \begin{pmatrix} 1 & 0 & 0 & 1 & 1 \\ 0 & 1 & 0 & 1 & 0 \\ 0 & 0 & 1 & 0 & 1 \end{pmatrix}$$

Hence all the code words are

$$\begin{array}{ll} 0\,0\,0 \longrightarrow 0\,0\,0\,0\,0 & 0\,1\,1 \longrightarrow 0\,1\,1\,1\,1 \\ 0\,0\,1 \longrightarrow 0\,0\,1\,0\,1 & 1\,0\,1 \longrightarrow 1\,0\,1\,1\,0 \\ 0\,1\,0 \longrightarrow 0\,1\,0\,1\,0 & 1\,1\,0 \longrightarrow 1\,1\,0\,0\,1 \\ 1\,0\,0 \longrightarrow 1\,0\,0\,1\,1 & 1\,1\,1 \longrightarrow 1\,1\,1\,0\,0 \end{array}$$

The minimum distance of the code is 2 and so it is single error detecting. It is not a single error correcting code; for example, for the received word 1 1 1 0 1 we have

$$1\,1\,1\,0\,0 + 0\,0\,0\,0\,1 = 1\,1\,0\,0\,1 + 0\,0\,1\,0\,0$$

it is at equal distance from two code words and so we are unable to decide about the code word transmitted.

Incidentally, the above also shows that the dual of a single error correcting code need not be single error correcting.

Exercise 1.3

1. Find generator and parity check matrices for dual codes to the codes in questions 2 and 3 of Exercise 1.2.
2. Find a code $E: \mathbb{B}^2 \to \mathbb{B}^6$ with minimum distance 4.
3. Describe the generator matrices **G** whose associated parity check matrices **H** will correctly decode all single errors. (We answer this question in the following theorem.)

Theorem 1.7
An (m, n)-code with generator matrix $\mathbf{G} = (\mathbf{I}_m \quad \mathbf{A})$ will decode all single errors correctly iff all the rows of **A** are distinct and the weight of each one of them is at least 2.

Proof
The parity check matrix of this code is $\mathbf{H} = (\mathbf{A}^t \quad \mathbf{I}_{n-m})$. But the parity check matrix **H** decodes all single errors correctly iff all the columns of **H** are non-zero and distinct. But this is so iff all the columns of $\mathbf{A}^t$ are distinct and each one of them is of weight at least 2. This is so iff the rows of **A** are all distinct and each one of them is of weight at least 2.

This theorem may be restated as follows:

Theorem 1.8
An (m, n)-code with generator matrix **G** will decode all single errors correctly iff the distance between any two rows of **G** is at least 3 and each one of them is of weight at least 3.

Let **H** be a parity check matrix and **C** be the (m, n)-code defined by **H**. Then, we known that C is a gr oup code. C is in fact a subgroup of $\mathbb{B}^n$. We then have the following proposition.

Proposition 1.3
Words $x, y \in \mathbb{B}^n$ are in the same coset of C iff they have the same syndrome.

Proof
x, y are in the same coset iff $y = x + c$ for some code word $c \in C$. But this is so iff $x + y = c \in C$. Now

$$
\begin{aligned}
x + y \in C \text{ iff } & \mathbf{H}(\mathbf{x} + \mathbf{y})^t = 0 \\
& \Leftrightarrow \mathbf{H}(\mathbf{x}^t + \mathbf{y}^t) = 0 \\
& \Leftrightarrow \mathbf{H}\mathbf{x}^t + \mathbf{H}\mathbf{y}^t = 0 \\
& \Leftrightarrow \mathbf{H}\mathbf{x}^t = \mathbf{H}\mathbf{y}^t
\end{aligned}
$$

Remark

Throughout this chapter, we restricted ourselves to the field $\mathbb{B}$ of two elements. This was done to keep the presentation simple and for having a concrete picture. However, most of the material could equally well have been developed over any finite field.

Let F be a finite field and $F^{(n)}$ be the set of all sequences of length n with entries in the field F. By the weight (or **Hamming weight**) wt(a) of a word a we mean the number of non-zero entries in a. Also for $a = a_1a_2\text{---}a_n$, $b = b_1b_2\text{---}b_n$ in $F^{(n)}$, we define the distance $d(a, b)$ between a and b to be the number of i's for which $a_i \neq b_i$. Lemma 1.1 then becomes $d(a, b) = \mathrm{wt}(a - b) \forall a, b \in F^{(n)}$ while its Corollary, Lemma 1.2, Theorems 1.1, 1.2, 1.3 remain valid without change (of course, slight changes in the proofs are needed). We could have, similarly, defined and studied matrix codes and group codes over F and almost all the results of section 1.2 and section 1.3 (up to Theorem 1.5) are valid in the general case. Statement of Theorem 1.6 needs slight change.

Exercise 1.4

Prove all the results stated in the above remark and others that are valid over any finite field.

2

Polynomial codes

We have considered earlier one algebraic technique for obtaining/studying codes namely that of **matrix techniques**. We now introduce another important algebraic technique (not totally unrelated to the matrix techniques) for studying codes. This technique is through **polynomial multiplication**. The codes thus obtained are called polynomial codes. Most of the important codes studied today are polynomial codes.

2.1 DEFINITION OF VECTOR SPACE AND POLYNOMIAL RING

Definition 2.1 – vector spaces

Let F be a field. A non-empty set V is called a **vector space over F** if:

(i) there is defined an addition in V, w.r.t. which V is an Abelian group;

(ii) for every $a \in F$, $v \in V$, there is defined a unique element $av \in V$ such that for $v, v_1, v_2 \in V$, $a, b \in F$ and 1 the identity of F,

  (a) $a(v_1 + v_2) = av_1 + av_2$

  (b) $(a + b)v = av + bv$

  (c) $(ab)v = a(bv)$

  (d) $1 \times v = v$

Let V be a vector space over a field F. A set $\{v_1, v_2, \ldots, v_n\}$ of elements of V is called **linearly independent** if whenever

$$a_1 v_1 + a_2 v_2 + \cdots + a_n v_n = 0$$

where $a_1, a_2, \ldots, a_n$ are in F, then

$$a_1 = a_2 = \cdots = a_n = 0$$

If the elements $v_1, v_2, \ldots, v_n \in V$ are linearly independent over F and every element v of V can be expressed in the form $a_1 v_1 + \cdots + a_n v_n$, $a_i \in F$, $1 \le i \le n$, then $\{v_1, \ldots, v_n\}$ is called a **basis of V**. Also then V is said to be of **dimension** n over F and we express it by $\dim V = n$.

Let V, W be two vector spaces over the same field F. A map $f: V \to W$ is called an **isomorphism** if the map f is one–one and onto and $\forall v, v_1, v_2 \in V, a \in F$,

$$f(v_1 + v_2) = f(v_1) + f(v_2) \qquad f(av) = af(v)$$

Proposition 2.1
If V, W are vector spaces of equal finite dimension over the same field F, then V and W are isomorphic.

Proof
Let $\dim V = \dim W = n < \infty$. Let $\{x_1, \ldots, x_n\}$ be a basis of V over F and $\{y_1, \ldots, y_n\}$ be a basis of W over F. Since every element of V can be uniquely written as $a_1x_1 + \cdots + a_nx_n$, with $a_i \in F$, the map $\theta: V \to W$ given by

$$\theta(a_1x_1 + \cdots + a_nx_n) = a_1y_1 + \cdots + a_ny_n \quad a_i \in F$$

is well defined. It is fairly easy to see that θ is a homomorphism. Also, if

$$\theta(a_1x_1 + \cdots + a_nx_n) = 0$$

then

$$a_1y_1 + \cdots + a_ny_n = 0$$

which implies that

$$a_1 = \cdots = a_n = 0$$

the elements $y_1, \ldots, y_n$ being linearly independent.

Thus $a_1x_1 + \cdots + a_nx_n = 0$ and θ is one–one. As every element of W is of the form

$$a_1y_1 + \cdots + a_ny_n = \theta(a_1x_1 + \cdots + a_nx_n)$$

for some $a_1, \ldots, a_n \in F$, θ is onto as well and, hence, an isomorphism. ■

Let F be a field and X be a variable. Let $F[X]$ denote the set of all polynomials in the variable X over F, i.e. $F[X]$ is the set of all finite formal sums of the form

$$a_0 + a_1X + \cdots + a_nX^n$$

where $a_i \in F$. The set $F[X]$ with the usual addition and multiplication, i.e. for

$$a(X) = a_0 + a_1X + \cdots + a_mX^m$$

$$b(X) = b_0 + b_1X + \cdots + b_nX^n$$

in $F[X]$

$$a(X) + b(X) = c_0 + c_1X + c_2X^2 + \cdots$$

and

$$a(X)b(X) = d_0 + d_1X + \cdots + d_{m+n}X^{m+n}$$

where $c_i = a_i + b_i \forall i$ and

$$d_j = a_0 b_j + a_1 b_{j-1} + \cdots + a_j b_0$$

it being understood that $a_i = 0$ for $i > m$ and $b_i = 0$ for $i > n$, becomes a commutative ring with identity. The elements

$$a(X) = a_0 + a_1 X + \cdots + a_m X^m$$

of $F[X]$ are called **polynomials** and if $a_m \neq 0$, we say that $a(X)$ is a polynomial of degree m (expressed as $\deg a(X) = m$). Since the product of non-zero elements in F is non-zero, it follows from the definition of multiplication in $F[X]$ that for non-zero $a(X)$, $b(X)$ in $F[X]$, $a(X)b(X) \neq 0$ and that

$$\deg(a(X)b(X)) = \deg a(X) + \deg b(X)$$

(Recall that a commutative ring R in which $ab = 0$, $a, b \in R$, implies $a = 0$ or $b = 0$ is called an **integral domain**.) Observe that a polynomial

$$a(X) = a_0 + a_1 X + \cdots + a_m X^m = 0$$

iff $a_0 = a_1 = \cdots = a_m = 0$. Thus for any $m \geq 1$, the elements $1, X, X^2, \ldots, X^m$ are linearly independent over F.

We could have defined a polynomial ring $R[X]$ in the variable X over any commutative ring R with identity but $R[X]$ is not, in general, an integral domain.

2.2 POLYNOMIAL CODES

Let F be any field and $F[X]$ be the polynomial ring in the variable X over F. Let n be a given positive integer. With a polynomial

$$a(X) = a_0 + a_1 X + \cdots + a_{n-1} X^{n-1}$$

of degree at most $n - 1$, we can associate an ordered n-tuple or a word-ordered sequence

$$a = (a_0, a_1, \ldots, a_{n-1})$$

of length n and conversely, with every word

$$a = (a_0, a_1, \ldots, a_{n-1}) \quad a_i \in F$$

we can associate a polynomial

$$a(X) = a_0 + a_1 X + \cdots + a_{n-1} X^{n-1}$$

of degree at most $n - 1$. Observe that for the polynomial

$$a_0 + a_1 X + \cdots + a_m X^m$$

of degree $m \leq n - 1$, we first rewrite $a(X)$, by introducing zero coefficients, as

$$a_0 + a_1 X + \cdots + a_m X^m + a_{m+1} X^{m+1} + \cdots + a_{n-1} X^{n-1}$$

with $a_i = 0$ for $m+1 \le i \le n-1$ and then associate with it the word

$$(a_0, a_1, \ldots, a_m, a_{m+1}, \ldots, a_{n-1})$$

of length n.

Let, as usual, $F^{(n)}$ denote the set of all words of length n with entries in F. Recall that $F^{(n)}$ is an n-dimensional vector space over F. Let $\mathscr{P}_n(X)$ denote the set of all polynomials in $F[X]$ which are of degree at most n. It is clear that the sum of two polynomials of degree at most n is again a polynomial of degree at most n and the multiplication of a polynomial of degree at most n with an element of F is again a polynomial of degree at most n. With these compositions (addition and scalar multiplication), $\mathscr{P}_n(X)$ becomes a vector space over F. The elements $1, X, \ldots, X^n$ of $\mathscr{P}_n(X)$ are such that any element of $\mathscr{P}_n(X)$ is a linear combination of these elements over F and that these elements are linearly independent over F. Thus, $\mathscr{P}_n(X)$ is a vector space of dimension $n+1$ over F. Two vector spaces over the same field and of equal dimensions are always isomorphic. We thus have the following theorem.

Theorem 2.1

The vector spaces $\mathscr{P}_{n-1}(X)$ and $F^{(n)}$ are isomorphic with the element

$$a(X) = a_0 + a_1 X + \cdots + a_{n-1} X^{n-1}$$

corresponding to the element

$$a = (a_0, a_1, \ldots, a_{n-1})$$

under this isomorphism.

Remark

In view of the above isomorphism, we use

$$a(X) = a_0 + a_1 X + \cdots + a_{n-1} X^{n-1}$$

and

$$a = (a_0, a_1, \ldots, a_{n-1}) \quad a_i \in F$$

interchangeably.

Definition 2.2

Let $g(X) = g_0 + g_1 X + \cdots + g_k X^k \in F[X]$ be a polynomial of degree at most k. The **polynomial code** with encoding (generating) polynomial $g(X)$ encodes each message word

$$a = (a_0, a_1, \ldots, a_{m-1})$$

of length m into the code word

$$b = (b_0, b_1, \ldots, b_{m+k-1})$$

which corresponds to the code polynomial

$$b(X) = b_0 + b_1 X + \cdots + b_{m+k-1} X^{m+k-1} = a(X)g(X)$$

where

$$a(X) = a_0 + a_1 X + \cdots + a_{m-1} X^{m-1}$$

Remark

If $g_0 = 0$ in the encoding polynomial $g(X)$, then the first entry in every code word is 0 – thus this entry gives no useful information about the code. Similarly, if $g_k = 0$ in the encoding polynomial $g(X)$, then the last entry in every code word in the code generated by $g(X)$ is 0 and thus the last code word digit is wasted. To avoid this waste, we shall assume throughout that $g_0 \neq 0$ and $g_k \neq 0$.

Proposition 2.2

The polynomial code of length $n = m + k$ generated by the polynomial $g(X) = g_0 + g_1 X + \cdots + g_k X^k$ is a subspace of $F^{(n)}$.

Proof

Let

$$a = (a_0, a_1, \ldots, a_{m-1}) \qquad b = (b_0, b_1, \ldots, b_{m-1})$$

be two message words. Let

$$a(X) = a_0 + a_1 X + \cdots + a_{m-1} X^{m-1}$$

$$b(X) = b_0 + b_1 X + \cdots + b_{m-1} X^{m-1}$$

be the corresponding message polynomials and $\alpha, \beta \in F$. Then the code polynomial corresponding to the message polynomial $\alpha a(X) + \beta b(X)$ is

$$[\alpha a(X) + \beta b(X)]g(X) = (\alpha a(X))g(X) + (\beta b(X))g(X) = \alpha(a(X)g(X)) + \beta(b(X)g(X))$$

■

Since a vector space is first a group under addition, a polynomial code is always a group code. For a group code, we have seen earlier (Proposition 1.2) that the minimum distance equals the minimum of the weights of non-zero code words. We thus have a similar proposition for polynomial codes.

Proposition 2.3

The minimum distance of a polynomial code with encoding polynomial $g(X)$ is the minimum weight $\mathrm{wt}(a(X)g(X))$ of the non-zero code polynomials $a(X)g(X)$.

Of course, by the weight of a code polynomial, we mean the number of non-zero terms in the polynomial. For example:

$$\mathrm{wt}(X + X^2 + X^3 + X^6) = 4 \qquad \mathrm{wt}(1 + X + X^3) = 3$$

The following simple observation will be useful later.

Proposition 2.4
A polynomial with coefficients in $\mathbb{B}$ is divisible by $1+X$ iff it has an even number of terms.

Proof
Let $f(X)=a_0+a_1X+\cdots+a_nX^n, a_i\in\mathbb{B}$ be a polynomial which is divisible by $1+X$. Then there exists a polynomial $b(X)\in\mathbb{B}[X]$ such that

$$f(X)\equiv(1+X)b(X)$$

Since this is an identity, taking $X=1$ on both sides of this identity gives

$$a_0+a_1+\cdots+a_n=2b(1)=0$$

in $\mathbb{B}$.

The field $\mathbb{B}$ being of characteristic 2, this is possible only if the number of non-zero a_i's is even.

Conversely, let

$$f(X)=a_0+a_1X+\cdots+a_nX^n \qquad a_i\in\mathbb{B}$$

be a polynomial having an even number of non-zero terms; say

$$f(X)=X^{i_1}+X^{i_2}+\cdots+X^{i_{2k}}$$

where $i_1<i_2<\cdots<i_{2k}$. We rewrite

$$f(X)=(X^{i_1}+X^{i_2})+(X^{i_3}+X^{i_4})+\cdots+(X^{i_{2k-1}}+X^{i_{2k}})$$

Since for $i<j$,

$$\begin{aligned}X^i+X^j&=X^i(1+X^{j-i})\\&=X^i(1+X)(1+X+\cdots+X^{j-i-1})\end{aligned}$$

$1+X\,|\,X^i+X^j$. Thus $1+X$ divides every bracketed term in $f(X)$ and, therefore, $f(X)$ is divisible by $1+X$.

Theorem 2.2
If $g(X)\in\mathbb{B}[X]$ divides no polynomial of the form X^k-1 for $k<n$, then the binary polynomial code of length n generated by $g(X)$ has minimum distance at least 3.

Proof
Let

$$g(X)=g_0+g_1X+\cdots+g_rX^r$$

with $g_i\in\mathbb{B}$ and $g_0\neq 0$, $g_r\neq 0$. Let $m=n-r$ so that every message polynomial is of degree at most $m-1$. From the definition of polynomial codes, it follows that every code polynomial is divisible by $g(X)$. Also polynomial codes being group codes, we have to prove that there is no non-zero code word of weight at most 2.

Suppose that the theorem fails. Then there exists a code polynomial $b(X)$ with at most two non-zero entries. Observe that $g(X) \nmid X^k - 1$ for any $k < n$ implies in particular that $g(X)$ is not a constant polynomial.

Case (i): $b(X) = X^i + X^j, i < j$

The code being of length n, we have $j < n$. Therefore, $0 < j - i < n$. But $g(X) | b(X)$ shows that $g(X) | X^i(1 + X^{j-i})$. Also $g_0 \neq 0$ implies that $X \nmid g(X)$. Strictly speaking, we can say that X and $g(X)$ are **relatively coprime**. Therefore $g(X) | 1 + X^{j-i}$ which contradicts the hypothesis.

Case (ii): $b(X) = X^i, i < n$

But, as seen above, $g(X) \nmid X^i$ and we have a contradiction. We have thus proved that there is no code polynomial with at most two non-zero terms and this establishes the theorem.

Examples 2.1

Case (i)

Consider first the binary polynomial code of length 6 generated by the polynomial $1 + X + X^3$. Then the message polynomials are of degree at most 2. For any message word (a_0, a_1, a_2) the corresponding message polynomial is

$$a(X) = a_0 + a_1 X + a_2 X^2$$

Therefore, the corresponding code polynomial is

$$\begin{aligned} b(X) &= (a_0 + a_1 X + a_2 X^2)(1 + X + X^3) \\ &= a_0 + (a_0 + a_1)X + (a_1 + a_2)X^2 + (a_0 + a_2)X^3 + a_1 X^4 + a_2 X^5 \end{aligned}$$

Thus, the code words of this code are

$$\begin{array}{ccc} (a_0, a_1, a_2) & \longrightarrow & (a_0, a_0 + a_1, a_1 + a_2, a_0 + a_2, a_1, a_2) \\ 0\ 0\ 0 & \longrightarrow & 0\ 0\ 0\ 0\ 0\ 0 \\ 0\ 0\ 1 & \longrightarrow & 0\ 0\ 1\ 1\ 0\ 1 \\ 0\ 1\ 0 & \longrightarrow & 0\ 1\ 1\ 0\ 1\ 0 \\ 1\ 0\ 0 & \longrightarrow & 1\ 1\ 0\ 1\ 0\ 0 \\ 0\ 1\ 1 & \longrightarrow & 0\ 1\ 0\ 1\ 1\ 1 \\ 1\ 0\ 1 & \longrightarrow & 1\ 1\ 1\ 0\ 0\ 1 \\ 1\ 1\ 0 & \longrightarrow & 1\ 0\ 1\ 1\ 1\ 0 \\ 1\ 1\ 1 & \longrightarrow & 1\ 0\ 0\ 0\ 1\ 1 \end{array}$$

The minimum distance of this code is 3.

Observe that the encoding polynomial $1 + X + X^3$ does not divide

$$X^4 + 1 = (X + 1)^4 \quad \text{and} \quad X^5 + 1 = (X + 1)(X^4 + X^3 + X^2 + X + 1)$$

and our computation of minimum distance confirms the result of Theorem 2.2.

Case (ii)
Next, we consider the binary polynomial code of length 7 generated by the polynomial $1 + X^2 + X^3$. The message words here are of length 4. The code words of this code are

$$(a_0, a_1, a_2, a_3) \longrightarrow (a_0, a_1, a_0 + a_2, a_0 + a_1 + a_3, a_1 + a_2, a_2 + a_3, a_3)$$

0 0 0 0 ⟶ 0 0 0 0 0 0 0
0 0 0 1 ⟶ 0 0 0 1 0 1 1
0 0 1 0 ⟶ 0 0 1 0 1 1 0
0 1 0 0 ⟶ 0 1 0 1 1 0 0
1 0 0 0 ⟶ 1 0 1 1 0 0 0
0 0 1 1 ⟶ 0 0 1 1 1 0 1
0 1 0 1 ⟶ 0 1 0 0 1 1 1
1 0 0 1 ⟶ 1 0 1 0 0 1 1
0 1 1 0 ⟶ 0 1 1 1 0 1 0
1 0 1 0 ⟶ 1 0 0 1 1 1 0
1 1 0 0 ⟶ 1 1 1 0 1 0 0
0 1 1 1 ⟶ 0 1 1 0 0 0 1
1 0 1 1 ⟶ 1 0 0 0 1 0 1
1 1 0 1 ⟶ 1 1 1 1 1 1 1
1 1 1 0 ⟶ 1 1 0 0 0 1 0
1 1 1 1 ⟶ 1 1 0 1 0 0 1

Observe that the minimum distance of this code is 3.

Case (iii)
Finally, we consider the binary polynomial code of length 7, generated by the polynomial $1 + X + X^3$. The message words are again of length 4. The code words of this code are

$$(a_0, a_1, a_2, a_3) \longrightarrow (a_0, a_0 + a_1, a_1 + a_2, a_0 + a_2 + a_3, a_1 + a_3, a_2, a_3)$$

0 0 0 0 ⟶ 0 0 0 0 0 0 0
0 0 0 1 ⟶ 0 0 0 1 1 0 1
0 0 1 0 ⟶ 0 0 1 1 0 1 0
0 1 0 0 ⟶ 0 1 1 0 1 0 0
1 0 0 0 ⟶ 1 1 0 1 0 0 0
0 0 1 1 ⟶ 0 0 1 0 1 1 1
0 1 0 1 ⟶ 0 1 1 1 0 0 1
1 1 1 0 ⟶ 1 0 0 0 1 1 0

(contd)

$$
\begin{array}{llll}
1\ 0\ 0\ 1 & \longrightarrow & 1\ 1\ 0\ 0\ 1\ 0\ 1 \\
0\ 1\ 1\ 0 & \longrightarrow & 0\ 1\ 0\ 1\ 1\ 1\ 0 \\
1\ 0\ 1\ 0 & \longrightarrow & 1\ 1\ 1\ 0\ 0\ 1\ 0 \\
1\ 1\ 0\ 0 & \longrightarrow & 1\ 0\ 1\ 1\ 1\ 0\ 0 \\
0\ 1\ 1\ 1 & \longrightarrow & 0\ 1\ 0\ 0\ 0\ 1\ 1 \\
1\ 0\ 1\ 1 & \longrightarrow & 1\ 1\ 1\ 1\ 1\ 1\ 1 \\
1\ 1\ 0\ 1 & \longrightarrow & 1\ 0\ 1\ 0\ 0\ 0\ 1 \\
1\ 1\ 1\ 1 & \longrightarrow & 1\ 0\ 0\ 1\ 0\ 1\ 1
\end{array}
$$

The minimum distance of this code is again 3. The encoding polynomial $1 + X + X^3$ does not divide $X^k + 1$ for $k < 7$ and the minimum distance is in conformity with Theorem 2.2.

In a group code, an error vector $\mathbf{e} = (e_0 \quad e_1 \quad \cdots \quad e_{n-1})$ goes undetected iff it is a code word. (Note that, as before, the notation e refers to a series and $\mathbf{e}$ refers to the vector formed using the elements in the series.) We thus have the following proposition.

Proposition 2.5

An error vector $\mathbf{e} = (e_0 \quad e_1 \quad \cdots \quad e_{n-1})$ of a polynomial (m, n) code with generator $g(X)$ is undetected iff the associated error polynomial

$$e(X) = e_0 + e_1 X + \cdots + e_{n-1} X^{n-1}$$

is a multiple of $g(X)$.

Definition 2.3

We say that the **exponent** of a polynomial $g(X) \in \mathbb{B}[X]$ is the least positive integer e such that $g(X) | X^e - 1$.

Definition 2.4

Two errors occurring at adjacent positions are called a **double error**.

Observe that $e(X) = X + X^3$ is not a double error, while $e(X) = X^2 + X^3$ or $e(X) = X + X^2$ are double errors.

Theorem 2.3

In a binary polynomial (m, n)-code, if the encoding polynomial

$$g(X) = (1 + X)h(X)$$

where $h(X)$ has exponent $e > n$, then any combination of two single or double errors will be detected.

Proof

Since the exponent of $h(X)$ is $e > n$, $h(X)$ is neither a constant polynomial nor a multiple of X.

If the error polynomial is

$$e(X) = X^i + X^j \quad i+1 \leq j \leq n-1$$

then $e(X) = X^i(1 + X^{j-i})$, $j - i < n$. Therefore, $h(X)$ being of exponent $e > n$ does not divide $X^i(1 + X^{j-i}) = e(X)$. Thus, $e(X)$ is not a code polynomial and, hence, this error pattern is detected.

If

$$e(X) = X^i + X^{i+1} + X^j \quad i+1 < j$$

or

$$e(X) = X^i + X^j + X^{j+1} \quad i < j$$

or

$$e(X) = X^j$$

then $1 + X \nmid e(X)$ (Proposition 2.4) and hence $g(X) \nmid e(X)$. Therefore, the three error patterns are detected. Finally, if the error polynomial is

$$e(X) = X^i + X^{i+1} + X^j + X^{j+1} = (1 + X)(X^i + X^j) \quad i < j < n$$

then $h(X) \nmid X^i + X^j$ (as seen above) and so $g(X) \nmid e(X)$. Thus, this error pattern is also detected.

Example 2.2

Consider the binary polynomial code of length 5 generated by the polynomial

$$g(X) = X^2 + 1 = (X + 1)(X + 1)$$

Observe that in this code, the following combinations of two single or double errors go undetected:

$$X, X^3 \quad 1, X^2 \quad X^2, X^4 \quad 1 + X, X^2 + X^3 \quad X + X^2, X^3 + X^4 \quad 1, X^4$$
$$1 + X, X^3 + X^4$$

However, any combination of one single and one double error is always detected. In fact, we can make the following general observation (in view of Proposition 2.4):

Remark

In a binary polynomial code with encoding polynomial $g(X) = (1 + X)h(X)$, any combination of one single and one double error is always detected.

Proposition 2.6

In a binary polynomial (m, n)-code with encoding polynomial $g(X)$, every code word is of even weight iff $1 + X \mid g(X)$.

Proof

If $g(X) = (1 + X)h(X)$, then $1 + X$ divides any code polynomial

$$b(X) = b_0 + b_1 X + \cdots + b_{n-1} X^{n-1}$$

and it follows from Proposition 2.4 that the corresponding code word $b = (b_0, b_1, \ldots, b_{n-1})$ is of even weight.

Conversely, suppose that $b(X)$ is a code polynomial corresponding to a code word b of even weight. Then, on pairing adjacent terms, we see that $b(X)$ is a sum of finite number of sums of the form $X^i + X^j, i < j$. But

$$X^i + X^j = X^i(1 + X^{j-i}) = X^i(1 + X)(1 + X + \cdots + X^{j-i-1})$$

Therefore, $b(X)$ is divisible by $1 + X$. In particular, the code polynomial $Xg(X)$ corresponding to the message polynomial

$$a(X) = a_0 + a_1X + \cdots + a_{m-1}X^{m-1} \quad a_i = 0 \forall i \neq 1, a_1 = 1$$

is also divisible by $1 + X$. But then it follows that $1 + X | g(X)$.

Exercise 2.1

1. Is the result of Proposition 2.4 true if we are working over a field of odd order? Justify.
2. Compute the minimum distance of the binary polynomial code of length 5 generated by $1 + X + X^2$.
3. Is the converse of Theorem 2.2 true? If the binary polynomial code of length n generated by $g(X)$ has minimum distance at least 3, is it always true that $g(X)$ divides no polynomial of the form $X^k - 1$ for $k < n$?
4. Find the minimum distance of the binary polynomial code of length 8 generated by the polynomials:
   (i) $1 + X + X^3$
   (ii) $1 + X^2 + X^3$
5. Compute all the code words of the polynomial code of length:
   (i) 3
   (ii) 4, generated by the polynomial $X^2 + 1$ over the field of 3 elements.
6. If F is a field of 3 elements and $g(X) \in F[X]$ divides no polynomial of the form $X^k - 1$ or $X^k + 1$ for $k < n$, prove that the polynomial code of length n generated by $g(X)$ over F has minimum distance at least 3.
7. Compute the exponent of the polynomial $a(X) \in \mathbb{B}[X]$ when
   (i) $a(X) = 1 + X + X^2$
   (ii) $a(X) = 1 + X + X^3$
   (iii) $a(X) = 1 + X^2 + X^3$
   (iv) $a(X) = X + X^3$
   (v) $a(X) = 1 + X^3$
   (vi) $a(X) = 1 + X^2 + X^4$

2.3 GENERATOR AND PARITY CHECK MATRICES – GENERAL CASE

Starting with an $m \times n$ matrix $\mathbf{G} = (\mathbf{I}_m \quad \mathbf{A})$, where $\mathbf{A}$ is $m \times (n - m)$ matrix, we defined a code $\mathscr{C} = \{\mathbf{aG} | a \in \mathbb{B}^m\}$ of length n and called $\mathbf{G}$ a generator matrix of

$\mathscr{C}$. Also we found that $\mathbf{H} = (\mathbf{A}^t \quad \mathbf{I}_{n-m})$ has the property that $\mathbf{H}\mathbf{b}^t = 0$ for every $\mathbf{b} \in \mathscr{C}$ and called it the parity check matrix of the code $\mathscr{C}$. However, it is not essential that we insist on the first m columns of $\mathbf{G}$ to form the identity matrix $\mathbf{I}_m$ or that the last $(n-m)$ columns of $\mathbf{H}$ to form the identity matrix $\mathbf{I}_{n-m}$. But, in that case, a great deal of information about the code defined by $\mathbf{G}$ will be lost. Also, the relation between the generator matrix $\mathbf{G}$ and parity check matrix $\mathbf{H}$ of the same code is not as clear as in the earlier case.

All codes considered here will be over $\mathbb{B}$, the field of two elements.

Recall that the rank of an $m \times n$ matrix $\mathbf{A}$ is the maximum number of linearly independent rows or the maximum number of linearly independent columns of the matrix $\mathbf{A}$.

Definition 2.5

Let $\mathscr{C}$ be an (m, n)-code. If there exists a $m \times n$ matrix $\mathbf{G}$ of rank m such that $\mathscr{C} = \{\mathbf{aG} \mid a \in \mathbb{B}^m\}$ then $\mathbf{G}$ is called a **generator matrix** of the code $\mathscr{C}$. Also then $\mathscr{C}$ is called a **matrix code** generated by $\mathbf{G}$.

Definition 2.6

Let $\mathscr{C}$ be an (m, n)-code. If there exists an $(n-m) \times n$ matrix $\mathbf{H}$ of rank $n-m$ such that $\mathbf{H}\mathbf{b}^t = 0$ for all $\mathbf{b} \in \mathscr{C}$, then $\mathbf{H}$ is called a parity check matrix of $\mathscr{C}$.

With this generalized definition of a matrix code, we have the following theorem.

Theorem 2.4

A polynomial code is a matrix code.

Proof

Let $\mathscr{C}$ be a polynomial (m, n)-code with encoding polynomial

$$g(X) = g_0 + g_1 X + \cdots + g_k X^k$$

Then $n = m + k$. Let $\mathbf{G}$ be the $m \times n$ matrix

$$\mathbf{G} = \begin{pmatrix} g_0 & g_1 & \cdots & g_k & 0 & 0 & \cdots & 0 \\ 0 & g_0 & \cdots & g_{k-1} & g_k & 0 & \cdots & 0 \\ \vdots & \vdots & & \vdots & \vdots & & \ddots & \vdots \\ 0 & 0 & \cdots & 0 & g_0 & g_1 & \cdots & g_k \end{pmatrix}$$

in which the first row has initial entries $g_0, g_1, \ldots, g_k$ and the rest of the rows are obtained by giving a cyclic shift (anticlockwise) to the entries of the previous row until we arrive at a row in which the last $k+1$ entries are $g_0, g_1, \ldots, g_k$.

The determinant of the submatrix formed by taking the first m columns is $g_0^m \neq 0$ as $g_0 \neq 0$. Therefore, the rank of $\mathbf{G}$ is m. It is straightforward to check that the code word in the code generated by the matrix $\mathbf{G}$ corresponding to the message word

$$a = (a_0, a_1, \ldots, a_{m-1})$$

equals the code word corresponding to the message word a in the polynomial code generated by $g(X)$. Thus, the two codes are identical.

Example 2.3

(i) The generator matrix of the $(3, 6)$ polynomial code with encoding polynomial

$$g(X) = 1 + X + X^3$$

is

$$\mathbf{G} = \begin{pmatrix} 1 & 1 & 0 & 1 & 0 & 0 \\ 0 & 1 & 1 & 0 & 1 & 0 \\ 0 & 0 & 1 & 1 & 0 & 1 \end{pmatrix}$$

(ii) The generator matrix of the $(4, 7)$ polynomial code with encoding polynomial

$$g(X) = 1 + X^2 + X^3$$

is

$$\mathbf{G} = \begin{pmatrix} 1 & 0 & 1 & 1 & 0 & 0 & 0 \\ 0 & 1 & 0 & 1 & 1 & 0 & 0 \\ 0 & 0 & 1 & 0 & 1 & 1 & 0 \\ 0 & 0 & 0 & 1 & 0 & 1 & 1 \end{pmatrix}$$

(iii) The generator matrix of the $(4, 7)$ polynomial code with encoding polynomial

$$g(X) = 1 + X + X^3$$

is

$$\mathbf{G} = \begin{pmatrix} 1 & 1 & 0 & 1 & 0 & 0 & 0 \\ 0 & 1 & 1 & 0 & 1 & 0 & 0 \\ 0 & 0 & 1 & 1 & 0 & 1 & 0 \\ 0 & 0 & 0 & 1 & 1 & 0 & 1 \end{pmatrix}$$

(iv) Consider the $(4, 7)$ polynomial code with encoding polynomial

$$g(X) = 1 + X + X^3$$

so that its generator matrix is as seen in (iii) above. The code word corresponding to the message word

$$a = (a_0, a_1, a_2, a_3) \text{ is } (a_0, a_0 + a_1, a_1 + a_2, a_0 + a_2 + a_3, a_1 + a_3, a_2, a_3)$$

Let $\alpha_i, 0 \leq i \leq 6$, be elements from $\mathbb{B}$ such that

$$\alpha_0 a_0 + \alpha_1(a_0 + a_1) + \alpha_2(a_1 + a_2) + \alpha_3(a_0 + a_2 + a_3) + \alpha_4(a_1 + a_3) + \alpha_5 a_2 + \alpha_6 a_3 = 0$$

Then

$$a_0(\alpha_0+\alpha_1+\alpha_3)+a_1(\alpha_1+\alpha_2+\alpha_4)+a_2(\alpha_2+\alpha_3+\alpha_5)+a_3(\alpha_3+\alpha_4+\alpha_6)=0$$

Since this holds $\forall a \in \mathbb{B}^4$, we have

$$\alpha_0 + \alpha_1 + \alpha_3 = 0 \quad \alpha_1 + \alpha_2 + \alpha_4 = 0$$

$$\alpha_2 + \alpha_3 + \alpha_5 = 0 \quad \alpha_3 + \alpha_4 + \alpha_6 = 0$$

We need to find out α_i which satisfy these equations. Suppose $\alpha_0 = 0$. Then $\alpha_1 = \alpha_3, \alpha_4 = \alpha_5$ and $\alpha_2 = \alpha_6$. We may take two sets of values of α's as

$$\alpha_0 = 0 \quad \alpha_1 = \alpha_3 = 1 \quad \alpha_4 = \alpha_5 = 0 \quad \alpha_2 = \alpha_6 = 1$$

and

$$\alpha_0 = 0 \quad \alpha_1 = \alpha_3 = 1 \quad \alpha_4 = \alpha_5 = 1 \quad \alpha_2 = \alpha_6 = 0$$

In order to avoid a column of zeros, let us next suppose $\alpha_0 = 1$. Then $\alpha_1 + \alpha_3 = 1$. Suppose that $\alpha_1 = 1, \alpha_3 = 0$. Then the above equations reduce to

$$1 + \alpha_2 + \alpha_4 = 0 \quad \alpha_2 + \alpha_5 = 0 \quad \alpha_4 + \alpha_6 = 0$$

If $\alpha_2 = 0$, then $\alpha_4 = 1 = \alpha_6$ and $\alpha_5 = 0$. Thus one set of values is

$$\alpha_0 = \alpha_1 = \alpha_4 = \alpha_6 = 1$$

$$\alpha_2 = \alpha_3 = \alpha_5 = 0$$

Therefore, a parity check matrix for this code is

$$\mathbf{H} = \begin{pmatrix} 0 & 1 & 1 & 1 & 0 & 0 & 1 \\ 0 & 1 & 0 & 1 & 1 & 1 & 0 \\ 1 & 1 & 0 & 0 & 1 & 0 & 1 \end{pmatrix}$$

Another parity check matrix of this code is

$$\mathbf{H}_1 = \begin{pmatrix} 0 & 1 & 1 & 1 & 0 & 0 & 1 \\ 0 & 0 & 1 & 0 & 1 & 1 & 1 \\ 1 & 1 & 0 & 0 & 1 & 0 & 1 \end{pmatrix}$$

There are quite a few other parity check matrices of the code.

Exercises 2.2

1. Find two parity check matrices for the code of Example 2.3(i).
2. Find two parity check matrices of the (4, 7) polynomial code of Example 2.3(ii).

Consider the 3×6 matrix

$$\mathbf{G} = \begin{pmatrix} 1 & 1 & 0 & 1 & 0 & 0 \\ 0 & 1 & 1 & 0 & 1 & 1 \\ 1 & 0 & 1 & 1 & 1 & 1 \end{pmatrix}$$

Then we find that

$$(000)\mathbf{G} = 000000 = (111)\mathbf{G}$$
$$(001)\mathbf{G} = 101111 = (110)\mathbf{G}$$

and

$$(100)\mathbf{G} = 110100 = (011)\mathbf{G}$$

Thus, the map $a \rightarrow \mathbf{aG}$, $a \in \mathbb{B}^3$ is not one–one. This is so because the rows of $\mathbf{G}$ are linearly dependent or that the rank $(\mathbf{G}) < 3$. To avoid this eventuality and the unnecessary extra work involved, we insist on the $m \times n$ generator matrix to be of rank m.

We shall come to generator matrices again when we discuss linear codes.

3

Hamming codes

3.1 BINARY REPRESENTATION OF NUMBERS

Recall that our ordinary 'decimal' number system is the denary number system which means that the number system uses 10 as base. In this system we know that the number 947 actually equals: $9 \times 10^2 + 4 \times 10^1 + 7$. Similarly

$$\begin{aligned} 3\,450\,671 &= 3 \times 10^6 + 4 \times 10^5 + 5 \times 10^4 + 0 \times 10^3 + 6 \times 10^2 + 7 \times 10 + 1 \\ &= 3 \times 10^6 + 4 \times 10^5 + 5 \times 10^4 + 6 \times 10^2 + 7 \times 10 + 1 \end{aligned}$$

We can in fact define a number system with any number r, $1 \leq r \leq 9$, as base. In such a number system each place will be occupied by one of the numbers, $0, 1, \ldots, r-1$. In view of this, the number system with base $r = 1$ is not of any interest. With the coming into existence of computers the number system using 2 as base has gained considerable importance. Such a system of numbers is called a **binary number system** or a binary representation of numbers. Observe that any place in this representation of the numbers is occupied by either 0 or 1. A number *abcde* in this system represents the number

$$a \times 2^4 + b \times 2^3 + c \times 2^2 + d \times 2 + e$$

For example

$$101101 = 1 \times 2^5 + 1 \times 2^3 + 1 \times 2^2 + 1$$

i.e. the number

$$32 + 8 + 4 + 1 = 45$$

in the denary system. In this way, it is quite easy to give denary representation of a number, binary representation of which is given. The reverse process is also equally simple and given any number in the denary system, we can get its binary representation. We explain this procedure through an example.

Example 3.1

Consider the number 67. Observe that

$$67 = 64 + 2 + 1 = 2^6 + 2^1 + 1$$

and so binary representation of the number 67 is 1000011. Similarly

$$43 = 32 + 11 = 32 + 8 + 2 + 1 = 2^5 + 2^3 + 2^1 + 1$$

and its binary representation is 101011.

Next consider the number 243. We have

$$\begin{aligned} 243 &= 128 + 115 \\ &= 128 + 64 + 51 \\ &= 128 + 64 + 32 + 19 \\ &= 128 + 64 + 32 + 16 + 2 + 1 \\ &= 2^7 + 2^6 + 2^5 + 2^4 + 2^1 + 1 \end{aligned}$$

and so the binary representation of 243 is 11110011. Observe that putting one or more zeros on the left of the binary representation of a number does not alter the number. Also, observe that a number which is at most 2^m can be represented by a string of 0s and 1s of length $m + 1$, and a string of 0s and 1s of length $m + 1$ will always represent a number at most 2^m. Moreover, a number has only one 1 (and the rest 0s) in its binary representation iff it is a power of 2.

We next describe a recursion algorithm for converting a number into a binary number.

Let a number n be given.

1. Divide by 2 with r_0 as remainder and n_1 the quotient.
2. Divide n_1 by 2. Let r_1 be the remainder and n_2 the quotient.

Continue the process until the quotient becomes 0. Let the remainders obtained in succession be $r_0, r_1, \ldots, r_k$. Then, the binary representation of the number n is $r_k r_{k-1} \cdots r_1 r_0$. We now illustrate this process with an example.

Example 3.2

Consider the number 483.

$$\begin{array}{rl} 2\,)\underline{483} & \\ 2\,)\underline{241} & r_0 = 1 \\ 2\,)\underline{120} & r_1 = 1 \\ 2\,)\underline{60} & r_2 = 0 \\ 2\,)\underline{30} & r_3 = 0 \\ 2\,)\underline{15} & r_4 = 0 \\ 2\,)\underline{7} & r_5 = 1 \\ 2\,)\underline{3} & r_6 = 1 \\ 2\,)\underline{1} & r_7 = 1 \\ 0 & r_8 = 1 \end{array}$$

Hence, in the binary system, 483 is represented by 111100011.

The algorithm for the conversion of a number from binary system to denary system is equally simple. Let a number n be represented in the binary system as $r_k r_{k-1} \cdots r_1 r_0$. To convert it into the decimal system:

1. Multiply r_k by 2 and add r_{k-1} to obtain $2r_k + r_{k+1}$.
2. Multiply $2r_k + r_{k-1}$ by 2 and add r_{k-2} to obtain

$$2(2r_k + r'_{k-1}) + r_{k-2} = 2^2 r_k + 2r_{k-1} + r_{k-2}$$

Continue this process to exhaust all the r_i's. The result so obtained is the number n in the decimal representation. The coefficient of r_k will be 2^k and so on.

Example 3.3

Consider a binary number 1011011001. We then obtain the decimal representation of this number as follows.

```
1  0  1  1  0  1  1  0  0  1
2
2+0=2
   4 + 1=5
      10 + 1=11
         22 + 0=22
            44 + 1=45
               90 + 1=91
                  182 + 0=182
                     364 + 0=364
                        728 + 1 = 729
```

The arrow at each step indicates multiplication of the number above it by 2.

3.2 HAMMING CODES

Hamming codes are single error, correcting codes. Given any positive integer r, we can always construct an $(m = 2^r - r - 1, n = 2^r - 1)$ code which corrects each single error that might occur and corrects no other errors. These codes are defined by the following procedure.

Procedure for forming a Hamming code

Step 1

Choose a positive integer r. Then the code words of the code will have $n = 2^r - 1$ digits and the message words will have $m = 2^r - r - 1$ digits. Thus, in each code word, there are r check digits. The number of check symbols present in any code word of a block code is called the **redundancy** of the code. So, the code being constructed has redundancy r.

Step 2

In each code word $b = (b_1, b_2, \ldots, b_n)$ use $b_{2^0}, b_{2^1}, \ldots, b_{2^{r-1}}$ as the r check digits and place the $2^r - r - 1$ message digits in the remaining b_j positions but in their original order. For example if we take $r = 3$, then in the code word $b = (b_1, \ldots, b_7)$ corresponding to the message word $a = (a_1, a_2, a_3, a_4)$, b_1, b_2, b_4 are the check symbols and $b_3 = a_1, b_5 = a_2, b_6 = a_3, b_7 = a_4$. Again, if we take $r = 4$, then in the code word $b = (b_1, \ldots, b_{15})$ corresponding to the message word $a = (a_1, \ldots, a_{11})$ b_1, b_2, b_4, b_8 are the check symbols and $b_3 = a_1, b_5 = a_2, b_6 = a_3, b_7 = a_4$, $b_9 = a_5$, $b_{10} = a_6$, $b_{11} = a_7$, $b_{12} = a_8$, $b_{13} = a_9$, $b_{14} = a_{10}$, $b_{15} = a_{11}$.

Step 3

Form a matrix $\mathbf{M}$ of $2^r - 1$ rows and r columns in which the ith row is the binary representation of the number i. For example:

(a) if $r = 2$ then $\mathbf{M}$ is the 3×2 matrix with

$$\mathbf{M} = \begin{pmatrix} 0 & 1 \\ 1 & 0 \\ 1 & 1 \end{pmatrix}$$

(b) if $r = 3$, then $\mathbf{M}$ is the 7×3 matrix with

$$\mathbf{M}' = \begin{pmatrix} 0 & 0 & 0 & 1 & 1 & 1 & 1 \\ 0 & 1 & 1 & 0 & 0 & 1 & 1 \\ 1 & 0 & 1 & 0 & 1 & 0 & 1 \end{pmatrix}$$

(c) if $r = 4$, then $\mathbf{M}$ is the 15×4 matrix with

$$\mathbf{M}' = \begin{pmatrix} 0 & 0 & 0 & 0 & 0 & 0 & 0 & 1 & 1 & 1 & 1 & 1 & 1 & 1 & 1 \\ 0 & 0 & 0 & 1 & 1 & 1 & 1 & 0 & 0 & 0 & 0 & 1 & 1 & 1 & 1 \\ 0 & 1 & 1 & 0 & 0 & 1 & 1 & 0 & 0 & 1 & 1 & 0 & 0 & 1 & 1 \\ 1 & 0 & 1 & 0 & 1 & 0 & 1 & 0 & 1 & 0 & 1 & 0 & 1 & 0 & 1 \end{pmatrix}$$

Step 4

Form the matrix equation $\mathbf{bM} = 0$ which gives r linear equations in the r unknowns $b_1, b_2, \ldots, b_{2^{r-1}}$. For example:

(a) if $r = 2$, then

$$\mathbf{bM} = 0 \Rightarrow b_2 + b_3 = 0 \quad \text{and} \quad b_1 + b_3 = 0$$

(b) if $r = 3$, then

$$\mathbf{bM} = 0 \Rightarrow b_4 + b_5 + b_6 + b_7 = 0 \quad b_2 + b_3 + b_6 + b_7 = 0$$

and

$$b_1 + b_3 + b_5 + b_7 = 0$$

(c) while if $r = 4$, then

$$\mathbf{bM} = 0 \Rightarrow b_8 + b_9 + b_{10} + b_{11} + b_{12} + b_{13} + b_{14} + b_{15} = 0$$
$$b_4 + b_5 + b_6 + b_7 + b_{12} + b_{13} + b_{14} + b_{15} = 0$$
$$b_2 + b_3 + b_6 + b_7 + b_{10} + b_{11} + b_{14} + b_{15} = 0$$
$$b_1 + b_3 + b_5 + b_7 + b_9 + b_{11} + b_{13} + b_{15} = 0$$

Observe that, in each of the above examples, the r linear equations are such that every one of them contains exactly one unknown variable. We shall prove this in the general case. In view of this, the r linear equations give a unique solution for the r unknown quantities so that the code word is uniquely determined by the given message word.

Step 5

To encode a message word, place the message digits in the correct b_j positions and then make the check digits $b_{2^i}, 0 \leq i \leq r - 1$ satisfy the r linear equations as in Step 4 above. There will be exactly one b_{2^i} in each equation and so the r equations lead to a unique solution for $b_{2^0}, b_{2^1}, \ldots, b_{2^{r-1}}$.

Lemma 3.1

In each of the r linear equations as in Step 4 above there is exactly one b_{2^i} present.

Proof

Suppose that b_{2^i} occurs in the equation which is obtained by multiplying $\mathbf{b} = (b_1 \cdots b_n)$ with the kth column of the matrix $\mathbf{M}$. Then the 2^ith entry in the kth column of $\mathbf{M}$ is 1, i.e. the $(2^i, k)$th entry of $\mathbf{M}$ is 1. This entry is in the 2^ith row of $\mathbf{M}$ and 2^ith row of $\mathbf{M}$ is the binary representation of the number 2^i which has only the $(i + 1)$th entry from the right equal to 1. Hence this entry occurs in the $(r - i)$th column and so $k = r - i$. Thus if b_{2^i} and b_{2^j} both occur in the equation which is obtained by multiplying $\mathbf{b}$ with the kth column of $\mathbf{M}$, then we have $k = r - i$ and $k = r - j$. Hence $i = j$ and at most one check symbol b_{2^i} occurs in any equation.

On the other hand, it is clear from the above reasoning that a given check symbol b_{2^i} occurs in the equation obtained by multiplying $\mathbf{b}$ with the $(r - i)$th column of $\mathbf{M}$.

The code as obtained by the constructive procedure (Steps 1–5 above) is called the $(2^r - r - 1, 2^r - 1)$ Hamming code.

Proposition 3.1
Hamming code is a group code.

Proof
Let $b = b_1 b_2 \cdots b_n$ be the code word corresponding to the message word $a = a_1 a_2 \cdots a_m$ and $b' = b'_1 b'_2 \cdots b'_n$ be the code word corresponding to the message word $a' = a'_1 a'_2 \cdots a'_m$. Let $\mathbf{M}$ be the $(2^r - 1) \times r$ matrix in which the ith row is the binary representation of the number i. Then $\mathbf{bM} = 0 = \mathbf{b'M}$ and, therefore, $(\mathbf{b} + \mathbf{b'})\mathbf{M} = 0$. Also the entries in $b + b'$ at the positions other than the $2^0, 2^1, \ldots, 2^{r-1}$ positions are $a_i + a'_i$ in the order in which they occur in $a + a'$. Also b_{2^i} is given by the equation when $\mathbf{b}$ is multiplied by the $(r - i)$th column of $\mathbf{M}$ and similarly for b'_{2^i}. Thus $b + b'$ is the code word corresponding to the message word $a + a'$ and $a \to b$ where b is the code word in the Hamming code corresponding to the message word a is a group homomorphism $\mathbb{B}^m \to \mathbb{B}^n$. Hence the Hamming code is a group code.

Theorem 3.1
The minimum distance of any Hamming code is 3.

Proof
Since Hamming code is a group code, it is enough to prove that the weight of any non-zero code word is at least 3 and there is at least one code word of weight exactly 3. Again, all the message symbols of a message word occur somewhere in the corresponding code word. Therefore for the first assertion, it is enough to prove that the weight of a non-zero code word which corresponds to a message word of weight at most 2 has weight at least 3.

Case (i)
Let a be a message word with $\text{wt}(a) = 1$ and $b = b_1 b_2 \cdots b_n$ be the corresponding code word. Suppose that the non-zero message symbol occurs in the ith position in b. Then $i \neq 2^j$ for any j and in the binary representation of i there are at least two non-zero entries. Let these be in the sth and tth position from the left.

Let $\mathbf{M}_1, \mathbf{M}_2, \ldots, \mathbf{M}_r$ denote the columns of the $(2^r - 1) \times r$ matrix $\mathbf{M}$ in which jth row is the binary representation of j. Consider the equations $\mathbf{bM}_s = 0$ and $\mathbf{bM}_t = 0$ out of the r linear equations $\mathbf{bM} = 0$. In the equation $\mathbf{bM}_s = 0$, there is exactly one check symbol and every other symbol is a message symbol. Let b_2k be the check symbol present in this equation. Every message symbol except b_i being 0, we have $b_2k + b_i = 0$ (by the choice of s) and so $b_2k \neq 0$. Similarly, we find that the check symbol occurring in the equation $\mathbf{bM}_t = 0$ is non-zero. Every one of the r linear equations $\mathbf{bM} = 0$ contains only one check symbol and check symbols in different equations are different. Thus, the two non-zero check symbols in the equations $\mathbf{bM}_s = 0$ and $\mathbf{bM}_t = 0$ are different. Therefore $\text{wt}(b) \geq 3$.

Case (ii)

Let a be a message word with $\text{wt}(a) = 2$ and let $b = b_1 b_2 \cdots b_n$ be the corresponding code word. Suppose that the two non-zero message symbols occur in the ith and jth positions. Then i and j are not powers of 2. Again $i \neq j$ implies that binary representations of i and j differ in at least one place. Suppose that these differ from each other in the sth position from the left. We may, therefore, suppose that the sth position (from the left) of i is 1 and that of j is 0. (We can reverse the roles of i and j otherwise.) Let b_{2^k} be the check symbol which appears in the equation $\mathbf{bM}_s = 0$. Since every other symbol in this equation is a message symbol and

$$b_\ell = 0 \forall \ell \text{ except } \ell = i \text{ and } \ell = j$$

and the ith entry in $\mathbf{M}_s$ is 1 while the jth entry in $\mathbf{M}_s$ is 0, this equation reduces to $b_{2^k} + b_i = 0$, i.e. $b_{2^k} \neq 0$. Thus, in b, there is at least one check symbol which is non-zero and $\text{wt}(b) \geq 3$.

Let $b = b_1 b_2 \cdots b_n$ be the code word corresponding to the message word $a = a_1 a_2 \cdots a_m$, where $a_1 = 1$ and $a_i = 0$ for $2 \leq i \leq m$. In the equation $\mathbf{bM}_s = 0$, the check symbol takes the value 1 iff the third entry in the column $\mathbf{M}_s$ is 1. Thus the number of non-zero check symbols in b equals the number of non-zero entries in the third row of $\mathbf{M}$. But the third row of $\mathbf{M}$ has exactly two non-zero entries. Hence, the number of non-zero check symbols in b is 2 and, so, $\text{wt}(b) = 3$. This proves that the minimum distance of the Hamming code is 3.

Corollary

Hamming code is a single error correcting and double error detecting code.

Proof

This follows from the above theorem and Theorems 1.1 and 1.3.

Remark

A Hamming code being a group code, it follows that error vectors that go undetected are precisely those which equal some code word.

Exercise 3.1

Compute all the code words of the (4, 7) Hamming code.

Remarks

The $(2^r - 1) \times r$ matrix $\mathbf{M}$ in the construction of Hamming code is of rank r. Also $\mathbf{bM} = 0$ iff $\mathbf{M}^t\mathbf{b}^t = 0$. Thus $\mathbf{M}' = \mathbf{M}^t$ is a parity check matrix of the

Hamming code. We may thus define a Hamming code as the matrix code defined by the parity check matrix **H**, the ith column of which is the binary representation of the number i, $1 \leq i \leq 2^r - 1$, subject also to the conditions given in Steps 1 and 2 of the procedure (p. 42).

In Chapter 1, we had required the last $n - m$ columns of a parity check matrix to form an identity matrix. However, Theorem 1.5 does not need this strong restriction and is valid in the general case, i.e. when a parity check matrix is required to be of rank $n - m$. Since the columns of the parity check matrix $\mathbf{M}'$ constructed for the Hamming code are non-zero and distinct, Theorem 1.5 shows that Hamming codes are single error correcting codes.

Exercise 3.2

Write down a parity check matrix of the (11, 15) Hamming code.

4

Finite fields and BCH codes

4.1 FINITE FIELDS

For the construction of BCH codes, we need to study construction of finite fields. However, the construction of finite fields depends heavily on some results on commutative rings. We first recall these results on rings, although briefly.

Definition 4.1
Let R be a ring and I be an ideal of R. Consider the set $R/I = \{x + I | x \in R\}$, where $x + I = \{x + a\colon a \in R\}$ are cosets of the ideal I in R. It is easy to check, using the ideal properties of I, that for $x, y \in R$,

$$(x + I) + (y + I) = (x + y) + I$$
$$(x + I)(y + I) = xy + I$$

are independent of the choice of the elements x, y in the cosets $x + I$ and $y + I$ respectively. Also w.r.t. these compositions R/I is a ring called the **ring of quotients** or the **difference ring** of R relative to the ideal I. If R is a commutative ring with identity, then so is R/I.

Definition 4.2
Let R be a commutative ring with identity and $a \in R$. Then $\langle a \rangle = aR = \{ar\colon r \in R\}$ is an ideal of R and is called a **principal ideal** generated by a. If every ideal of R is of this form, R is called a **principal ideal ring**.

Theorem 4.1
The polynomial ring $F[X]$, where F is a field, is a principal ideal ring.

Proof

The ring $F[X]$ is already a commutative ring with identity. Let I be an ideal of $F[X]$. If $I=0$, then I is a principal ideal generated by 0 and we, therefore, suppose that $I \neq 0$. Choose a $0 \neq f(X) \in I$ such that

$$\deg f(X) \leq \deg g(X) \forall 0 \neq g(X) \in I$$

This choice is possible because the set of non-negative integers is well ordered. Let $g(X) \in F[X]$, $g(X) \neq 0$. We claim that

$$g(X) = q(X) f(X) + r(X)$$

for some $q(X)$, $r(X) \in F[X]$ where either $r(X)=0$ or $\deg r(X) < \deg f(X)$. If $\deg g(X) < \deg f(X)$, we can take $q(X)=0$ and $r(X)=f(X)$. Suppose that

$$n = \deg f(X) \leq \deg g(X)$$

Let

$$f(X) = a_0 X^n + a_1 X^{n-1} + \cdots + a_n$$

$$g(X) = b_0 X^m + b_1 X^{m-1} + \cdots + b_m$$

Then $a_0 \neq 0$ and

$$g(X) = a_0^{-1} b_0 X^{m-n} f(X) + g_1(X)$$

where $\deg g_1(X) \leq m-1$. Induction on the degree of $g(X)$ then shows that there exist $q_1(X)$, $r(X) \in F[X]$ such that

$$g_1(X) = q_1(X) f(X) + r(X)$$

where either $r(X)=0$ or $\deg r(X) < \deg f(X)$. Then

$$g(X) = q(X) f(X) + r(X)$$

where

$$q(X) = a_0^{-1} b_0 X^{m-n} + q_1(X) \in F[X]$$

This proves the claim. Now, if $r(X) \neq 0$, then

$$r(X) = g(X) - q(X) f(X) \in F[X]$$

and $\deg r(X) < \deg f(X)$. This contradicts the choice of $f(X)$. Hence $g(X) = q(X) f(X)$ and I is a principal ideal generated by $f(X)$. ■

Definition 4.3

A non-constant polynomial $f(X) \in R[X]$, where R is a commutative ring with identity, is called **irreducible** if

$$f(X) = g(X)h(X) \quad \text{for } g(X), h(X) \in R[X]$$
$$\Rightarrow \deg g(X) = 0 \quad \text{or} \quad \deg h(X) = 0$$

Otherwise, we say that $f(X)$ is **reducible**.

Theorem 4.2
Let F be a field and $f(X) \in F[X]$ be an irreducible polynomial. Then $F[X]/\langle f(X) \rangle$ is a field.

Proof
Let I denote the ideal $\langle f(X) \rangle$ of $F[X]$ generated by $f(X)$. If $I = F[X]$, then $1 = f(X)g(X)$ for some $g(X) \in F[X]$. Comparing the degrees of the two sides of this relation implies that $f(X)$ is a constant polynomial which is a contradiction. Hence $F[X]/I$ has at least two elements. Already $F(X)/I$ is a commutative ring with identity.

Let $g(X) \in F[X]$, $g(X) \notin I$. Then

$$J = \{a(X)f(X) + b(X)g(X) \colon a(X), b(X) \in F[X]\}$$

is an ideal of $F[X]$ and there exists $h(X) \in F[X]$ such that $J = \langle h(X) \rangle$. Now

$$f(X) = 1 \times f(X) + 0 \times g(X)$$

is in J and, therefore,

$$f(X) = a(X)h(X) \quad \text{for some } a(X) \in F[X]$$

Irreducibility of $f(X)$ shows that $\deg h(X) = 0$ or $\deg a(X) = 0$. If $\deg a(X) = 0$, then $a(X)$ is a unit in $F[X]$ so that $h(X) \in I$ which implies that $J = I$. This is a contradiction because $g(X) \in J$ but $g(X) \notin I$. Hence $h(X)$ is a unit in $F[X]$ and $J = F[X]$. Therefore

$$1 = a(X)f(X) + b(X)g(X) \quad \text{for some } a(X), b(X) \in F[X]$$

and

$$1 + I = (b(X) + I)(g(X) + I)$$

Thus $g(X) + I$ is invertible and $F[X]/I$ is a field.

Definition 4.4
Let K be a field and F be a subfield of K. Then K is called an **extension** of the field F. The fact that K is an extension of F is expressed by writing

$$K|_F$$

Observe that K can be regarded as a vector space over F by using the multiplication in K. The dimension of the vector space K over F is called the **degree** of the extension K of F and is denoted by $[K{:}F]$. The extension $K|_F$ is called a **finite extension** if the degree $[K{:}F]$ is finite.

The intersection of all subfields of a field F is called a **prime subfield** of F. It is unique and is, in fact, the smallest subfield of F. In general, a field which has no proper subfield is called a **prime field**.

Let $K|_F$ be an extension of a field F. An element $\alpha \in K$ is said to be **algebraic over** F if there exists a polynomial $f(X) \in F[X]$ which has α as a root. Let $\alpha \in K$

be algebraic over F and consider

$$A = \{f(X) \in F[X] : f(\alpha) = 0\}$$

Then A is an ideal of $F[X]$ and $F[X]$ is a principal ideal domain. Let $m_1(X) \in F[X]$ be a generator of the ideal A. If a is the coefficient of the highest power of X in $m_1(X)$, then $m(X) = a^{-1}m_1(X)$ is a monic polynomial with $\deg m(X) = \deg m_1(X)$ which is also a generator of A. If

$$m(X) = r(X)s(X) \quad \text{for some } r(X), s(X) \in F[X]$$

then either $r(\alpha) = 0$ or $s(\alpha) = 0$, i.e. either $m(X) | r(X)$ or $m(X) | s(X)$. But

$$\deg m(X) = \deg r(X) + \deg s(X)$$

and, therefore, either $\deg r(X) = 0$ or $\deg s(X) = 0$. Thus $m(X)$ is irreducible. Also, it is clear from the choice of $m(X)$ that it is a monic, irreducible polynomial of the least possible degree which has α as a root. It is easily seen that $m(X)$ with these properties is unique and is called the **minimal polynomial** of α over F. Observe that the minimal polynomial of α depends on the subfield F. For example, the minimal polynomial of $\alpha = \sqrt{2}$ over Q is $X^2 - 2$ while over $\mathbb{R}$ – the field of real numbers – it is $X - \sqrt{2}$. Also observe that every element of the field F is algebraic over F.

Proposition 4.1
The order of a finite field is p^n for some prime p and some positive integer n.

Proof
Let F be a finite field. Since F is a finite group w.r.t. addition, it follows from Lagrange's theorem on finite groups (order of every subgroup of a finite group divides the order of the group) that $O(F)a = 0 \forall a \in F$. Choose m to be the smallest positive integer with $ma = 0$ for every $a \in F$. Let e denote the identity of F. Suppose that $m = rs$, where r, s are integers with $r > 1$, $s > 1$. Then

$$0 = me = (rs)e = (re)(se)$$

Therefore either $re = 0$ or $se = 0$. For any $a \in F$ and positive integer k, $ka = (ke)a$. Therefore, either $ra = 0 \forall a \in F$ or $sa = 0 \forall a \in F$ which contradicts the choice of m. Hence m must be a prime p (say). It is easy to check that

$$K = \{re : 0 \leq r < p\}$$

is a subfield of F. Since F is finite, F is a finite dimensional vector space over K. If $\dim_K F = n$, and $x_1, x_2, \ldots, x_n$ is a basis of F over K, then every element of F can be uniquely written as

$$a_1x_1 + a_2x_2 + \cdots + a_nx_n, a_i \in K$$

Since each a_i has exactly p choices and choice of any a_j is independent of the choice of other a_i, there are exactly p^n elements in F, i.e. $O(F) = p^n$.

Let F be a finite field of order p^n. Then $F^* = F \setminus \{0\}$ is a multiplicative group of order $p^n - 1$. Therefore, it follows from the Lagrange theorem, that

$$a^{p^n - 1} = 1 \forall a \in F^*$$

or that

$$a^{p^n} = a \forall a \in F^*$$

Since this result is trivially true for the element $0 \in F$, we have

$$a^{p^n} = a \forall a \in F \tag{4.1}$$

Observe that every element of F is algebraic over the prime subfield of F.

Proposition 4.2

Let F be a finite field of order p^n with k as its prime subfield. Then α and α^p have the same minimal polynomial over k for every $\alpha \in F$.

Proof

Define a map $\theta: F \to F$ by $\theta(a) = a^p$, $a \in F$. Since

$$pa = 0 \forall a \in F$$

$$ab = ba \forall a, b \in F$$

and p divides the binomial coefficient

$$\binom{p}{i} \forall i \quad 1 \leq i \leq p - 1$$

it follows that θ is a homomorphism of fields. Since $a^p \neq 0$ for $a \neq 0$, θ is a monomorphism. For any $a \in k$, $a^p = a$ (by (4.1)) and therefore θ keeps the elements of the prime subfield k fixed. Thus $\theta: F \to F$ is a monomorphism of finite dimensional vector space F over k into itself and hence θ is onto as well.

Let $a \in F$ and $m(X)$ be the minimal polynomial of α. Then $m(\alpha) = 0$ and θ being a field-homomorphism $m(\alpha^p) = 0$, i.e. α^p is also a root of $m(X)$. If $m_1(X)$ is the minimal polynomial of α^p, then $m_1(X) | m(X)$. But $m(X)$ is an irreducible, monic polynomial. Therefore $m_1(X) = m(X)$.

A finite field is called a **Galois field** and if F is a field of order p^n, we write $F = \mathrm{GF}(p^n)$.

For the proof of the next theorem, we need the following result about Abelian groups.

Lemma 4.1

Let G be a non-cyclic Abelian group of finite order m. Then there is a proper divisor k of m such that $x^k = 1 \forall x \in G$.

Theorem 4.3
In any finite field $F = \mathrm{GF}(p^n)$, the multiplicative group F^* of all non-zero elements is cyclic.

Proof
The multiplicative group F^* of F is an Abelian group of order $q-1$, where $q = p^n$. If F^* is not cyclic, there exists an integer r, $1 < r < q-1$ such that $a^r = 1 \forall a \in F^*$. Thus, every $a \in F$ is a root of the polynomial $X^{r+1} - X$ and hence

$$X - a | X^{r+1} - X \forall a \in F$$

Also $X - a, X - b$ are relatively coprime for $a, b \in F$, $a \neq b$. Therefore

$$\prod_{a \in F} (X - a) | X^{r+1} - X$$

But

$$\deg \prod_{a \in F} (X - a) = \mathrm{O}(F) = q$$

and

$$r + 1 < q - 1 + 1 = q$$

Thus, a polynomial of degree q divides a polynomial of degree $< q$ – which is a contradiction. This proves that there is no r with $1 < r < q-1$ such that $a^r = 1 \forall a \in F^*$. Hence F^* is cyclic.

Definition 4.5
A generator of the cyclic group F^* of the finite field F is called a **primitive element** of F.

We know that there are

$$\phi(\mathrm{O}(F^*)) = \phi(\mathrm{O}(F) - 1) = \phi(q-1) \quad q = p^n = \mathrm{O}(F)$$

elements in F^*, every one of which generates F^*. Therefore, in a field of order p^n there are $\phi(p^n - 1)$ primitive elements. Here ϕ is Euler's ϕ-function.

Proposition 4.3
Let K be a finite extension of F and L be a finite extension of K. Then L is a finite extension of F and

$$[L:F] = [L:K][K:F]$$

Proof
Let $[L:K] = m$ and $[K:F] = n$. Let $\alpha_1, \ldots, \alpha_n$ be a basis of the vector space K over F and $\beta_1, \ldots, \beta_m$ be a basis of the vector space L over K. Let $x \in L$. Then there exist elements $y_1, \ldots, y_m \in K$ such that

$$x = \sum_{i=1}^{m} y_i \beta_i$$

For every y_i, $1 \leq i \leq m$, there exist elements $a_{ij} \in F$, $1 \leq j \leq n$, such that

$$y_i = \sum_{j=1}^{n} a_{ij}\alpha_j$$

Substituting these values of y_i in the expression for x, we find that x can be expressed as a linear combination of the elements $\alpha_j\beta_i$, $1 \leq i \leq m$, $1 \leq j \leq n$, with coefficients in F. Thus, the elements $\{\alpha_j \beta_i\}$ $1 \leq i \leq m$, $1 \leq j \leq n$, generate L over F.

Suppose that elements $a_{ij} \in F$, $1 \leq i \leq m$, $1 \leq j \leq n$ are such that

$$\sum_{i=1}^{m} \sum_{j=1}^{n} a_{ij}\alpha_j\beta_i = 0$$

Then

$$\sum_{i=1}^{m} \left(\sum_{j=1}^{n} a_{ij}\alpha_j \right) \beta_i = 0$$

and $\{\beta_i\}$, $1 \leq i \leq m$, being a basis of L over K, we have

$$\sum_{j=1}^{n} a_{ij}\alpha_j = 0 \quad \forall i, 1 \leq i \leq m$$

But then $\{\alpha_j\}$, $1 \leq j \leq n$ being a basis of K over F, we have $a_{ij} = 0 \forall j$, $1 \leq j \leq n$ and $\forall i$, $1 \leq i \leq m$. This proves that $\{\alpha_j\beta_i\}$, $1 \leq i \leq m$, $1 \leq j \leq n$ is a basis of L over F and

$$[L:F] = mn = [L:K][K:F]$$ ■

Let F be a prime field of characteristic $p \neq 0$, $f(X) \in F[X]$ be an irreducible polynomial of degree n and $I = \langle f(X) \rangle$ be the ideal generated by $f(X)$. Then $K = F[X]/I$ is a field and clearly an arbitrary element of K is of form $g(X) + I$, where $g(X) \in F[X]$ is a polynomial of degree at most $n - 1$. Also it follows that such an expression of an element of K is uniquely determined. Therefore $O(K) = p^n$. Thus, in order to construct a field K of order p^n, p a prime and n a positive integer, we need to find an irreducible polynomial $f(X)$ of degree n over the field of p elements. Also, observe that every element of $K = F[X]/I$ is a root of the polynomial

$$X^{p^n} - X$$

and, so, the irreducible polynomial $f(X)$ must be a divisor of $X^{p^n} - X$.

Proposition 4.4

Let $f(X)$ be a polynomial of degree 2 or 3 over a field F. Then $f(X)$ is irreducible iff none of the elements of F is a root of $f(X)$.

Proof

Suppose that $f(X)$ is reducible. Then $f(X)$ has a linear factor $aX + b$, say, where $a, b \in F$, $a \neq 0$. Then $-b/a \in F$ is a root of $f(X)$. Conversely, suppose that

$\alpha \in F$ is a root of $f(X)$. Then $X - \alpha \mid f(X)$. For example, if

$$f(X) = aX^3 + bX^2 + cX + d \qquad a, b, c, d \in F$$

then

$$\begin{aligned} f(X) &= aX^3 + bX^2 + cX + d - (a\alpha^3 + b\alpha^2 + c\alpha + d) \\ &= (X - \alpha)[a(X^2 + \alpha X + \alpha^2) + b(X + \alpha) + c] \end{aligned}$$

and

$$a(X^2 + \alpha X + \alpha^2) + b(X + \alpha) + c \in F[X]$$

This proves that $f(X)$ is reducible.

Example 4.1

Let F be the field of 5 elements. None of the elements of F is a root of the polynomial $f(X) = X^2 + 2$ and so $f(X)$ is irreducible in $F[X]$. Hence

$$K = F[X]/\langle f(X) \rangle$$

is a field of order 25. An arbitrary element of K is of the form

$$aX + b + \langle f(X) \rangle \quad a, b \in F$$

Write $X + \langle f(X) \rangle = \alpha$. The powers of α are then determined as follows: $\alpha^2 = 3$, $\alpha^3 = 3\alpha$, $\alpha^4 = 4$, $\alpha^5 = 4\alpha$, $\alpha^6 = 2$, $\alpha^7 = 2\alpha$ and $\alpha^8 = 1$. Thus α is not a primitive element of K.

Taking $\beta = \alpha + 4$ gives

$$\begin{aligned} \beta^2 &= \alpha^2 + 3\alpha + 1 = 3\alpha + 4 \\ \beta^3 &= (3\alpha + 4)(\alpha + 4) = \alpha \\ \beta^4 &= 4\alpha + \alpha^2 = 4\alpha + 3 \\ \beta^5 &= (\alpha + 4)(4\alpha + 3) = 4\alpha + 4 \\ \beta^6 &= (4\alpha + 4)(\alpha + 4) = 3 \\ \beta^7 &= 3\alpha + 2 \\ \beta^8 &= (3\alpha + 2)(\alpha + 4) = 4\alpha + 2 \\ \beta^9 &= (4\alpha + 2)(\alpha + 4) = 3\alpha \\ \beta^{10} &= 3\alpha^2 + 2\alpha = 2\alpha + 4 \\ \beta^{11} &= (2\alpha + 4)(\alpha + 4) = 2\alpha + 2 \\ \beta^{12} &= (2\alpha + 2)(\alpha + 4) = 4 \end{aligned}$$

Thus the order of β in the multiplicative group of K is greater than 12 and hence β is a primitive element of K.

Example 4.2

Let F be the field of 3 elements. None of the elements of F is a root of the polynomial $X^3 + 2X + 2 \in F[X]$ and so it is irreducible in $F[X]$. Therefore

$$K = F[X]/\langle X^3 + 2X + 2 \rangle$$

is a field of order $3^3 = 27$. Let the element $X + \langle X^3 + 2X + 2\rangle$ be denoted by α. Then

$$\begin{aligned}
\alpha^3 &= \alpha + 1 \\
\alpha^6 &= \alpha^2 + 2\alpha + 1 \\
\alpha^{12} &= \alpha^4 + 4\alpha^2 + 1 + 4\alpha^3 + 2\alpha^2 + 4\alpha \\
&= \alpha^4 + \alpha^3 + \alpha + 1 \\
&= \alpha^2 + \alpha + \alpha + 1 + \alpha + 1 \\
&= \alpha^2 + 2
\end{aligned}$$

and then

$$\alpha^{13} = \alpha^3 + 2\alpha = \alpha + 1 + 2\alpha = 1$$

Thus α is not a primitive element of K. Taking $\beta = \alpha^2 + 1$, we find that

$$\begin{aligned}
&\beta^2 = \alpha + 1 \neq 1 \\
&\beta^3 = \alpha^6 + 1 = (\alpha + 1)^2 + 1 = \alpha^2 + 2\alpha + 2 \\
&\beta^6 = \alpha^4 + \alpha^2 + 1 + \alpha^3 + \alpha^2 + 2\alpha = \alpha^2 + \alpha + \alpha^2 + 1 + \alpha + 1 + \alpha^2 + 2\alpha = \alpha + 2 \\
&\beta^{12} = \alpha^2 + \alpha + 1 \\
&\beta^{13} = (\alpha^2 + \alpha + 1)(\alpha^2 + 1) = \alpha^4 + \alpha^3 + 2\alpha^2 + \alpha + 1 = 2 \neq 1
\end{aligned}$$

Hence β is a primitive element of K.

Example 4.3

As in Chapter 1, let $\mathbb{B}$ be the field of 2 elements.

Case (i)

Neither of the two elements of $\mathbb{B}$ is a root of the polynomial $X^3 + X + 1 \in \mathbb{B}[X]$ and, so, the polynomial $X^3 + X + 1$ is irreducible over $\mathbb{B}$. Therefore,

$$K = \mathbb{B}[X]/\langle X^3 + X + 1\rangle$$

is a field of order 8. Let the element $X + \langle X^3 + X + 1\rangle$ be denoted by α. The multiplicative group of K is of order 7 and, so, any non-zero, non-identity element of K is primitive. In particular, so is the element α.

Case (ii)

Consider the polynomial $X^4 + X + 1 \in \mathbb{B}[X]$. Neither of the elements of $\mathbb{B}$ is a root of this polynomial and, so, $f(X) = X^4 + X + 1$ does not have a linear factor in $\mathbb{B}[X]$. Therefore, if $f(X)$ is reducible in $\mathbb{B}[X]$, it must be a product of only quadratic polynomials. But the only polynomial in $\mathbb{B}[X]$ of degree 2 which is irreducible is $X^2 + X + 1$ and

$$(X^2 + X + 1)^2 = X^4 + X^2 + 1 \neq X^4 + X + 1$$

Thus $f(X)$ is an irreducible polynomial and $K = \mathbb{B}[X]/\langle f(X)\rangle$ is a field of order 16. Let

$$\alpha = X + \langle f(X)\rangle$$

Then

$$\alpha^4 = \alpha + 1 \neq 1$$

$$\alpha^5 = \alpha^2 + \alpha \neq 1$$

and since $O(\alpha)$ as an element of the multiplicative group of K divides 15, α is a primitive element of K. All the elements of K then are 0, 1, α, α^2, α^3, $\alpha + 1$, $\alpha^2 + \alpha$, $\alpha^3 + \alpha^2$, $\alpha^3 + \alpha + 1$, $\alpha^2 + 1$, $\alpha^3 + \alpha$, $\alpha^2 + \alpha + 1$, $\alpha^3 + \alpha^2 + \alpha$, $\alpha^3 + \alpha^2 + \alpha + 1, \alpha^3 + \alpha^2 + 1, \alpha^3 + 1$. These non-zero elements are the powers of α in order.

Case (iii)

We can prove as in Case (ii) above that $X^4 + X^3 + 1$ is another polynomial of degree 4 which is irreducible over $\mathbb{B}$. Hence

$$K = \mathbb{B}[X]/\langle X^4 + X^3 + 1\rangle$$

is a field of order 16. Setting

$$\alpha = X + \langle X^4 + X^3 + 1\rangle$$

we find that $\alpha^4 = \alpha^3 + 1$. Then

$$\alpha^5 = \alpha^3 + \alpha + 1 \quad \alpha^6 = \alpha^3 + \alpha^2 + \alpha + 1 \quad \alpha^7 = \alpha^2 + \alpha + 1$$

$$\alpha^8 = \alpha^3 + \alpha^2 + \alpha \quad \alpha^9 = \alpha^2 + 1 \quad \alpha^{10} = \alpha^3 + \alpha \quad \alpha^{11} = \alpha^3 + \alpha^2 + 1$$

$$\alpha^{12} = \alpha + 1 \quad \alpha^{13} = \alpha^2 + \alpha \quad \alpha^{14} = \alpha^3 + \alpha^2 \quad \alpha^{15} = 1$$

Thus α is a primitive element of K (this could have been concluded from $\alpha^3 \neq 1$, $\alpha^5 \neq 1$ but the above illustrates all the powers of α).

Case (iv)

Yet another polynomial of degree 4 which is irreducible over $\mathbb{B}$ is

$$X^4 + X^3 + X^2 + X + 1$$

Thus

$$\mathbb{B}[X]/\langle X^4 + X^3 + X^2 + X + 1\rangle$$

is a field of order 16. However, in this case, the element

$$\alpha = X + \langle X^4 + X^3 + X^2 + X + 1\rangle$$

is not a primitive element of the field because

$$\alpha^4 = \alpha^3 + \alpha^2 + \alpha + 1$$

and then

$$\alpha^5 = \alpha^4 + \alpha^3 + \alpha^2 + \alpha = 1$$

Case (v)

Next consider the polynomial $X^6 + X^5 + 1$ over $\mathbb{B}$. It is clear that neither 0 nor 1 is a root of this polynomial. Therefore a possible factor of degree 2 of this polynomial is $X^2 + X + 1$. Let

$$X^6 + X^5 + 1 = (X^2 + X + 1)(X^4 + aX^3 + bX^2 + cX + 1)$$

Comparing the coefficients of various powers of X, gives

$$a + 1 = 1 \quad a + b + 1 = 0 \quad a + b + c = 0 \quad b + c + 1 = 0$$

The first three equations imply that $a = 0$, $b = c = 1$ and then the fourth gives $1 = 0$ – a contradiction. Thus the polynomial can have only cubic factors. Irreducible polynomials of degree 3 are

$$X^3 + X + 1 \qquad \text{and} \qquad X^3 + X^2 + 1$$

Now

$$(X^3 + X + 1)^2 = X^6 + X^2 + 1$$

$$(X^3 + X^2 + 1)^2 = X^6 + X^4 + 1$$

and

$$(X^3 + X^2 + 1)(X^3 + X + 1) = X^6 + X^5 + X^4 + X^2 + X + 1$$

This proves that $X^6 + X^5 + 1$ is irreducible over $\mathbb{B}$. Then

$$K = \mathbb{B}[X]/\langle X^6 + X^5 + 1 \rangle$$

is a field of order $2^6 = 64$. Let

$$\alpha = X + \langle X^6 + X^5 + 1 \rangle$$

Then

$$\alpha^6 = \alpha^5 + 1 \quad \alpha^7 = \alpha^5 + \alpha + 1 \quad \alpha^8 = \alpha^5 + \alpha^2 + \alpha + 1$$

$$\alpha^9 = \alpha^5 + \alpha^3 + \alpha^2 + \alpha + 1 \quad \alpha^{10} = \alpha^5 + \alpha^4 + \alpha^3 + \alpha^2 + \alpha + 1$$

$$\alpha^{11} = \alpha^4 + \alpha^3 + \alpha^2 + \alpha + 1 \quad \alpha^{12} = \alpha^5 + \alpha^4 + \alpha^3 + \alpha^2 + \alpha$$

$$\alpha^{13} = \alpha^4 + \alpha^3 + \alpha^2 + 1 \quad \alpha^{14} = \alpha^5 + \alpha^4 + \alpha^3 + \alpha \quad \alpha^{15} = \alpha^4 + \alpha^2 + 1$$

$$\alpha^{16} = \alpha^5 + \alpha^3 + \alpha \quad \alpha^{17} = \alpha^5 + \alpha^4 + \alpha^2 + 1 \quad \alpha^{18} = \alpha^3 + \alpha + 1$$

$$\alpha^{19} = \alpha^4 + \alpha^2 + \alpha \quad \alpha^{20} = \alpha^5 + \alpha^3 + \alpha^2 \quad \alpha^{21} = \alpha^5 + \alpha^4 + \alpha^3 + 1$$

Since the order of α divides 63 – the order of the multiplicative group K^* of K – it follows from the above computations that $O(\alpha) = 63$ and α is a primitive element of K.

Recall that a polynomial

$$f(X) = a_0 + a_1X + a_2X^2 + \cdots + a_nX^n$$

with integer coefficients $a_0, a_1, \ldots, a_n$ is called **primitive** if

$$\text{GCD}(a_0, a_1, \ldots, a_n) = 1$$

However, while working with polynomials over a field of p elements (p a prime), we deviate from this accepted terminology and call an irreducible polynomial $f(X) \in F_p[X]$ of degree n, where F_p is the field of p elements, primitive if:

(i) $f(X)$ divides $X^{p^n-1} - 1$; and
(ii) $f(X)$ does not divide $X^k - 1$ for any $k < p^n - 1$.

In our applications of finite fields to coding theory we are concerned with:

(i) the construction of a field K of order p^n for a given prime p and natural number n; and
(ii) finding a primitive element in K.

We have already proved above that if $f(X) \in F_p[X]$, where F_p is a field of p elements, is an irreducible polynomial of degree n, then $F_p[X]/\langle f(X) \rangle$ is a field of order p^n. Also as seen in Examples 4.3 Cases (i), (ii), (iii) and (v) there are situations in which $X + \langle f(X) \rangle$ is a primitive element of $K = F_p[X]/\langle f(X) \rangle$. That this is not always the case is shown by the Examples 4.1, 4.2 and 4.3 Case (iv). In fact we have the following proposition.

Proposition 4.5
Given an irreducible polynomial $f(X) \in F_p[X]$, the element

$$\alpha = X + \langle f(X) \rangle \in F_p[X]/\langle f(X) \rangle$$

($= K$, say) is primitive iff $f(X)$ is a primitive polynomial.

Proof
Let $\deg f(X) = n$. Then $\text{O}(K) = p^n = m$ (say) and $\text{O}(\alpha) = t \leq m - 1$. Therefore

$$X^{m-1} - 1 + \langle f(X) \rangle = 0$$

i.e. $f(X) | X^{m-1} - 1$.

Observe that

$$f(X) | X^r - 1 \quad \text{iff} \quad \text{O}(\alpha) | r$$

Thus $t = \text{O}(\alpha)$ is the smallest value of r for which $f(X) | X^r - 1$. This proves that α is primitive iff the smallest value of r for which $f(X) | X^r - 1$ is $m - 1$, i.e. iff $f(X)$ is a primitive polynomial.

Deciding whether a given irreducible polynomial over F_p is primitive is not an easy problem. More information on this problem is found when we consider factorization of the polynomials $X^n - 1$ over F_p as a product of irreducible polynomials over F_p.

Definition 4.6

Let K be a field, F a subfield of K and $\alpha \in K$. By $F(\alpha)$ we denote the smallest subfield of K which contains both F and α. We also call $F(\alpha)$ a **simple extension** of F.

Recall that if α is algebraic over F and the degree of the minimal polynomial of α over F is n then an arbitrary element of $F(\alpha)$ is of the form

$$a_0 + a_1\alpha + \cdots + a_m\alpha^m$$

where $a_i \in F$, $0 \leq i \leq m$ and $m \leq n-1$. If, on the other hand, α is transcendental over F, then an arbitrary element of $F(\alpha)$ is of the form $f(\alpha)/g(\alpha)$, where $f(X)$, $g(X) \in F[X]$ and $g(\alpha) \neq 0$.

Again, if $\alpha_1, \alpha_2, \ldots, \alpha_m$ are elements of the extension K of F, we write $F(\alpha_1, \alpha_2, \ldots, \alpha_m)$ for $F(\alpha_1)(\alpha_2)\ldots(\alpha_m)$.

Definition 4.7

Let F be a field and $f(X) \in F[X]$. An extension field K of F is called a **splitting field** of $f(X)$ if

(i) $f(X)$ factors as a product of linear factors over K; and
(ii) if $\alpha_1, \alpha_2, \ldots, \alpha_m$ are the roots of $f(X)$ then $K = F(\alpha_1, \alpha_2, \ldots, \alpha_m)$.

The procedure adopted in Examples 4.1–3 for the construction of finite fields also yields the following result which besides being of independent interest is needed for the construction of a splitting field of a given polynomial.

Proposition 4.6

Let $f(X)$ be an irreducible polynomial over a field F. Then there exists an extension K of F in which $f(X)$ has a root.

Proof

The polynomial $f(X)$ being irreducible, $K = F[X]/\langle f(X) \rangle$ is a field (Theorem 4.2). The element $\alpha = X + \langle f(X) \rangle$ of K is then a root of $f(X)$. The map

$$a \to a + \langle f(X) \rangle \quad a \in F$$

is a homomorphism: $F \to K$ which is clearly a monomorphism. Identifying the element a of F with the corresponding element $a + \langle f(X) \rangle$ of K we can regard K as an extension of F.

Corollary

Given any non-constant polynomial $f(X)$ over a field F, there exists an extension K of F in which $f(X)$ has a root.

Proof

Let $g(X) \in F[X]$ be an irreducible factor of $f(X)$. Then there exists an extension K of F in which $g(X)$ has a root α (say). But then α is also a root of $f(X)$.

Observe that the field K constructed in Proposition 4.6 is the smallest extension of F in which $f(X)$ has a root.

Examples 4.4

Case (i)

Consider the polynomial $X^2 + 1$ over the field Q of rational numbers. This is an irreducible polynomial over Q and so

$$K = Q[X]/\langle X^2 + 1 \rangle$$

is a field. Let α denote the element $X + \langle X^2 + 1 \rangle$ of K. Then $\alpha^2 = -1$ and an arbitrary element of K is of the form $a + b\alpha$, $a, b \in Q$. In the usual terminology of complex numbers, α is the complex number i $(= \sqrt{-1})$ and so

$$K = \{a + b\mathrm{i}/a, b \in Q\} = Q(\mathrm{i})$$

is a subfield of the field $\mathbb{C}$ of complex numbers.

Case (ii)

As another example, consider the polynomial $X^2 - X + 1$ over Q. This is irreducible over Q and so $\langle X^2 - X + 1 \rangle$ is a maximal ideal in $Q[X]$. Hence

$$K = Q[X]/\langle X^2 - X + 1 \rangle$$

is a field. Let

$$\alpha = X + \langle X^2 - X + 1 \rangle$$

Since any element of K is of the form

$$aX + b + \langle X^2 - X + 1 \rangle$$

where $a, b \in Q$, and uniquely so, arbitrary element of K can be written as $a\alpha + b$, $a, b \in Q$. Also clearly α is a root of the polynomial $X^2 - X + 1$. The map

$$a \to a + \langle X^2 - X + 1 \rangle \quad a \in Q$$

is a ring monomorphism from Q into K and, therefore, its image in K is a subfield of K isomorphic to Q. We may identify $a \in Q$ with the corresponding element

$$a + \langle X^2 - X + 1 \rangle$$

of K and so Q may be regarded as a subfield of K.

In the usual terminology of complex numbers we may take α to be

$$\frac{1 + \sqrt{3}\mathrm{i}}{2} \quad \text{or} \quad \frac{1 - \sqrt{3}\mathrm{i}}{2}$$

If we take

$$\alpha = \frac{1+\sqrt{3}\mathrm{i}}{2}$$

then

$$\frac{1-\sqrt{3}\mathrm{i}}{2} = 1-\alpha \in Q(\alpha)$$

and $K = Q(\alpha)$.

Given an irreducible polynomial $f(X) \in F[X]$, there can exist distinct but isomorphic field extensions of F in which $f(X)$ has a root.

Example 4.5
Consider the polynomial $X^4 - 2 \in Q[X]$. The Eisenstein's irreducibility criterion shows that $X^4 - 2$ is irreducible over Q. The roots of this polynomial are $\alpha, -\alpha, \alpha\mathrm{i}, -\alpha\mathrm{i}$, where α is the real positive fourth root of 2 and $\mathrm{i} = \sqrt{-1}$. The polynomial $X^4 - 2$ is the minimal polynomial of both α and $\alpha\mathrm{i}$. Therefore, every element of $Q(\alpha)$ can be uniquely written as

$$a + b\alpha + c\alpha^2 + d\alpha^3$$

where $a, b, c, d \in Q$ and every element of $Q(\alpha\mathrm{i})$ can be uniquely written as

$$a + b\alpha\mathrm{i} - c\alpha^2 - d\alpha^3\mathrm{i}$$

where $a, b, c, d \in Q$. The map

$$\theta: Q(\alpha) \to Q(\alpha\mathrm{i})$$

defined by

$$\theta(a + b\alpha + c\alpha^2 + d\alpha^3) = a + b\alpha\mathrm{i} - c\alpha^2 - d\alpha^3\mathrm{i} \quad a, b, c, d \in Q$$

is an isomorphism of the two vector spaces. Also

$$\begin{aligned}
&\theta((a + b\alpha + c\alpha^2 + d\alpha^3)(a' + b'\alpha + c'\alpha^2 + d'\alpha^3)) \\
&\quad = \theta(aa' + (ab' + a'b)\alpha + (ac' + bb' + ca')\alpha^2 + (ad' + bc' + cb' + da')\alpha^3 \\
&\qquad + (bd' + cc' + db')2 + (cd' + dc')2\alpha + dd'2\alpha^2) \\
&\quad = \theta(aa' + 2(bd' + cc' + db') + (ab' + a'b + 2(cd' + dc'))\alpha \\
&\qquad + (ac' + bb' + ca' + 2dd')\alpha^2 + (ad' + bc' + cb' + da')\alpha^3) \\
&\quad = aa' + 2(bd' + cc' + db') + (ab' + a'b + 2(cd' + dc'))\alpha\mathrm{i} \\
&\qquad - (ac' + bb' + ca' + 2dd')\alpha^2 - (ad' + bc' + cb' + da')\alpha^3\mathrm{i} \\
&\quad = (a + b\alpha\mathrm{i} - c\alpha^2 - d\alpha^3\mathrm{i})(a' + b'\alpha\mathrm{i} - c'\alpha^2 - d'\alpha^3\mathrm{i}) \\
&\quad = \theta(a + b\alpha + c\alpha^2 + d\alpha^3)\theta(a' + b'\alpha + c'\alpha^2 + d'\alpha^3)
\end{aligned}$$

for $a, a', b, b', c, c', d, d' \in Q$.

Thus θ is an isomorphism of fields. Also $Q(\alpha)$ contains the root α of X^4-2 and $Q(\alpha i)$ contains the root αi of X^4-2. Clearly $Q(\alpha)$ is a subfield of $\mathbb{R}$, the field of real numbers while $Q(\alpha i)$ is not contained in $\mathbb{R}$.

Theorem 4.4
Let F be a field and $f(X) \in F[X]$. Then there exists a splitting field of $f(X)$ over F.

Proof
We prove the theorem by induction on the degree of $f(X)$. If $\deg f(X)=1$, then it is clear that F itself is a splitting field of $f(X)$. Suppose that $\deg f(X)=n \geq 2$. By Proposition 4.6 there exists an extension of F in which $f(X)$ has a root α_1 (say). Let

$$F_1 = F(\alpha_1)$$

Then

$$f(X) = (X - \alpha_1)g(X)$$

where $g(X) \in F_1[X]$ and $\deg g(X) = n-1$. By induction hypothesis $g(X)$ has a splitting field K over F_1, i.e. $g(X)$ factors as a product of linear factors over K and

$$K = F_1(\alpha_2, \ldots, \alpha_n)$$

where $\alpha_2, \ldots, \alpha_n$ are the roots of $g(X)$. But then $\alpha_1, \alpha_2, \ldots, \alpha_n$ are the roots of $f(X)$, $K = F(\alpha_1, \alpha_2, \ldots, \alpha_n)$ and $f(X)$ factors as a product of linear factors over K. Hence K is a splitting field of $f(X)$.

4.2 SOME EXAMPLES OF PRIMITIVE POLYNOMIALS

Examples 4.6
Let $F = F_3$ – the field of 3 elements. Then

$$aX^2 + bX + 1 \quad a, b \in F$$

are the only possible irreducible polynomials of degree 2 over F. But then we must also have

$$a + b + 1 \neq 0 \quad \text{and} \quad a - b + 1 \neq 0$$

Therefore the possible values of a and b are: $a = 1$, $b = 0$ or $a = -1$, $b = 1$ or $a = -1 = b$. Thus all the irreducible polynomials over F of degree 2 are:

$$X^2 + 1 \quad X^2 - X - 1 \quad X^2 + X - 1$$

The polynomial $X^2 + 1$ divides $X^4 - 1$ over F and so is not primitive. Again

$$X^3 + 1 = (X + 1)(X^2 - X + 1) = (X + 1)(X + 1)^2 = (X + 1)^3$$

and neither of $X^2 - X - 1$ and $X^2 + X - 1$ is a divisor of $X^3 + 1$. It is also clear that neither of these two polynomials is a divisor of

$$X^4 - 1 = (X - 1)(X + 1)(X^2 + 1)$$

Also

$$X^6 - 1 = (X^3 - 1)(X^3 + 1) = (X - 1)^3(X + 1)^3$$

and so neither of the two polynomials is a divisor of $X^6 - 1$. Suppose that

$$X^2 + X - 1 | X^5 - 1$$

Then $X^2 + X - 1$ must divide $X^4 + X^3 + X^2 + X + 1$. Then

$$X^4 + X^3 + X^2 + X + 1 = (X^2 + X - 1)(X^2 + aX - 1)$$

and $1 = a + 1$ and $-a - 1 = 1$. This gives a contradiction. Similarly

$$X^4 + X^3 + X^2 + X + 1 = (X^2 - X - 1)(X^2 + aX - 1)$$

gives $-1 + a = 1$ and $-a - 1 = 1$ which again lead to a contradiction. Thus, neither of the two polynomials under consideration divides $X^5 - 1$ over F. If

$$X^6 + X^5 + X^4 + X^3 + X^2 + X + 1 = (X^2 + X - 1)(X^4 + aX^3 + bX^2 + cX - 1)$$

then comparing coefficients of powers of X gives $1 + a = 1$, $a + b - 1 = 1$, $-a + b + c = 1$, $-b + c - 1 = 1$, $-c - 1 = 1$ in F which lead to a contradiction. Also

$$X^6 + X^5 + X^4 + X^3 + X^2 + X + 1 = (X^2 - X - 1)(X^4 + aX^3 + bX^2 + cX - 1)$$

gives $-1 + a = 1$, $-1 - a + b = 1$, $-a - b + c = 1$, $-b - c - 1 = 1$, $-c + 1 = 1$ which again lead to a contradiction.

Thus, both the polynomials $X^2 + X - 1$ and $X^2 - X - 1$ are primitive over F. We have thus proved that $X^2 + X - 1$ and $X^2 - X - 1$ are the only primitive polynomials of degree 2 over $F_3 = F$.

Examples 4.7

Here we consider some primitive polynomials over $\mathbb{B}$.

Case (i)

The polynomial $X + 1$ over $\mathbb{B}$ is trivially the only primitive polynomial of degree 1.

Case (ii)

Neither 0 nor 1 is a root of the polynomial $X^2 + X + 1$ over $\mathbb{B}$ and so it is an irreducible polynomial. It is trivially a primitive polynomial.

Case (iii)

We have already proved in Example 4.3 Case (i) that $X^3 + X + 1$ is a primitive polynomial of degree 3 because the element $X + \langle X^3 + X + 1 \rangle$ of the field $\mathbb{B}[X]/\langle X^3 + X + 1 \rangle$ is primitive.

Observe that if $f(X)$ is any irreducible polynomial over $\mathbb{B}$ of degree 3, then the field

$$K = \mathbb{B}[X]/\langle f(X)\rangle$$

is of order 8 and the multiplicative group K^* of K being of prime order, every non-zero, non-identity element of K is a primitive element. In particular so is the element $X + \langle f(X)\rangle$. This proves that $f(X)$ is a primitive polynomial.

Clearly $X^3 + X^2 + 1$ and $X^3 + X + 1$ are the only cubic polynomials over $\mathbb{B}$ which are irreducible. Thus $X^3 + X + 1$ and $X^3 + X^2 + 1$ are the only cubic primitive polynomials over $\mathbb{B}$.

Case (iv)

The argument in Case (iii) above can also be used to prove that every irreducible polynomial of degree 5 over $\mathbb{B}$ is primitive. Observe that

$$(X^2 + X + 1)(X^3 + X + 1) = X^5 + X^4 + 1$$

and

$$(X^2 + X + 1)(X^3 + X^2 + 1) = X^5 + X + 1$$

Also neither 0 nor 1 is a root of the polynomial $X^5 + X^2 + 1$ or $X^5 + X^3 + 1$. Hence $X^5 + X^2 + 1$ and $X^5 + X^3 + 1$ are two primitive polynomials.

Similarly $X^5 + X^4 + X^3 + X^2 + 1$, $X^5 + X^4 + X^3 + X + 1$, $X^5 + X^4 + X^2 + X + 1$ and $X^5 + X^3 + X^2 + X + 1$ are all the other irreducible polynomials of degree 5 over $\mathbb{B}$. Hence all the primitive polynomials of degree 5 over $\mathbb{B}$ are

$$X^5 + X^2 + 1 \quad X^5 + X^3 + 1 \quad X^5 + X^4 + X^3 + X^2 + 1$$

$$X^5 + X^4 + X^3 + X + 1 \quad X^5 + X^4 + X^2 + X + 1 \quad X^5 + X^3 + X^2 + X + 1$$

Case (v)

Since the multiplicative group of a field K of order 2^7 is of order 127 (a prime), every non-zero, non-identity element of K is primitive and, therefore, every irreducible polynomial of degree 7 over $\mathbb{B}$ is primitive. By direct computation we find that

$$(X^2 + X + 1)(X^5 + X^2 + 1) = X^7 + X^6 + X^5 + X^4 + X^3 + X + 1$$

$$(X^2 + X + 1)(X^5 + X^3 + 1) = X^7 + X^6 + X^4 + X^3 + X^2 + X + 1$$

$$(X^2 + X + 1)(X^5 + X^4 + X^3 + X^2 + 1) = X^7 + X^5 + X^4 + X + 1$$

$$(X^2 + X + 1)(X^5 + X^4 + X^3 + X + 1) = X^7 + X^5 + 1$$

$$(X^2 + X + 1)(X^5 + X^4 + X^2 + X + 1) = X^7 + X^2 + 1$$

$$(X^2 + X + 1)(X^5 + X^3 + X^2 + X + 1) = X^7 + X^6 + X^3 + X^2 + 1$$

$$(X^3 + X + 1)(X^4 + X^3 + 1) = X^7 + X^6 + X^5 + X + 1$$

$$(X^3 + X + 1)(X^4 + X + 1) = X^7 + X^5 + X^3 + X^2 + 1$$

$$(X^3+X+1)(X^4+X^3+X^2+X+1)=X^7+X^6+X^5+X^4+X^3+1$$
$$(X^3+X^2+1)(X^4+X^3+1)=X^7+X^5+X^4+X^2+1$$
$$(X^3+X^2+1)(X^4+X+1)=X^7+X^6+X^2+X+1$$
$$(X^3+X^2+1)(X^4+X^3+X^2+X+1)=X^7+X^4+X^3+X+1$$

From these computations, we can read of all irreducible and hence primitive polynomials of degree 7 over $\mathbb{B}$. These are

X^7+X^6+1	$X^7+X^6+X^3+X+1$
X^7+X^4+1	$X^7+X^5+X^4+X^3+1$
X^7+X^3+1	$X^7+X^5+X^3+X+1$
X^7+X+1	$X^7+X^4+X^3+X^2+1$
$X^7+X^6+X^5+X^2+1$	$X^7+X^4+X^2+X+1$
$X^7+X^6+X^5+X^3+1$	$X^7+X^3+X^2+X+1$
$X^7+X^6+X^5+X^4+1$	$X^7+X^5+X^2+X+1$
$X^7+X^6+X^4+X+1$	$X^7+X^6+X^5+X^4+X^3+X^2+1$
$X^7+X^6+X^4+X^2+1$	$X^7+X^6+X^5+X^4+X^2+X+1$
$X^7+X^6+X^4+X^3+1$	$X^7+X^6+X^5+X^3+X^2+X+1$

4.3 BOSE–CHAUDHURI–HOCQUENGHEM CODES

Hocquenghem (1959) and Bose and Ray-Chaudhuri (1960) independently proved a remarkable theorem which enables us to systematically construct one of the most powerful multiple error-correcting codes for random independent errors. These are polynomial codes and are now called Bose–Chaudhuri–Hocquenghem codes (BCH codes for short!). Recall that a polynomial code is determined as soon as the generator polynomial is determined. Procedure for constructing a BCH code is as follows.

Suppose that a BCH code with code word length n, minimum distance d and with symbols in $F=\mathrm{GF}(q)$, a field of order q (=a power of a prime p) is required. We choose the least positive integer r which satisfies $q^r \geq n+1$. Let K be an extension of F of degree r and let α be a primitive element of K. Let $m_i(X)$ be the minimal polynomial of α^i, $1 \leq i \leq d-1$ and set

$$g(X)=\mathrm{LCM}(m_1(X),\ldots,m_{d-1}(X))$$

Theorem 4.5
The polynomial code with symbols in F and encoding polynomial $g(X)$ has minimum distance at least d.

Proof
Let $h(X)$ be any polynomial over F which has $\alpha, \alpha^2, \ldots, \alpha^{d-1}$ among its roots. Then

$$m_i(X) \mid h(X) \forall i,\ 1 \leq i \leq d-1$$

and hence $g(X)|h(X)$. Thus $g(X)$ is the polynomial of least possible degree (up to a constant factor) with roots $\alpha, \alpha^2, \ldots, \alpha^{d-1}$.

If $c(X)$ is a code polynomial in the polynomial code generated by $g(X)$, then $c(X) = a(X)g(X)$ for some $a(X) \in F[X]$ and, therefore, $\alpha, \alpha^2, \ldots, \alpha^{d-1}$ are roots of $c(X)$.

We know that in a group code the minimum distance of the code equals the minimum of the weights of non-zero code words. Since polynomial codes are group codes, it follows that the code generated by $g(X)$ has minimum distance at least d if there is no code word $c_0 c_1 \ldots c_{n-1}$ with less than d non-zero entries. Suppose, to the contrary, that a code word has less than d non-zero entries. Then the corresponding code polynomial is of the form

$$c(X) = b_1 X^{n_1} + b_2 X^{n_2} + \cdots + b_{d-1} X^{n_{d-1}}$$

where $b_1, b_2, \ldots, b_{d-1} \in F$ and also, we may assume that

$$n_1 > n_2 > \cdots > n_{d-1} \geq 0$$

Since the code is of length n, every code polynomial is of degree at most $n-1$ and, therefore, $n_1 \leq n-1 \; (\leq q^r - 2)$. As already pointed out, $\alpha, \alpha^2, \ldots, \alpha^{d-1}$ are roots of $c(X)$ and we have

$$\begin{aligned} b_1 \alpha^{n_1} + b_2 \alpha^{n_2} + \cdots + b_{d-1} \alpha^{n_{d-1}} &= 0 \\ b_1 \alpha^{2n_1} + b_2 \alpha^{2n_2} + \cdots + b_{d-1} \alpha^{2n_{d-1}} &= 0 \\ \cdots \qquad \cdots \qquad & \cdots \\ b_1 \alpha^{(d-1)n_1} + b_2 \alpha^{(d-1)n_2} + \cdots + b_{d-1} \alpha^{(d-1)n_{d-1}} &= 0 \end{aligned}$$

or

$$\mathbf{A} \begin{pmatrix} b_1 \\ b_2 \\ \vdots \\ b_{d-1} \end{pmatrix} = 0 \tag{4.2}$$

where

$$\mathbf{A} = \begin{pmatrix} \alpha^{n_1} & \alpha^{n_2} & \cdots & \alpha^{n_{d-1}} \\ \alpha^{2n_1} & \alpha^{2n_2} & \cdots & \alpha^{2n_{d-1}} \\ \cdots & & \ddots & \\ \alpha^{(d-1)n_1} & \alpha^{(d-1)n_2} & \cdots & \alpha^{(d-1)n_{d-1}} \end{pmatrix}$$

The determinant of $\mathbf{A}$ is a **Vandermonde determinant** and we know that

$$\det \mathbf{A} = \prod_{i>j} (\alpha^{n_i} - \alpha^{n_j})$$

The element α of K being primitive and

$$q^r - 1 > n_1 > n_2 > \cdots > n_{d-1} \geq 0$$

we have $\alpha^{n_i} - \alpha^{n_j} \neq 0$ for $i \neq j$ and, therefore, $\det \mathbf{A} \neq 0$. Now (4.2) is a system of $d-1$ homogeneous linear equations in $d-1$ variables $b_1, \ldots, b_{d-1}$ and $\det \mathbf{A} \neq 0$. Therefore the system of equations admits only the zero solution and $c(X) = 0$. Hence there is no non-zero code word with less than d non-zero entries and the code has minimum distance at least d.

Examples 4.8

Case (i)

Construct a binary BCH code of length 7 and minimum distance 3.

Here $n = 7$ and so we need to construct an extension of $\mathbb{B}$ of degree r where $2^r \geq 7 + 1 = 8$. Thus we take $r = 3$. We know from Example 4.7 (Case (iii)) that $X^3 + X + 1$ is a primitive polynomial of degree 3 over $\mathbb{B}$. Therefore

$$K = B[X]/\langle X^3 + X + 1 \rangle$$

is a field of order 8 and

$$\alpha = X + \langle X^3 + X + 1 \rangle$$

is a primitive element of K. Then α satisfies the relations $\alpha^3 + \alpha + 1 = 0$, and $\alpha^7 = 1$ and $X^3 + X + 1$ is the minimal polynomial of α. Since α and α^2 have the same minimal polynomial (Proposition 4.2), the generator polynomial of the required BCH code is $X^3 + X + 1$.

The message polynomials are of degree at most 3. If

$$a(X) = a_0 + a_1 X + a_2 X^2 + a_3 X^3$$

is an arbitrary message polynomial, the corresponding code polynomial is $a(X)(X^3 + X + 1)$ and so the corresponding code word is

$$(a_0, a_1 + a_0, a_2 + a_1, a_3 + a_2 + a_0, a_3 + a_1, a_2, a_3)$$

The encoding polynomial has 3 non-zero terms and, therefore, the code has minimum distance 3.

If we had started with the primitive polynomial $X^3 + X^2 + 1$, the corresponding BCH code with code word length 7 and minimum distance at least 3 is the polynomial code with encoding polynomial $X^3 + X^2 + 1$.

Case (ii)

Next we construct a binary BCH code of length 15 and minimum distance 5.

Here $n = 15 \geq 2^4 - 1$ and so we need to construct an extension K of $\mathbb{B}$ of degree 4. We have seen earlier (Example 4.3 Case (ii)) that $X^4 + X + 1$ is a primitive polynomial and so

$$\alpha = X + \langle X^4 + X + 1 \rangle$$

is a primitive element of

$$K = \mathbb{B}[X]/\langle X^4 + X + 1 \rangle$$

The minimal polynomial of α is $m_1(X) = X^4 + X + 1$. Also

$$m_2(X) = m_4(X) = m_1(X)$$

(Proposition 4.2). We next have to find the minimal polynomial of α^3. The elements $\alpha^3, \alpha^6, \alpha^{12}, \alpha^9$ have the same minimal polynomial and so

$$\begin{aligned} m_3(X) &= (X-\alpha^3)(X-\alpha^6)(X-\alpha^9)(X-\alpha^{12}) \\ &= X^4 + X^3(\alpha^3+\alpha^6+\alpha^9+\alpha^{12}) + X^2(\alpha^9+\alpha^{12}+\alpha^{15}+\alpha^{15}+\alpha^{18}+\alpha^{21}) \\ &\quad + X(\alpha^{18}+\alpha^{21}+\alpha^{24}+\alpha^{27}) + \alpha^{30} \\ &= X^4 + X^3(\alpha^3+\alpha^3+\alpha^2+\alpha^3+\alpha+\alpha^3+\alpha^2+\alpha+1) \\ &\quad + X^2(\alpha^9+\alpha^{12}+\alpha^3+\alpha^6) + X(\alpha^3+\alpha^6+\alpha^9+\alpha^{12}) + 1 \\ &= X^4+X^3+X^2+X+1 \end{aligned}$$

Therefore, the encoding polynomial of the BCH code with minimum distance at least 5 is

$$\begin{aligned} g(X) &= (X^4+X+1)(X^4+X^3+X^2+X+1) \\ &= X^8+X^7+X^6+X^4+1 \end{aligned}$$

Since the encoding polynomial has 5 non-zero terms, the minimum distance of the code is exactly 5. The code being of length 15, a message polynomial is of degree at most 6. Let

$$a(X) = a_0 + a_1X + a_2X^2 + a_3X^3 + a_4X^4 + a_5X^5 + a_6X^6$$

be an arbitrary message polynomial. The code word corresponding to the code polynomial $a(X)g(X)$ is

$$\begin{aligned} &(a_0, a_1, a_2, a_3, a_4+a_0, a_5+a_1, a_6+a_2+a_0, a_0+a_1+a_3, a_0+a_1+a_2+a_4, \\ &\quad a_1+a_2+a_3+a_5, a_2+a_3+a_4+a_6, a_3+a_4+a_5, a_4+a_5+a_6, a_5+a_6, a_6) \end{aligned}$$

Case (iii)
Find a generator polynomial of the binary BCH code of length 31 and minimum distance 5, it being given that $X^5 + X^2 + 1$ is an irreducible polynomial over $\mathbb{B}$.

Solution
The polynomial $X^5 + X^2 + 1$ being irreducible,

$$F = \mathbb{B}[X]/\langle X^5 + X^2 + 1\rangle$$

is a field of order 32. Since $F^* = F\setminus\{0\}$ is a cyclic group of order 31 – a prime – every non-identity element of F^* is a primitive element of F. In particular

$$\alpha = X + \langle X^5 + X^2 + 1\rangle$$

is a primitive element of F and $X^5 + X^2 + 1$ is the minimal polynomial of α.

Let $m_i(X)$ be the minimal polynomial of α^i, $1 \le i \le 4$. Observe that α, α^2, α^4 have the same minimal polynomial. So,

$$m_1(X) = m_2(X) = m_4(X) = X^5 + X^2 + 1$$

We now have to find $m_3(X)$. As α^3, α^6, α^{12}, α^{24}, α^{17} have the same minimal polynomial and $\{3, 6, 12, 24, 17\}$ is the complete cyclotomic class modulo 31 relative to 2,

$$m_3(X) = (X - \alpha^3)(X - \alpha^6)(X - \alpha^{12})(X - \alpha^{17})(X - \alpha^{24})$$

In $m_3(X)$, the coefficient of X^4 is

$$\begin{aligned}
\alpha^3 + \alpha^6 + \alpha^{12} + \alpha^{17} + \alpha^{24} &= \alpha^3 + \alpha^3 + \alpha + \alpha^{12} + (\alpha^3 + \alpha)^4 + (\alpha^2 + 1)(\alpha^6 + \alpha^2) \\
&= \alpha + \alpha^{12} + \alpha^{12} + \alpha^4 + \alpha^8 + \alpha^6 + \alpha^4 + \alpha^2 \\
&= \alpha + \alpha^2 + \alpha^6 + \alpha^8 \\
&= \alpha + \alpha^2 + \alpha^3 + \alpha + \alpha^3(\alpha^2 + 1) \\
&= \alpha^5 + \alpha^2 \\
&= 1
\end{aligned}$$

The coefficient of X^3 is

$$\begin{aligned}
&\alpha^9 + \alpha^{15} + \alpha^{20} + \alpha^{27} + \alpha^{18} + \alpha^{23} + \alpha^{30} + \alpha^{29} + \alpha^5 + \alpha^{10} \\
&\quad = \alpha^9 + \alpha^{17} + \alpha^7(\alpha^2 + 1)^4 + \alpha^{20} + (\alpha^4 + 1)(\alpha^8 + 1) + \alpha^9(\alpha^8 + 1) + \alpha^7 \\
&\quad = \alpha^9 + \alpha^{17} + \alpha^{15} + \alpha^7 + \alpha^8 + 1 + \alpha^{12} + \alpha^8 + \alpha^4 + 1 + \alpha^{17} + \alpha^9 + \alpha^7 \\
&\quad = \alpha^{15} + \alpha^{12} + \alpha^4 \\
&\quad = (\alpha^2 + 1)^3 + \alpha^2(\alpha^4 + 1) + \alpha^4 \\
&\quad = \alpha^6 + \alpha^4 + \alpha^2 + 1 + \alpha^6 + \alpha^2 + \alpha^4 \\
&\quad = 1
\end{aligned}$$

The coefficient of X^2 is

$$\begin{aligned}
&\alpha^{21} + \alpha^{26} + \alpha^2 + \alpha + \alpha^8 + \alpha^{13} + \alpha^4 + \alpha^{11} + \alpha^{16} + \alpha^{22} \\
&\quad = \alpha^{23} + \alpha^2 + \alpha + \alpha^{10} + \alpha^4 + \alpha^{13} + \alpha^2(\alpha^8 + 1) \\
&\quad = \alpha^3(\alpha^8 + 1) + \alpha + \alpha^4 + \alpha^3(\alpha^4 + 1) \\
&\quad = \alpha^{11} + \alpha + \alpha^4 + \alpha^7 \\
&\quad = \alpha(\alpha^4 + 1) + \alpha + \alpha^4 + \alpha^2(\alpha^2 + 1) \\
&\quad = \alpha^5 + \alpha^2 \\
&\quad = 1
\end{aligned}$$

The coefficient of X is

$$\begin{aligned}\alpha^7+\alpha^{14}+\alpha^{19}+\alpha^{25}+\alpha^{28} &= \alpha^7+\alpha^{16}+\alpha^5(\alpha^8+1)+\alpha^8(\alpha^8+1)\\ &= \alpha^7+\alpha^{13}+\alpha^5+\alpha^8\\ &= \alpha^2(\alpha^2+1)+\alpha^{10}+\alpha^5\\ &= \alpha^4+\alpha^2+\alpha^4+1+\alpha^2+1\\ &= 0\end{aligned}$$

The constant term is

$$\alpha^{3+6+12+17+24}=\alpha^{62}=1$$

Hence

$$m_3(X)=X^5+X^4+X^3+X^2+1$$

Therefore the generator polynomial of the BCH code of length 31 over $\mathbb{B}$ is

$$\begin{aligned}g(X) &= \mathrm{LCM}\{m_1(X), m_2(X), m_3(X), m_4(X)\}\\ &= \mathrm{LCM}\{m_1(X), m_3(X)\}\\ &= m_1(X)m_3(X)\\ &= (X^5+X^2+1)(X^5+X^4+X^3+X^2+1)\\ &= X^{10}+X^9+X^8+X^6+X^5+X^3+1\end{aligned}$$

Case (iv)
Find a generator polynomial for a 5-error-correcting binary BCH code of length 63, it being given that X^6+X+1 is a primitive polynomial over $\mathbb{B}$.

Solution
The length of the code is $63=2^6-1$ and we need to find an extension of $\mathbb{B}$ of degree 6. It being given that X^6+X+1 is a primitive polynomial over $\mathbb{B}$,

$$K=\mathbb{B}[X]/\langle X^6+X+1\rangle$$

is a field of order 2^6,

$$\alpha=X+\langle X^6+X+1\rangle$$

is a primitive element of K and X^6+X+1 is the minimal polynomial of α.

Since the code we are looking for is to be 5-error-correcting, the minimum distance of the code is at least $2\times 5+1=11$ (Theorem 1.2). We thus have to find the minimal polynomials $m_i(X)$ of α^i for $1\leq i\leq 10$. It follows from

Proposition 4.2 that

$$m_1(X) = m_2(X) = m_4(X) = m_8(X) = m_{16}(X) = m_{32}(X)$$
$$m_3(X) = m_6(X) = m_{12}(X) = m_{24}(X) = m_{48}(X) = m_{33}(X)$$
$$m_5(X) = m_{10}(X) = m_{20}(X) = m_{40}(X) = m_{17}(X) = m_{34}(X)$$
$$m_7(X) = m_{14}(X) = m_{28}(X) = m_{56}(X) = m_{49}(X) = m_{35}(X)$$
$$m_9(X) = m_{18}(X) = m_{36}(X)$$

Thus $m_1(X), m_3(X), m_5(X), m_7(X)$ are of degree 6 each while $m_9(X)$ is of degree 3. Also then the encoding polynomial of the BCH code we are looking for is

$$g(X) = m_1(X)m_3(X)m_5(X)m_7(X)m_9(X)$$

which is of degree 27. Now

$$\alpha^6 = \alpha + 1 \Rightarrow \alpha^{12} = \alpha^2 + 1$$
$$\alpha^{24} = \alpha^4 + 1$$
$$\alpha^{48} = \alpha^8 + 1 = \alpha^3 + \alpha^2 + 1$$
$$\alpha^{33} = (\alpha^4 + 1)(\alpha^4 + \alpha^3) = \alpha^8 + \alpha^7 + \alpha^4 + \alpha^3 = \alpha^3 + \alpha^2 + \alpha^2 + \alpha + \alpha^4 + \alpha^3$$
$$= \alpha^4 + \alpha$$

Therefore

$$\begin{aligned}
m_3(X) &= (X + \alpha^3)(X + \alpha + 1)(X + \alpha^2 + 1)(X + \alpha^4 + 1)(X + \alpha^3 + \alpha^2 + 1) \\
&\quad \times (X + \alpha^4 + \alpha) \\
&= [X^2 + X(\alpha^3 + \alpha + 1) + \alpha^4 + \alpha^3][X^2 + X(\alpha^4 + \alpha^2) + \alpha^4 + \alpha^2 + \alpha] \\
&\quad \times [X^2 + X(\alpha^4 + \alpha^3 + \alpha^2 + \alpha + 1) + \alpha^3 + \alpha^2 + \alpha + 1] \\
&= [X^4 + X^3(\alpha^4 + \alpha^3 + \alpha^2 + \alpha + 1) + X^2(\alpha^4 + \alpha^2) + X(\alpha^5 + \alpha^2 + 1) \\
&\quad + \alpha^4 + \alpha^3 + 1][X^2 + X(\alpha^4 + \alpha^3 + \alpha^2 + \alpha + 1) + \alpha^3 + \alpha^2 + \alpha + 1] \\
&= X^6 + X^4(\alpha^8 + \alpha^6 + \alpha^4 + \alpha^2 + 1 + \alpha^4 + \alpha^2 + \alpha^3 + \alpha^2 + \alpha + 1) \\
&\quad + X^3(\alpha^8 + \alpha^2 + \alpha^3) + X^2(\alpha^9 + \alpha^8 + \alpha^6 + \alpha^4 + \alpha^2 + \alpha) \\
&\quad + X(\alpha^2 + \alpha^6 + \alpha^7) + (\alpha^7 + \alpha^2 + \alpha + 1) \\
&= X^6 + X^4 + X^3 + X^2 + X + 1
\end{aligned}$$

Again

$$\alpha^{10} = \alpha^5 + \alpha^4 \quad \alpha^{20} = \alpha^5 + \alpha^4 + \alpha^3 + \alpha^2$$
$$\alpha^{40} = \alpha^5 + \alpha^3 + \alpha^2 + \alpha + 1$$
$$\alpha^{17} = (\alpha^5 + \alpha + 1)(\alpha + 1) = \alpha^5 + \alpha^2 + \alpha$$
$$a^{34} = \alpha^{10} + \alpha^4 + \alpha^2 = \alpha^5 + \alpha^2$$

Therefore

$$\begin{aligned}
m_5(X) = &(X + \alpha^5 + \alpha^4)(X + \alpha^5 + \alpha^4 + \alpha^3 + \alpha^2)(X + \alpha^5 + \alpha^3 + \alpha^2 + \alpha + 1) \\
&\times (X + \alpha^5 + \alpha^2 + \alpha)(X + \alpha^5 + \alpha^2)(X + \alpha^5) \\
= &[X^2 + X(\alpha^3 + \alpha^2) + \alpha^5 + \alpha^4 + \alpha + 1] \\
&\times [X^2 + X(\alpha^3 + 1) + \alpha^5 + \alpha^4 + \alpha^3 + \alpha^2 + \alpha] \\
&\times [X^2 + \alpha^2 X + \alpha^5 + \alpha^4 + \alpha^2 + \alpha] \\
= &[X^4 + X^3(\alpha^2 + 1) + X^2(\alpha^3 + \alpha^2 + 1 + \alpha^5 + \alpha^3 + \alpha^2 + \alpha + 1) \\
&+ X(\alpha^5 + \alpha^2 + 1) + \alpha^4 + 1][X^2 + \alpha^2 X + \alpha^5 + \alpha^4 + \alpha^2 + \alpha] \\
= &X^6 + X^5(\alpha^2 + 1 + \alpha^2) + X^4(\alpha^5 + \alpha^4 + \alpha^2 + \alpha + \alpha^4 + \alpha^2 + \alpha^5 + \alpha) \\
&+ X^3[(\alpha^2 + 1)(\alpha^5 + \alpha^4 + \alpha^2 + \alpha) + \alpha^2(\alpha^5 + \alpha) + \alpha^5 + \alpha^2 + 1] \\
&+ X^2[\alpha^4 + 1 + \alpha^2(\alpha^5 + \alpha^2 + 1) + (\alpha^5 + \alpha)(\alpha^5 + \alpha^4 + \alpha^2 + \alpha)] \\
&+ X[\alpha^2(\alpha^4 + 1) + (\alpha^5 + \alpha^2 + 1)(\alpha^5 + \alpha^4 + \alpha^2 + \alpha)] \\
&+ (\alpha^4 + 1)(\alpha^5 + \alpha^4 + \alpha^2 + \alpha) \\
= &X^6 + X^5 + X^2 + X + 1
\end{aligned}$$

For finding $m_7(X)$:

$$\begin{aligned}
\alpha^7 &= \alpha^2 + \alpha \\
a^{14} &= \alpha^4 + \alpha^2 \\
\alpha^{28} &= \alpha^4 + \alpha^3 + \alpha^2 \\
\alpha^{56} &= \alpha^4 + \alpha^3 + \alpha^2 + \alpha + 1 \\
\alpha^{49} &= \alpha^4 + \alpha^3 + \alpha \\
a^{35} &= \alpha^3 + \alpha + 1
\end{aligned}$$

Then

$$\alpha^7 + \alpha^{14} + \alpha^{28} + \alpha^{56} + \alpha^{49} + \alpha^{35} = 0$$

$$\alpha^7 \times \alpha^{14} \times \alpha^{28} \times \alpha^{56} \times \alpha^{49} \times \alpha^{35} = \alpha^{189} = 1$$

Sum of the products of these 6 powers of α taken 5 at a time

$$\begin{aligned}
&= \alpha^{63-7} + \alpha^{63-14} + \alpha^{63-28} + \alpha^{63-56} + \alpha^{63-49} + \alpha^{63-35} \\
&= \alpha^{56} + \alpha^{49} + \alpha^{35} + \alpha^7 + \alpha^{14} + \alpha^{28} \\
&= 0
\end{aligned}$$

Since α is an element of a field K and

$$\alpha^7 \times \alpha^{56} = \alpha^{14} \times \alpha^{49} = \alpha^{28} \times \alpha^{35} = 1$$

sum of products of these elements taken 2 at a time

$$= \text{sum of products of these elements taken 4 at a time}$$
$$= \alpha^7(\alpha^{14} + \alpha^{28} + \alpha^{35} + \alpha^{49} + \alpha^{56}) + \alpha^{14}(\alpha^{28} + \alpha^{35} + \alpha^{49} + \alpha^{56})$$
$$+ \alpha^{28}(\alpha^{35} + \alpha^{49} + \alpha^{56}) + \alpha^{35}(\alpha^{49} + \alpha^{56}) + \alpha^{49} \times \alpha^{56}$$
$$= \alpha^7 \times \alpha^7 + \alpha^{14}(\alpha^7 + \alpha^{14}) + (1 + \alpha^{14} + \alpha^{21}) + (\alpha^{21} + \alpha^{28}) + \alpha^{42}$$
$$= \alpha^{14} + \alpha^{21} + \alpha^{28} + 1 + \alpha^{14} + \alpha^{21} + \alpha^{21} + \alpha^{28} + \alpha^{42}$$
$$= 1 + \alpha^{21} + \alpha^{42}$$
$$= 1 + (\alpha^2 + \alpha)^3 + (\alpha^2 + \alpha)^6$$
$$= 1 + (\alpha^5 + \alpha^4 + \alpha^3 + \alpha + 1) + (\alpha^5 + \alpha^4 + \alpha^3 + \alpha + 1)^2$$
$$= 1 + (\alpha^5 + \alpha^4 + \alpha^3 + \alpha + 1) + (\alpha^5 + \alpha^4 + \alpha^3 + \alpha^2 + \alpha + 1 + \alpha^2 + 1)$$
$$= 0$$

Sum of the products of these powers of α taken 3 at a time

$$= (\alpha^{49} + \alpha^{56} + \alpha^7 + \alpha^{14}) + (\alpha^7 + \alpha^{21} + \alpha^{28}) + (\alpha^{28} + \alpha^{35}) + \alpha^{49}$$
$$+ (\alpha^{14} + \alpha^{28} + \alpha^{35}) + (\alpha^{35} + \alpha^{42}) + \alpha^{56} + (\alpha^{49} + \alpha^{56}) + \alpha^7 + \alpha^{14}$$
$$= \alpha^7 + \alpha^{14} + \alpha^{21} + \alpha^{28} + \alpha^{35} + \alpha^{42} + \alpha^{49} + \alpha^{56}$$
$$= \alpha^{21} + \alpha^{42}$$
$$= 1$$

Therefore

$$m_7(X) = X^6 + X^3 + 1$$

We are now left with computing $m_9(X)$.

$$m_9(X) = (X + \alpha^9)(X + \alpha^{18})(X + \alpha^{36})$$
$$= X^3 + X^2(\alpha^9 + \alpha^{18} + \alpha^{36}) + X(\alpha^{27} + \alpha^{45} + \alpha^{54}) + 1$$
$$= X^3 + X^2[(\alpha^4 + \alpha^3) + (\alpha^4 + \alpha^3)^2 + (\alpha^4 + \alpha^3)^4]$$
$$+ X[\alpha^{27} + \alpha^{45} + \alpha^{54}] + 1$$

Now

$$\alpha^{18} = (\alpha^4 + \alpha^3)^2 = \alpha^8 + \alpha^6 = \alpha^3 + \alpha^2 + \alpha + 1$$
$$\alpha^{36} = (\alpha^3 + \alpha^2 + \alpha + 1)^2 = \alpha^4 + \alpha^2 + \alpha$$
$$\alpha^{27} = (\alpha^4 + \alpha^3)(\alpha^3 + \alpha^2 + \alpha + 1) = \alpha^7 + \alpha^3 = \alpha^3 + \alpha^2 + \alpha$$
$$\alpha^{54} = (\alpha^3 + \alpha^2 + \alpha)^2 = \alpha^4 + \alpha^2 + \alpha + 1$$
$$\alpha^{45} = (\alpha^3 + \alpha^2 + \alpha + 1)(\alpha^3 + \alpha^2 + \alpha) = \alpha^4 + \alpha^3 + 1$$

Therefore

$$m_9(X) = X^3 + X^2 + 1$$

The encoding polynomial $g(X)$ of the code with minimum distance at least 11 is

$$\begin{aligned}
g(X) &= m_1(X)m_3(X)m_5(X)m_7(X)m_9(X) \\
&= (X^6 + X + 1)(X^6 + X^4 + X^3 + X^2 + X + 1)(X^6 + X^5 + X^2 + X + 1) \\
&\quad \times (X^6 + X^3 + 1)(X^3 + X^2 + 1) \\
&= (X^{12} + X^{10} + X^9 + X^8 + X^7 + X^6 + X^7 + X^5 + X^4 + X^3 + X^2 + X \\
&\quad + X^6 + X^4 + X^3 + X^2 + X + 1)(X^6 + X^5 + X^2 + X + 1) \\
&\quad \times (X^9 + X^8 + X^6 + X^6 + X^5 + X^3 + X^3 + X^2 + 1) \\
&= (X^{12} + X^{10} + X^9 + X^8 + X^5 + 1)(X^6 + X^5 + X^2 + X + 1) \\
&\quad \times (X^9 + X^8 + X^5 + X^2 + 1) \\
&= (X^{12} + X^{10} + X^9 + X^8 + X^5 + 1)(X^{15} + X^{14} + X^{11} + X^{10} + X^9 \\
&\quad + X^{14} + X^{13} + X^{10} + X^9 + X^8 + X^{11} + X^{10} + X^7 + X^6 + X^5 + X^8 \\
&\quad + X^7 + X^4 + X^3 + X^2 + X^6 + X^5 + X^2 + X + 1) \\
&= (X^{12} + X^{10} + X^9 + X^8 + X^5 + 1)(X^{15} + X^{13} + X^{10} + X^4 + X^3 \\
&\quad + X + 1) \\
&= (X^{27} + X^{25} + X^{24} + X^{23} + X^{20} + X^{15} + X^{25} + X^{23} + X^{22} + X^{21} \\
&\quad + X^{18} + X^{13} + X^{22} + X^{20} + X^{19} + X^{18} + X^{15} + X^{10} + X^{16} + X^{14} \\
&\quad + X^{13} + X^{12} + X^9 + X^4 + X^{15} + X^{13} + X^{12} + X^{11} + X^8 + X^3 + X^{13} \\
&\quad + X^{11} + X^{10} + X^9 + X^6 + X + X^{12} + X^{10} + X^9 + X^8 + X^5 + 1 \\
&= X^{27} + X^{24} + X^{21} + X^{19} + X^{16} + X^{15} + X^{14} + X^{12} + X^{10} + X^9 \\
&\quad + X^6 + X^5 + X^4 + X^3 + X + 1
\end{aligned}$$

Case (v)
To construct the minimal polynomial of a 2-error-correcting binary BCH code of length 21.

Solution
The length of the code is $21 \leq 2^5 - 1$ and so we need to construct an extension of $\mathbb{B}$ of degree 5. We know from Example 4.7 Case (iv) that $X^5 + X^2 + 1$ is a primitive polynomial of degree 5 over $\mathbb{B}$. Therefore

$$K = \mathbb{B}[X]/\langle X^5 + X^2 + 1 \rangle$$

is a field of order 32,

$$\alpha = X + \langle X^5 + X^2 + 1 \rangle$$

is a primitive element in K (Proposition 4.5) and $X^5 + X^2 + 1$ is the minimal polynomial of α.

The code is 2-error-correcting, therefore the minimum distance of the code is at least $2 \times 2 + 1 = 5$ (Theorem 1.2). We then need the minimal polynomials $m_i(X)$ of α^i, $1 \leq i \leq 4$. But it follows from Proposition 4.2 that

$$m_1(X) = m_2(X) = m_4(X)$$

and

$$m_3(X) = m_6(X) = m_{12}(X) = m_{24}(X) = m_{17}(X)$$

Thus $m_3(X)$ is the polynomial with roots α^3, α^6, α^{12}, α^{24} and α^{17}. Now $\alpha^5 = \alpha^2 + 1$ and so

$$\alpha^6 = \alpha^3 + \alpha \quad \alpha^{12} = \alpha^6 + \alpha^2 = \alpha^3 + \alpha^2 + \alpha$$

$$\alpha^{24} = \alpha^6 + \alpha^4 + \alpha^2 = \alpha^4 + \alpha^3 + \alpha^2 + \alpha$$

$$\alpha^{17} = \alpha^8 + \alpha^6 + \alpha^4 + \alpha^2 = \alpha^3 + \alpha^2 + 1 + \alpha^3 + \alpha + \alpha^4 + \alpha^2 = \alpha^4 + \alpha + 1$$

Then sum of the roots of $m_3(X) = 1$.

Sum of the roots of $m_3(X)$ taken 2 at a time

$$\begin{aligned}
&= \alpha^3(\alpha^6 + \alpha^{12} + \alpha^{24} + \alpha^{17}) + \alpha^6(\alpha^{12} + \alpha^{24} + \alpha^{17}) + \alpha^5 + \alpha^{29} + \alpha^{10} \\
&= \alpha^3(1 + \alpha^3) + \alpha^6(1 + \alpha^3 + \alpha^6) + \alpha^5 + \alpha^{10} + \alpha^{29} \\
&= \alpha^3 + \alpha^5 + \alpha^{10} + \alpha^9 + \alpha^{12} + \alpha^{29} \\
&= \alpha^3 + \alpha^2 + 1 + \alpha^4 + 1 + (\alpha^4 + \alpha^3 + \alpha) + (\alpha^3 + \alpha^2 + \alpha) \\
&\quad + (\alpha^4 + \alpha^3 + \alpha^2 + \alpha)(\alpha^2 + 1) \\
&= \alpha^3 + \alpha^6 + \alpha^5 + \alpha^4 + \alpha^3 + \alpha^4 + \alpha^3 + \alpha^2 + \alpha \\
&= 1
\end{aligned}$$

Sum of the roots of $m_3(X)$ taken 3 at a time

$$\begin{aligned}
&= \alpha^{21} + \alpha^2 + \alpha^{26} + \alpha^8 + \alpha + \alpha^{11} + \alpha^4 + \alpha^{22} + \alpha^{13} + \alpha^{16} \\
&= \alpha + \alpha^2 + \alpha^4 + \alpha^4(\alpha^4 + \alpha + 1) + \alpha^8 + (\alpha^2 + \alpha + 1) + (\alpha^2 + \alpha + 1)^2 \\
&\quad + \alpha^2(\alpha^2 + \alpha + 1) + \alpha^{16} + \alpha^{26} \\
&= 1 + \alpha^5 + (\alpha^4 + \alpha^2 + 1) + (\alpha^4 + \alpha^3 + \alpha^2) + \alpha^4(\alpha^3 + \alpha^2 + \alpha) \\
&\quad + \alpha^2(\alpha^4 + \alpha^3 + \alpha^2 + \alpha) \\
&= \alpha^4 + \alpha^5 + \alpha^7 = \alpha^4 + \alpha^2 + 1 + \alpha^4 + \alpha^2 \\
&= 1
\end{aligned}$$

Sum of the roots of $m_3(X)$ taken 4 at a time

$$= \alpha^{14} + \alpha^7 + \alpha^{19} + \alpha^{25} + \alpha^{28}$$

$$= \alpha^4 + \alpha^2 + \alpha^2(\alpha^3 + \alpha^2 + \alpha) + \alpha^2(\alpha^4 + \alpha + 1) + \alpha(\alpha^4 + \alpha^3 + \alpha^2 + \alpha) + (\alpha^4 + \alpha + 1)(\alpha^2 + \alpha + 1)$$

$$= \alpha^2 + \alpha^3 + \alpha^4 + \alpha^6 + \alpha^6 + \alpha^5 + \alpha^4 + \alpha^3 + \alpha^2 + \alpha + \alpha^2 + \alpha + 1$$

$$= 1 + \alpha^2 + \alpha^5 = 0$$

Therefore

$$m_3(X) = X^5 + X^4 + X^3 + X^2 + 1$$

Hence the encoding polynomial of the required BCH code is

$$g(X) = (X^5 + X^2 + 1)(X^5 + X^4 + X^3 + X^2 + 1)$$
$$= X^{10} + X^9 + X^8 + X^6 + X^5 + X^3 + 1$$

Case (vi)

So far we have only constructed binary BCH codes. We now want to find the encoding polynomial of a 2-error-correcting BCH code of length 8 over GF(3).

The length of the code being $8 = 3^2 - 1$, we have to construct a field extension of GF(3) of degree 2. We have already proved (Example 4.6) that $X^2 + X + 2$ is a primitive polynomial of degree 2 over $\mathrm{GF}(3) = F_3$. Therefore

$$K = F_3[X]/\langle X^2 + X + 2\rangle$$

is a field of order 9,

$$\alpha = X + \langle X^2 + X + 2\rangle$$

is a primitive element of K (Proposition 4.5) and $X^2 + X + 2$ is the minimal polynomial of α over F_3. The code is 2-error-correcting and so the minimum distance of the code is at least $2 \times 2 + 1 = 5$ (Theorem 1.2). We therefore need to find the minimal polynomials $m_i(X)$ of α^i, $1 \le i \le 4$. We know from Proposition 4.2 that

$$m_1(X) = m_3(X) \quad \text{and} \quad m_2(X) = m_6(X)$$

Therefore

$$m_2(X) = (X - \alpha^2)(X - \alpha^6) = X^2 - (\alpha^2 + \alpha^6)X + 1$$
$$= X^2 - [(2\alpha + 1) + (2\alpha + 1)^3]X + 1$$
$$= X^2 - [2\alpha + 1 + 2\alpha^3 + 1]X + 1$$
$$= X^2 - [2\alpha + 2 + 2\alpha(2\alpha + 1)]X + 1$$
$$= X^2 + 1$$

$$\begin{aligned} m_4(X) &= X - \alpha^4 \\ &= X - (2\alpha + 1)^2 \\ &= X - 4\alpha^2 - 4\alpha - 1 \\ &= X - (2\alpha + 1) - \alpha - 1 \\ &= X - 2 \\ &= X + 1 \end{aligned}$$

The generating polynomial of the BCH code is

$$\begin{aligned} g(X) &= m_1(X)m_2(X)m_4(X) \\ &= (X^2 + X + 2)(X^2 + 1)(X + 1) \\ &= (X^4 + X^3 + 2X^2 + X^2 + X + 2)(X + 1) \\ &= X^5 + 2X^4 + X^3 + X^2 + 2 \end{aligned}$$

The generating polynomial contains 5 non-zero terms and, therefore, the minimum distance of the code is 5.

Case (vii)
We end this set of examples by constructing a single-error-correcting BCH code of length 10 over $GF(5) = F_5$, it being given that $X^2 + X + 2$ is primitive over F_5.

The length of the code is $10 \leq 5^2 - 1$ and so we have to construct an extension of F_5 of degree 2. It being given that $X^2 + X + 2$ is primitive over F_5,

$$K = F_5[X]/\langle X^2 + X + 2\rangle$$

is an extension of F_5 of degree 2,

$$\alpha = X + \langle X^2 + X + 2\rangle$$

is a primitive element of K (Proposition 4.5) and $X^2 + X + 2$ is the minimal polynomial of α over F_5. The code being single-error-correcting has to have minimum distance at least 3. Therefore, we have to find the minimal polynomials of α and α^2. But α^2 and α^{10} have the same minimal polynomial (Proposition 4.2) $m_2(X)$. In fact, these are the only two roots of $m_2(X)$ and so

$$\begin{aligned} m_2(X) &= (X - \alpha^2)(X - \alpha^{10}) \\ &= X^2 - X(\alpha^2 + \alpha^{10}) + \alpha^{12} \end{aligned}$$

Now $\alpha^2 = -\alpha - 2$ and so

$$\begin{aligned}\alpha^{10} &= -\alpha^5 - 2^5\\ &= -\alpha^5 - 2\\ &= -\alpha(\alpha^2 + 4\alpha + 4) - 2\\ &= -\alpha(3\alpha + 2) - 2\\ &= 3(\alpha + 2) - 2\alpha - 2\\ &= \alpha - 1\end{aligned}$$

Also then

$$\begin{aligned}\alpha^{12} &= \alpha^2(\alpha - 1)\\ &= -(\alpha + 2)(\alpha - 1)\\ &= -\alpha^2 - \alpha + 2\\ &= \alpha + 2 - \alpha + 2\\ &= 4\end{aligned}$$

Hence

$$m_2(X) = X^2 + 3X + 4$$

Therefore the generating polynomial of the BCH code is

$$\begin{aligned}g(X) &= m_1(X)m_2(X)\\ &= (X^2 + X + 2)(X^2 + 3X + 4)\\ &= X^4 + 4X^3 + 4X^2 + 3\end{aligned}$$

Theorem 4.6
A binary BCH code with code word length $n = 2^m - 1$ and minimum distance at least $d = 2t + 1$ can always be constructed with check digits at most mt.

Proof
The code needed is of length $n = 2^m - 1$. We therefore construct an extension K of $\mathbb{B}$ of degree m and let α be a primitive element of K. The degree $[K:\mathbb{B}] = m$ is finite and so every element of K is algebraic over $\mathbb{B}$. Moreover if β is any element of K, then $\mathbb{B}(\beta)$ is a subfield of K and the degree relation

$$[K:\mathbb{B}] = [K:\mathbb{B}(\beta)][\mathbb{B}(\beta):\mathbb{B}]$$

(Proposition 4.3) then shows that $[\mathbb{B}(\beta):\mathbb{B}] \leq m$. This then shows that the degree of the minimal polynomial of β which equals $[\mathbb{B}(\beta):\mathbb{B}]$ is at most m.

For the construction of the code, we need the minimal polynomials $m_i(X)$ of α^i, $1 \leq i \leq 2t$. Let $g(X)$ be the generating polynomial of the code. Then

$$g(X) = \text{LCM}\{m_1(X), m_2(X), \ldots, m_{2t}(X)\}$$

We prove by induction on t that $\deg g(X) \leq mt$. For $t = 1$

$$g(X) = \text{LCM}\{m_1(X), m_2(X)\}$$

But α and α^2 have the same minimal polynomial. Therefore $g(X) = m_1(X)$ and we are through. Suppose that $t \geq 1$ and that we have proved the claim for t. Let:

$$g_1(X) = \text{LCM}\{m_1(X), \ldots, m_{2t}(X), m_{2t+1}(X), m_{2t+2}(X)\}$$

Then, since $m_{2t+2}(X) = m_{t+1}(X)$

$$\begin{aligned} g_1(X) &= \text{LCM}\{m_1(X), \ldots, m_{2t}(X), m_{2t+1}(X)\} \\ &= \text{LCM}[\text{LCM}\{m_1(X), \ldots, m_{2t}(X)\}, m_{2t+1}(X)] \\ &= \text{LCM}\{g(X), m_{2t+1}(X)\} \end{aligned}$$

Then

$$\begin{aligned} \deg g_1(X) &\leq \deg g(X) + \deg m_{2t+1}(X) \\ &\leq mt + m \\ &= m(t+1) \end{aligned}$$

which completes induction.

Since in a polynomial code, the number of check symbols equals the degree of the generating polynomial, the theorem follows.

Remark

If $g(X)$ is the generating polynomial of the binary BCH code with minimum distance at least $2t$, then in the notations of the proof of the above theorem

$$\begin{aligned} g(X) &= \text{LCM}\{m_1(X), \ldots, m_{2t-1}(X)\} \\ &= \text{LCM}\{m_1(X), \ldots, m_{2t-1}(X), m_{2t}(X)\} \end{aligned}$$

and, therefore, $g(X)$ is the generating polynomial of a BCH code with minimum distance at least $2t + 1$.

We have proved Theorem 4.6 from the point of view of practical utility. We can in fact prove the following theorem for BCH codes over a field of any prime order exactly on the lines of the proof of Theorem 4.6.

Theorem 4.7

A BCH code over GF(p) with code word length $n = p^m - 1$ and minimum distance at least $pt + 1$ can always be constructed with at most $(p - 1)mt$ check digits.

Exercises 4.1

1. Prove that $X^2 + X + 1$ is the only irreducible polynomial of degree 2 in $\mathbb{B}[X]$.

2. Prove that X^4+X^3+1, X^4+X+1 and $X^4+X^3+X^2+X+1$ are the only irreducible polynomials of degree 4 over $\mathbb{B}$.
3. Prove that X^3+X+1 and X^3+X^2+1 are the only irreducible polynomials of degree 3 over $\mathbb{B}$.
4. Prove that the polynomial $X^6+X^5+X^3+X^2+1$ is irreducible over $\mathbb{B}$. Prove that this polynomial is primitive.
5. If R is a commutative ring, a, $b\in R$ and n is a positive integer, prove that

$$(a+b)^n = a^n + \binom{n}{1}a^{n-1}b + \cdots + \binom{n}{i}a^{n-i}b^i + \cdots + \binom{n}{n-1}ab^{n-1} + b^n$$

where $\binom{n}{i}$ are the binomial coefficients.

6. Is GF(4) a subfield of GF(8)? Justify your answer.
7. Find a primitive element α of $F_3[X]/\langle X^2+1\rangle$. (Observe that if we put $\beta = X + \langle X^2+1\rangle$, then $1+\beta, 1-\beta, -1-\beta, -1+\beta$ are all the primitive elements of the field.)
8. Is the polynomial X^2-2 primitive over $\mathrm{GF}(3)=F_3$? Justify your answer.
9. Let F be a field of order p^n, p a prime and let α be a primitive element of F. Prove that α^r is primitive iff $\mathrm{GCD}(r, p^n-1)=1$.
10. Prove that the number of primitive elements in a field of order p^n, p a prime is $\phi(p^n-1)$. (Here ϕ stands for Euler's ϕ-function.)
11. Can the words 'at least' be omitted from the statement of Theorem 4.5?
12. Compute all the code words of the code of Example 4.8 Case (i). Compare this code with the (4, 7) Hamming code.
13. Construct binary BCH code of length 7 with minimum distance 3 by using the primitive polynomial X^3+X^2+1. Compare this code with
    (i) the (4, 7) Hamming code
    (ii) the code of Example 4.8 Case (i).

5

Linear codes

We have earlier studied codes which are Abelian groups. These codes were called group codes. Also these codes were considered over $\mathbb{B}$ the field of two elements. In this chapter, we study codes over a finite field GF(q) of q elements which generalize the concept of group codes studied earlier. Codes studied here are called **linear codes**.

Let $F = \mathrm{GF}(q)$, where q is a prime power, be a field of q elements. Let $V(n, q)$ denote the set of all vectors or sequences of length n over F. Then $V(n, q)$ is a vector space of dimension n over F. (Observe the change of notation here).

Definition 5.1
A subspace $\mathscr{C}$ of $V(n, q)$ is called a **linear code** of length n over F.

A vector space is first of all an Abelian group w.r.t. addition. It therefore follows that a linear code is always a group code. In view of this, we have the following proposition.

Proposition 5.1
The minimum distance d of a linear code $\mathscr{C}$ equals the minimum among the weights of non-zero code words.

5.1 GENERATOR AND PARITY CHECK MATRICES

Let $\mathscr{C}$ be a linear code of length n over F. Let $k(\leq n)$ be the dimension of $\mathscr{C}$ over F and choose a basis

$$X^{(1)}, X^{(2)}, \ldots, X^{(k)}$$

of $\mathscr{C}$ over F. Then any element of (or code word in) $\mathscr{C}$ is of the form

$$a_1 X^{(1)} + a_2 X^{(2)} + \cdots + a_k X^{(k)} = (a_1 \quad a_2 \quad \cdots \quad a_k)\mathbf{G}$$

where

$$\mathbf{G} = \begin{pmatrix} X^{(1)} \\ X^{(2)} \\ \vdots \\ X^{(k)} \end{pmatrix}$$

is a $k \times n$ matrix over F. The rows of the matrix $\mathbf{G}$ being linearly independent, the matrix $\mathbf{G}$ is of rank k. Thus $\mathscr{C}$ is a matrix code with a generator matrix $\mathbf{G}$. By choosing a different basis of $\mathscr{C}$ over F, we produce another generator matrix of $\mathscr{C}$.

Let $\mathbf{A} = (a_{ij})$ be a non-singular square matrix of order k over F. Then $\mathbf{AG}$ is a $k \times n$ matrix over F and $\forall i$, $1 \leq i \leq k$, the ith row of $\mathbf{AG}$ is

$$a_{i1}X^{(1)} + a_{i2}X^{(2)} + \cdots + a_{ik}X^{(k)}$$

Therefore, the rows of $\mathbf{AG}$ generate a subspace of $\mathscr{C}$.

Proposition 5.2
$\mathbf{AG}$ is a generator matrix of $\mathscr{C}$.

Proof
We know that

$$\text{rank } \rho(\mathbf{AG}) \leq \min\{\text{rank}(\mathbf{A}), \text{rank}(\mathbf{G})\} \leq \text{rank } \mathbf{G} = \rho(\mathbf{G}) = k$$

Again,

$$k = \rho(\mathbf{G}) = \rho(\mathbf{A}^{-1}\mathbf{AG}) \leq \min\{\rho(\mathbf{A}^{-1}), \rho(\mathbf{AG})\} \leq \rho(\mathbf{AG})$$

Hence $\rho(\mathbf{AG}) = \rho(\mathbf{G}) = k$ and the rows of $\mathbf{AG}$ are linearly independent. Thus the rows of $\mathbf{AG}$ generate a subspace of dimension k of $\mathscr{C}$ which itself is of dimension k. Then any element of $\mathscr{C}$ is of the form $a(\mathbf{AG})$ for some $a \in V(k, q)$. Hence $\mathbf{AG}$ is a generator matrix of $\mathscr{C}$.

If $\mathscr{C}$ is a linear code of length n, the dimension of $\mathscr{C}$ is k, and d is the minimum distance of $\mathscr{C}$, we then say that $\mathscr{C}$ is a linear $[n, k, d]$ code over F.

Definition 5.2
Two codes $\mathscr{C}$ and $\mathscr{C}'$ of length n are said to be **equivalent** if there exists a permutation σ of the n-symbols $\{1, 2, \ldots, n\}$ such that $c' = (c'_1, c'_2, \ldots, c'_n) \in \mathscr{C}'$ iff $c' = \sigma(c)$ for some $c \in \mathscr{C}$, where

$$\sigma(c) = \sigma(c_1, \ldots, c_n) = (c_{\sigma(1)}, c_{\sigma(2)}, \ldots, c_{\sigma(n)})$$

Observe that equivalent codes have the same minimum distance and, therefore, the same error detection/correction capability. Therefore, for studying error detection/correction, we may work with equivalent code if that helps our study.

Let σ be a permutation of the set $\{1, 2, \ldots, n\}$. Suppose that $\sigma(j) = i_j$, $1 \leq j \leq n$, so that

$$\{i_1, \ldots, i_n\} = \{1, 2, \ldots, n\}$$

Let $\mathbf{P}$ be a square matrix of order n in which (i_j, j) entry is 1 for every j, $1 \leq j \leq n$, and every other entry is 0. The matrix $\mathbf{P}$ has exactly one non-zero entry (which is in fact 1) in every row and in every column. Such a matrix is called a **permutation matrix** and is clearly non-singular. Let $\mathbf{P} = (p_{ij})$. For any vector $\mathbf{c} = (c_1 \quad \cdots \quad c_n)$, the jth entry of $\mathbf{cP}$ is

$$c_1 p_{1j} + c_2 p_{2j} + \cdots + c_n p_{nj} = c_{i_j}$$

as $p_{kj} = 0$ for $k \neq i_j$. Therefore,

$$\mathbf{cP} = (c_{i_1} \quad c_{i_2} \quad \cdots \quad c_{i_n}) = \sigma(\mathbf{c})$$

Conversely, given a permutation matrix $\mathbf{P}$ of order n, we can define a permutation σ of the set $\{1, 2, \ldots, n\}$ such that for any vector $\mathbf{c} \in V(n, q)$, $\mathbf{cP} = \sigma(\mathbf{c})$. (This justifies the name permutation matrix.) We thus have the following proposition.

Proposition 5.3
Two codes $\mathscr{C}$ and $\mathscr{C}'$ of length n are equivalent iff there exists a permutation matrix $\mathbf{P}$ of order n such that $\mathscr{C} = \{\mathbf{cP}/\mathbf{c} \in \mathscr{C}\}$.

Corollary
If $\mathscr{C}$ is a linear $[n, k, d]$ code, then so is its equivalent code $\mathscr{C}'$.

Theorem 5.1
Given a linear $[n, k, d]$ code $\mathscr{C}$ over F, there exists an equivalent code $\mathscr{C}'$ having a generator matrix, the first k columns of which form the identity matrix $\mathbf{I}_k$.

Proof
Let $\mathbf{G}$ be a generator matrix of $\mathscr{C}$. Let $\mathbf{G}_1, \mathbf{G}_2, \ldots, \mathbf{G}_n$ denote the columns of $\mathbf{G}$ and suppose that

$$\mathbf{G}_{i_1}, \mathbf{G}_{i_2}, \ldots, \mathbf{G}_{i_k} \quad \text{for} \quad 1 \leq i_1 < i_2 < \cdots < i_k \leq n$$

are linearly independent. Let σ be a permutation of the set $\{1, 2, \ldots, n\}$ with $\sigma(j) = i_j$, $1 \leq i \leq k$. Let $\mathbf{P} = (p_{ij})$ be the permutation matrix of order n associated with the permutation σ. Then, for $1 \leq j \leq k$, $p_{\ell j} = 0$ for $\ell \neq i_j$ and $p_{i_j j} = 1$. In the matrix

$$\mathbf{M} = \mathbf{GP} = (\mathbf{G}_1 \quad \cdots \quad \mathbf{G}_n) \begin{pmatrix} p_{11} & p_{12} & \cdots & p_{1n} \\ p_{21} & p_{22} & \cdots & p_{2n} \\ \vdots & \vdots & & \vdots \\ p_{n1} & p_{n2} & \cdots & p_{nn} \end{pmatrix}$$

$$= \left(\sum_j \mathbf{G}_j p_{j1} \quad \sum_j \mathbf{G}_j p_{j2} \quad \cdots \quad \sum_j \mathbf{G}_j p_{jk} \quad \cdots \quad \sum_j \mathbf{G}_j p_{jn} \right)$$

$$= (\mathbf{G}_{i_1} \quad \mathbf{G}_{i_2} \quad \cdots \quad \mathbf{G}_{i_k} \quad \cdots)$$

the first k columns are linearly independent. Let $\mathscr{C}'$ be the code of length n with $\mathbf{M}$ as a generator matrix. Now

$$\begin{aligned} \mathscr{C}' &= \{\mathbf{aM} \mid \mathbf{a} \in V(k, q)\} \\ &= \{\mathbf{aGP} \mid \mathbf{a} \in V(k, q)\} \\ &= \{\mathbf{cP} \mid \mathbf{c} \in \mathscr{C}\} \end{aligned}$$

and, therefore, $\mathscr{C}'$ is equivalent to $\mathscr{C}$. The code $\mathscr{C}'$ is linear.

Let

$$\mathbf{A} = (\mathbf{G}_{i_1} \quad \mathbf{G}_{i_2} \quad \cdots \quad \mathbf{G}_{i_k})$$

The columns $\mathbf{G}_{i_1}, \ldots, \mathbf{G}_{i_k}$ being linearly independent, $\mathbf{A}$ is a non-singular square matrix of order k. Therefore $\mathbf{A}^{-1}\mathbf{M}$ is also a generator matrix of $\mathscr{C}'$. Writing $\mathbf{M} = (\mathbf{A} \quad \mathbf{B})$, where $\mathbf{B}$ is a $k \times (n-k)$ matrix, we find that a generator matrix (of $\mathscr{C}'$) is

$$\mathbf{A}^{-1}\mathbf{M} = \mathbf{A}^{-1}(\mathbf{A} \quad \mathbf{B}) = (\mathbf{A}^{-1}\mathbf{A} \quad \mathbf{A}^{-1}\mathbf{B}) = (\mathbf{I}_k \quad \mathbf{A}^{-1}\mathbf{B})$$

Proposition 5.4

If $\mathscr{C}$ is a linear $[n, k, d]$ code over F, then $d \leq n - k + 1$.

Proof

Since equivalent codes have the same minimum distance, we may assume that $\mathscr{C}$ is a linear code with a generator matrix $\mathbf{G}$ such that the first k columns of $\mathbf{G}$ form the identity matrix. Let e^i be a message word with 1 in the ith position and zero everywhere else. The corresponding code word $\mathbf{e}^i\mathbf{G}$ is the ith row of $\mathbf{G}$ which has at least $k-1$ zero entries. So the weight of this non-zero code word is at most $n-(k-1)$, i.e. $n-k+1$. Hence $d \leq n-k+1$.

In the case of binary codes with generator matrix $\mathbf{G} = (\mathbf{I}_k \quad \mathbf{A})$ and parity check matrix $\mathbf{H} = (\mathbf{B} \quad \mathbf{I}_{n-k})$ we have seen that $\mathbf{B} = \mathbf{A}^t$. However, in the case of codes over an arbitrary finite field F, the relationship between the generator matrix and parity check matrix is given by the following proposition.

Proposition 5.5

If $\mathscr{C}$ is a linear $[n, k, d]$ code over F with parity check matrix $\mathbf{H} = (\mathbf{A} \quad \mathbf{I}_{n-k})$ where $\mathbf{A}$ is an $(n-k) \times k$ matrix over F, then a generator matrix of $\mathscr{C}$ is given by $\mathbf{G} = (\mathbf{I}_k \quad -\mathbf{A}^t)$.

Proof

Let $u = u_1 \text{---} u_n$ be the code word corresponding to the message word $a = a_1 \text{---} a_k$. Let $\mathbf{u} = (\mathbf{v} \quad \mathbf{w})$, where $\mathbf{v}$ is $1 \times k$ and $\mathbf{w}$ is $1 \times (n-k)$ matrix. From the definition of the code given by parity check matrix $\mathbf{H}$, $\mathbf{v}=\mathbf{a}$ and then $\mathbf{Hu}^{\mathrm{t}}=0$ implies that

$$(\mathbf{A} \quad \mathbf{I}_{n-k})\begin{pmatrix}\mathbf{a}^{\mathrm{t}}\\ \mathbf{w}^{\mathrm{t}}\end{pmatrix}=0 \quad \text{or} \quad \mathbf{Aa}^{\mathrm{t}}+\mathbf{w}^{\mathrm{t}}=0$$

Thus $\mathbf{w}=-\mathbf{aA}^{\mathrm{t}}$ and hence $\mathbf{u}=\mathbf{a}(\mathbf{I}_k \quad -\mathbf{A}^{\mathrm{t}})$. Thus u is a code word corresponding to the message word a in the code defined by the generator matrix $\mathbf{G}=(\mathbf{I}_k \quad -\mathbf{A}^{\mathrm{t}})$.

Conversely, consider the code word u with

$$\mathbf{u}=\mathbf{aG}=(\mathbf{a} \quad -\mathbf{aA}^{\mathrm{t}})$$

corresponding to the message word a in the code defined by $\mathbf{G}=(\mathbf{I}_k \quad -\mathbf{A}^{\mathrm{t}})$. Then

$$\mathbf{Hu}^{\mathrm{t}}=(\mathbf{A} \quad \mathbf{I}_{n-k})\begin{pmatrix}\mathbf{a}^{\mathrm{t}}\\ -\mathbf{Aa}^{\mathrm{t}}\end{pmatrix}=\mathbf{Aa}^{\mathrm{t}}-\mathbf{Aa}^{\mathrm{t}}=0$$

Therefore, u is the code word corresponding to a in the code given by $\mathbf{H}$.

Examples 5.1

Case (i)
Every binary group code is a linear code.

Case (ii)
Every polynomial code and every matrix code is a linear code. In particular, BCH codes are linear codes and so are Hamming codes.

Case (iii)
Every group code over the field of three elements is a linear code. (A code over the field of three elements is called a **ternary code**.)

Case (iv)
The following binary codes are linear codes:

$$\text{(a)}\ \begin{array}{cccc} 1&1&0&0\\ 0&1&1&1\\ 1&0&1&1\\ 1&0&1&0\\[1em] 0&1&1&0\\ 1&1&0&1\\ 0&0&0&1\\ 0&0&0&0 \end{array} \qquad \text{(b)}\ \begin{array}{cccc} 1&1&0&0\\ 1&0&1&1\\ 0&1&1&1\\ 0&1&1&0\\[1em] 1&0&1&0\\ 1&1&0&1\\ 0&0&0&1\\ 0&0&0&0 \end{array} \qquad \text{(c)}\ \begin{array}{cccc} 0&1&1&0\\ 1&1&0&1\\ 1&0&1&1\\ 0&0&1&1\\[1em] 0&1&0&1\\ 1&1&1&0\\ 1&0&0&0\\ 0&0&0&0 \end{array}$$

The generator matrices of these codes are respectively

$$\mathbf{A} = \begin{pmatrix} 1 & 1 & 0 & 0 \\ 0 & 1 & 1 & 1 \\ 1 & 0 & 1 & 0 \end{pmatrix} \qquad \mathbf{B} = \begin{pmatrix} 1 & 1 & 0 & 0 \\ 1 & 0 & 1 & 1 \\ 0 & 1 & 1 & 0 \end{pmatrix} \qquad \mathbf{C} = \begin{pmatrix} 0 & 1 & 1 & 0 \\ 1 & 1 & 0 & 1 \\ 0 & 0 & 1 & 1 \end{pmatrix}$$

Case (v)
The matrices

$$\mathbf{P} = \begin{pmatrix} 0 & 1 & 0 & 0 \\ 1 & 0 & 0 & 0 \\ 0 & 0 & 1 & 0 \\ 0 & 0 & 0 & 1 \end{pmatrix} \quad \text{and} \quad \mathbf{Q} = \begin{pmatrix} 0 & 0 & 0 & 1 \\ 0 & 1 & 0 & 0 \\ 1 & 0 & 0 & 0 \\ 0 & 0 & 1 & 0 \end{pmatrix}$$

are permutation matrices and with **A**, **B**, **C** as in Case (iv) above we have

$$\mathbf{B} = \mathbf{AP} \quad \mathbf{A} = \mathbf{CQ}$$

Thus, the codes (a) and (b) of Case (iv) are equivalent and so are the codes (a) and (c). But then the codes (b) and (c) are also equivalent. This could also be observed from the relation $\mathbf{B} = \mathbf{C}(\mathbf{QP})$, as the product of permutation matrices is again a permutation matrix.

Case (vi)
Observe the following:

(a) The permutation matrix **P** corresponding to the permutation $\sigma = (1, 2, 3)$ of the set $\{1, 2, 3, 4\}$ is given by

$$\mathbf{P} = \begin{pmatrix} 0 & 0 & 1 & 0 \\ 1 & 0 & 0 & 0 \\ 0 & 1 & 0 & 0 \\ 0 & 0 & 0 & 1 \end{pmatrix}$$

(b) That corresponding to $\sigma = (1, 2)(3, 4)$ of the set $\{1, 2, 3, 4\}$ is given by

$$\mathbf{P} = \begin{pmatrix} 0 & 1 & 0 & 0 \\ 1 & 0 & 0 & 0 \\ 0 & 0 & 0 & 1 \\ 0 & 0 & 1 & 0 \end{pmatrix}$$

(c) The permutation matrix **P** corresponding to the permutation $\sigma = (1, 3)(2, 4)$ of the set $\{1, 2, 3, 4\}$ is given by

$$\mathbf{P} = \begin{pmatrix} 0 & 0 & 1 & 0 \\ 0 & 0 & 0 & 1 \\ 1 & 0 & 0 & 0 \\ 0 & 1 & 0 & 0 \end{pmatrix}$$

Exercises 5.1

1. Prove that the inverse of a permutation matrix is a permutation matrix.
2. Find a code equivalent to the code of Case (iv)(a) above such that, in the generating matrix of the equivalent code, the first three columns form the identity matrix.
3. Prove that the relation of two codes being equivalent is an equivalence relation.
4. Give a proof of the corollary to Proposition 5.3.
5. Find the permutation matrix which corresponds to
   (i) the permutation $\sigma = (1, 3, 2)$ of the set $\{1, 2, 3\}$.
   (ii) the permutation $\sigma = (1, 3, 2, 5)$ of the set $\{1, 2, 3, 4, 5\}$.
6. Find the permutation of the appropriate set corresponding to these permutation matrices.

   (i)

$$\mathbf{P} = \begin{pmatrix} 0 & 0 & 1 \\ 1 & 0 & 0 \\ 0 & 1 & 0 \end{pmatrix}$$

   (ii)

$$\mathbf{P} = \begin{pmatrix} 0 & 0 & 1 \\ 0 & 1 & 0 \\ 1 & 0 & 0 \end{pmatrix}$$

   (iii)

$$\mathbf{P} = \begin{pmatrix} 1 & 0 & 0 & 0 \\ 0 & 0 & 1 & 0 \\ 0 & 0 & 0 & 1 \\ 0 & 1 & 0 & 0 \end{pmatrix}$$

   (iv)

$$\mathbf{P} = \begin{pmatrix} 0 & 0 & 1 & 0 & 0 \\ 0 & 0 & 0 & 0 & 1 \\ 0 & 1 & 0 & 0 & 0 \\ 1 & 0 & 0 & 0 & 0 \\ 0 & 0 & 0 & 1 & 0 \end{pmatrix}$$

5.2 DUAL CODE OF A LINEAR CODE

Definition 5.3
Let $\mathbf{x} = (x_1 \quad x_2 \quad \cdots \quad x_n)$, $\mathbf{y} = (y_1 \quad y_2 \quad \cdots \quad y_n)$ be two vectors of length n over a field F. Then, by the intersection $\mathbf{x} * \mathbf{y}$ of $\mathbf{x}$ and $\mathbf{y}$, we mean the vector

$$\mathbf{x} * \mathbf{y} = (x_1 y_1 \quad x_2 y_2 \quad \cdots \quad x_n y_n)$$

while by their scalar product $\mathbf{x}\cdot\mathbf{y}$ we mean the element

$$\mathbf{x}\cdot\mathbf{y} = x_1y_1 + x_2y_2 + \cdots + x_ny_n \text{ of } F$$

Thus

$$\mathbf{x}\cdot\mathbf{y} = \mathbf{x}\mathbf{y}^t = {}^{\mathbf{y}\cdot\mathbf{x}} = \mathbf{y}\mathbf{x}^t$$

For example, if $F = \mathbb{B}$ and

$$\mathbf{x} = (1 \quad 1 \quad 0 \quad 1) \qquad \mathbf{y} = (1 \quad 1 \quad 1 \quad 1)$$
$$\mathbf{z} = (1 \quad 0 \quad 1 \quad 0 \quad 1 \quad 1) \qquad \mathbf{t} = (1 \quad 1 \quad 0 \quad 1 \quad 0 \quad 1)$$

then

$$\mathbf{x} * \mathbf{y} = (1 \quad 1 \quad 0 \quad 1)$$
$$\mathbf{x}\cdot\mathbf{y} = 1 + 1 + 0 + 1 = 1$$

and

$$\mathbf{z} * \mathbf{t} = (1 \quad 0 \quad 0 \quad 0 \quad 0 \quad 1)$$
$$\mathbf{z}\cdot\mathbf{t} = 1 + 0 + 0 + 0 + 0 + 1 = 0$$

Definition 5.4

Two vectors $\mathbf{x}$ and $\mathbf{y}$ of the same length n over F are called **orthogonal** if $\mathbf{x}\cdot\mathbf{y} = 0$ or equivalently

$$\mathbf{x}\mathbf{y}^t = \mathbf{y}\cdot\mathbf{x} = \mathbf{y}\mathbf{x}^t = 0$$

Exercises 5.2

1. For binary vectors $\mathbf{x}$ and $\mathbf{y}$ of the same length, prove that $\mathbf{x}\cdot\mathbf{y} = 0$ iff $\text{wt}(\mathbf{x} * \mathbf{y})$ is even and $\mathbf{x}\cdot\mathbf{y} = 1$ iff $\text{wt}(\mathbf{x} * \mathbf{y})$ is odd.
2. Prove that a ternary vector $\mathbf{x}$ is orthogonal to itself iff its weight is divisible by 3.

We have defined dual of a code $\mathscr{C}$ as the code generated by a parity check matrix of the code $\mathscr{C}$. We now define the dual in a more general way – but finally it amounts to the same thing.

Definition 5.5

If $\mathscr{C}$ is an $[n, k, d]$ linear code over F, its **dual code** or **orthogonal code** $\mathscr{C}^\perp$ is the set of all vectors of length n that are orthogonal to all code words of $\mathscr{C}$, i.e.

$$\mathscr{C}^\perp = \{\mathbf{u} \in \mathbf{V}(n, q) \mid \mathbf{u}\cdot\mathbf{v} = 0 \ \forall \mathbf{v} \in \mathscr{C}\}$$

Examples 5.2

Case (i)

Consider the linear code as in Case (iv)(a) of Examples 5.1. Let $(x_1 \quad x_2 \quad x_3 \quad x_4)$ be a vector orthogonal to all (those vectors isomorphic to) the code words. In particular $(x_1 \quad x_2 \quad x_3 \quad x_4)$ is orthogonal to code words 1100, 0111 and 1010.

Therefore,

$$x_1 + x_2 = 0 = x_2 + x_3 + x_4 = x_1 + x_3$$

These relations then show that

$$x_1 = x_2 = x_3 \quad \text{and} \quad x_4 = 0$$

Therefore

$$\mathscr{C}^{\perp} = \{0000, 1110\}$$

Observe that $\mathscr{C}^{\perp}$ is also a linear code and $\dim \mathscr{C}^{\perp} + \dim \mathscr{C} = 4$.

Case (ii)

Consider the (4, 7) binary Hamming code $\mathscr{C}$:

$$\begin{array}{ccccccc c ccccccc}
0 & 0 & 0 & 0 & 0 & 0 & 0 & \quad & 0 & 0 & 0 & 1 & 0 & 1 & 1 \\
1 & 1 & 1 & 0 & 1 & 0 & 0 & & 1 & 0 & 0 & 0 & 1 & 0 & 1 \\
0 & 1 & 1 & 1 & 0 & 1 & 0 & & 1 & 1 & 0 & 0 & 0 & 1 & 0 \\
0 & 0 & 1 & 1 & 1 & 0 & 1 & & 0 & 1 & 1 & 0 & 0 & 0 & 1 \\
1 & 0 & 0 & 1 & 1 & 1 & 0 & & 1 & 0 & 1 & 1 & 0 & 0 & 0 \\
0 & 1 & 0 & 0 & 1 & 1 & 1 & & 0 & 1 & 0 & 1 & 1 & 0 & 0 \\
1 & 0 & 1 & 0 & 0 & 1 & 1 & & 0 & 0 & 1 & 0 & 1 & 1 & 0 \\
1 & 1 & 0 & 1 & 0 & 0 & 1 & & 1 & 1 & 1 & 1 & 1 & 1 & 1
\end{array}$$

It is a $[7, 4, 3]$ linear code with 1110100, 0111010, 0011101 and 0001011 as a basis. As a consequence of this, it follows that the code words of the (4, 7) Hamming code are in one-to-one correspondence with the code words of the code generated by the matrix

$$\begin{pmatrix} 1 & 1 & 1 & 0 & 1 & 0 & 0 \\ 0 & 1 & 1 & 1 & 0 & 1 & 0 \\ 0 & 0 & 1 & 1 & 1 & 0 & 1 \\ 0 & 0 & 0 & 1 & 0 & 1 & 1 \end{pmatrix}$$

A vector $(x_1 \quad x_2 \quad x_3 \quad x_4 \quad x_5 \quad x_6 \quad x_7)$ is orthogonal to $\mathscr{C}$ iff it is orthogonal to the basis vectors. For this, we have

$$x_1 + x_2 + x_3 + x_5 = 0$$

$$x_2 + x_3 + x_4 + x_6 = 0$$

$$x_3 + x_4 + x_5 + x_7 = 0$$

$$x_4 + x_6 + x_7 = 0$$

These imply

$$x_5 = x_1 + x_2 + x_3$$

$$x_7 = x_2 + x_3$$

$$x_4 = x_1 + x_3$$

$$x_6 = x_1 + x_2$$

In matrix form, these equations can be rewritten as

$$\begin{pmatrix} 1 & 0 & 1 & 1 & 0 & 0 & 0 \\ 1 & 1 & 1 & 0 & 1 & 0 & 0 \\ 1 & 1 & 0 & 0 & 0 & 1 & 0 \\ 0 & 1 & 1 & 0 & 0 & 0 & 1 \end{pmatrix} \begin{pmatrix} x_1 \\ x_2 \\ \vdots \\ x_7 \end{pmatrix} = 0$$

Thus the generator matrix of the dual code $\mathscr{C}^\perp$ is

$$\mathbf{G} = \begin{pmatrix} 1 & 0 & 0 & 1 & 1 & 1 & 0 \\ 0 & 1 & 0 & 0 & 1 & 1 & 1 \\ 0 & 0 & 1 & 1 & 1 & 0 & 1 \end{pmatrix}$$

In view of this, it follows that $\mathscr{C}^\perp$ is a linear code of dimension 3. All the code words of $\mathscr{C}^\perp$ are:

$$\begin{matrix} 0 & 0 & 0 & 0 & 0 & 0 & 0 & \quad & 1 & 1 & 0 & 1 & 0 & 0 & 1 \\ 1 & 0 & 0 & 1 & 1 & 1 & 0 & \quad & 1 & 0 & 1 & 0 & 0 & 1 & 1 \\ 0 & 1 & 0 & 0 & 1 & 1 & 1 & \quad & 0 & 1 & 1 & 1 & 0 & 1 & 0 \\ 0 & 0 & 1 & 1 & 1 & 0 & 1 & \quad & 1 & 1 & 1 & 0 & 1 & 0 & 0 \end{matrix}$$

Case (iii)

Consider the (4, 7) polynomial code $\mathscr{C}$ generated by the polynomial $1 + X + X^3$. A generator matrix of this code is

$$\mathbf{G} = \begin{pmatrix} 1 & 1 & 0 & 1 & 0 & 0 & 0 \\ 0 & 1 & 1 & 0 & 1 & 0 & 0 \\ 0 & 0 & 1 & 1 & 0 & 1 & 0 \\ 0 & 0 & 0 & 1 & 1 & 0 & 1 \end{pmatrix}$$

A word $x = (x_1 x_2 x_3 x_4 x_5 x_6 x_7)$ is orthogonal to every code word in $\mathscr{C}$ iff $\mathbf{G}\mathbf{x}^t = 0$, i.e. iff

$$x_1 + x_2 + x_4 = 0$$

$$x_2 + x_3 + x_5 = 0$$

$$x_3 + x_4 + x_6 = 0$$

$$x_4 + x_5 + x_7 = 0$$

These relations give

$$x_4 = x_1 + x_2$$

$$x_5 = x_2 + x_3$$
$$x_6 = x_1 + x_2 + x_3$$
$$x_7 = x_1 + x_3$$

In matrix notation, it follows that

$$\begin{aligned}\mathbf{x} &= (x_1 \quad x_2 \quad x_3 \quad x_4 \quad x_5 \quad x_6 \quad x_7) \\ &= (x_1 \quad x_2 \quad x_3)\begin{pmatrix} 1 & 0 & 0 & 1 & 0 & 1 & 1 \\ 0 & 1 & 0 & 1 & 1 & 1 & 0 \\ 0 & 0 & 1 & 0 & 1 & 1 & 1 \end{pmatrix}\end{aligned}$$

Thus the dual code is generated by the matrix

$$\mathbf{H} = \begin{pmatrix} 1 & 0 & 0 & 1 & 0 & 1 & 1 \\ 0 & 1 & 0 & 1 & 1 & 1 & 0 \\ 0 & 0 & 1 & 0 & 1 & 1 & 1 \end{pmatrix}$$

and hence is a (3, 7) code. All the code words of this code are:

$$\begin{array}{ccccccc} 0 & 0 & 0 & 0 & 0 & 0 & 0 \\ 1 & 0 & 0 & 1 & 0 & 1 & 1 \\ 0 & 1 & 0 & 1 & 1 & 1 & 0 \\ 0 & 0 & 1 & 0 & 1 & 1 & 1 \end{array} \qquad \begin{array}{ccccccc} 1 & 1 & 0 & 0 & 1 & 0 & 1 \\ 0 & 1 & 1 & 1 & 0 & 0 & 1 \\ 1 & 0 & 1 & 1 & 1 & 0 & 0 \\ 1 & 1 & 1 & 0 & 0 & 1 & 0 \end{array}$$

If the dual code were a polynomial code generated by

$$f(X) = 1 + bX + cX^2 + dX^3 + eX^4$$

then it follows from the above that

$$\begin{aligned}(a_0 + a_1X + a_2X^2)f(X) = a_0 + a_1X + a_2X^2 + (a_0 + a_1)X^3 + (a_1 + a_2)X^4 \\ + (a_0 + a_1 + a_2)X^5 + (a_0 + a_2)X^6 \ \forall a_0, a_1, a_2 \in \mathbb{B}\end{aligned}$$

Comparing the coefficients of X, X^2, X^3 gives

$$a_1 + a_0b = a_1$$
$$a_2 + a_1b + a_0c = a_2$$
$$a_0d + a_1c + a_2b = a_0 + a_1$$

for all $a_0, a_1, a_2 \in \mathbb{B}$. These relations then imply that

$$b = c = 0 \quad \text{and} \quad a_0d = a_0 + a_1 \ \forall a_0, a_1$$

But whatever the value of d in $\mathbb{B}$, the relation

$$a_0d = a_0 + a_1 \ \forall a_0, a_1 \in \mathbb{B}$$

is not possible. This proves that the dual code $\mathscr{C}^{\perp}$ is not a polynomial code. We thus have the following observation: The dual of a polynomial code need not be a polynomial code.

Theorem 5.2
Let $\mathscr{C}$ be a linear $[n,k,d]$ code with generator matrix $\mathbf{G}$ and parity check matrix $\mathbf{H}$. Then $\mathscr{C}^{\perp}$ is a linear $[n,n-k,-]$ code with generator matrix $\mathbf{H}$ and parity check matrix $\mathbf{G}$.

Proof
By definition

$$\mathscr{C}^{\perp}=\{\mathbf{u}\in V(n,q)\,|\,\mathbf{u}\mathbf{v}^{\mathrm{t}}=0\ \forall\,\mathbf{v}\in\mathscr{C}\}$$

Let $\mathbf{u}$, $\mathbf{w}\in\mathscr{C}^{\perp}$ and $\alpha,\beta\in\mathrm{GF}(q)=F$. For any $\mathbf{v}\in\mathscr{C}$,

$$\begin{aligned}(\alpha\mathbf{u}+\beta\mathbf{w})\mathbf{v}^{\mathrm{t}}&=(\alpha\mathbf{u})\mathbf{v}^{\mathrm{t}}+(\beta\mathbf{w})\mathbf{v}^{\mathrm{t}}\\&=\alpha(\mathbf{u}\mathbf{v}^{\mathrm{t}})+\beta(\mathbf{w}\mathbf{v}^{\mathrm{t}})=0\end{aligned}$$

Therefore $\mathscr{C}^{\perp}$ is a linear code.

Suppose that $\mathbf{G}$ is an $r\times n$ matrix. Then

$$\mathscr{C}=\{\mathbf{a}\mathbf{G}\,|\,\mathbf{a}\in V(r,q)\}$$

But the map $\phi\colon V(r,q)\rightarrow\mathscr{C}$ given by

$$\phi(\mathbf{a})=\mathbf{a}\mathbf{G},\quad \mathbf{a}\in V(r,q)$$

is a vector space homomorphism which is clearly onto. Let $\mathbf{G}_1,\mathbf{G}_2,\ldots,\mathbf{G}_n$ denote the columns of $\mathbf{G}$. We may suppose that the columns $\mathbf{G}_1,\mathbf{G}_2,\ldots,\mathbf{G}_r$ are linearly independent. Then $\mathbf{B}=(\mathbf{G}_1\quad\mathbf{G}_2\quad\cdots\quad\mathbf{G}_r)$ is a non-singular square matrix of order r.

Let $\mathbf{a}\in V(r,q)$ such that $\phi(\mathbf{a})=0$, i.e. $\mathbf{a}\mathbf{G}=0$. Then also $\mathbf{a}\mathbf{B}=0$ which implies that $\mathbf{a}=(\mathbf{a}\mathbf{B})\mathbf{B}^{-1}=0\mathbf{B}^{-1}=0$. Hence ϕ is an isomorphism. Therefore

$$k=\dim\mathscr{C}=\dim V(r,q)=r$$

Hence $\mathbf{G}$ is a $k\times n$ matrix.

Let $\mathbf{u}\in\mathscr{C}^{\perp}$. Then

$$\mathbf{v}\mathbf{u}^{\mathrm{t}}=0\,\forall\,\mathbf{v}\in\mathscr{C}$$

i.e.

$$\mathbf{a}(\mathbf{G}\mathbf{u}^{\mathrm{t}})=(\mathbf{a}\mathbf{G})\mathbf{u}^{\mathrm{t}}=0\,\forall\,\mathbf{a}\in V(k,q)$$

This easily implies that $\mathbf{G}\mathbf{u}^{\mathrm{t}}=0$.

Conversely, if $\mathbf{u}\in V(n,q)$ such that $\mathbf{G}\mathbf{u}^{\mathrm{t}}=0$, then

$$(\mathbf{a}\mathbf{G})\mathbf{u}^{\mathrm{t}}=0\,\forall\,\mathbf{a}\in V(k,q)$$

Hence

$$\mathscr{C}^{\perp}=\{\mathbf{u}\in V(n,q)\,|\,\mathbf{G}\mathbf{u}^{\mathrm{t}}=0\}$$

which proves that $\mathbf{G}$ is a parity check matrix for $\mathscr{C}^{\perp}$.

Next define a map θ: $V(n,q) \to V(n,q)$ by

$$\theta(\mathbf{x}) = \mathbf{G}\mathbf{x}^t, \quad \mathbf{x} \in V(n,q)$$

θ is clearly a linear transformation. Therefore

$$\text{rank}(\theta) + \text{nullity of } \theta = n - \text{the dimension of } V(n,q)$$

But

$$\text{rank}(\theta) = \text{rank } \mathbf{G} = k$$

Therefore

$$\text{nullity of } \theta = n - k$$

Also, the nullity of θ is the dimension of $\ker\theta$ and $\ker\theta = \mathscr{C}^\perp$. Hence, $\dim \mathscr{C}^\perp = n - k$.

It is clear from the definition of $\mathscr{C}^\perp$ that $\mathscr{C} \subseteq (\mathscr{C}^\perp)^\perp$. Also

$$\dim(\mathscr{C}^\perp)^\perp = n - \dim \mathscr{C}^\perp = n - (n-k) = k = \dim \mathscr{C}$$

Hence $\mathscr{C} = (\mathscr{C}^\perp)^\perp$.

Let $\mathscr{C}'$ be the code generated by $\mathbf{H}$. Every code word in $\mathscr{C}$ is orthogonal to every row of $\mathbf{H}$ and hence is orthogonal to every code word in $\mathscr{C}'$. Therefore $\mathscr{C}' \subseteq \mathscr{C}^\perp$.

The matrix $\mathbf{H}$ being an $(n-k) \times n$ matrix with $n-k$ columns linearly independent, $\mathscr{C}'$ is a vector space of dimension $n-k$. But $\mathscr{C}^\perp$ is already proved to be of dimension $n-k$. Hence $\mathscr{C}' = \mathscr{C}^\perp$ and so $\mathbf{H}$ is a generator matrix of $\mathscr{C}^\perp$.

Remarks 5.1

1. The above reasoning has also proved that $(\mathscr{C}^\perp)^\perp = \mathscr{C}$.
2. We have defined dual only of a linear code. However, we may define the dual of a fixed length code which is not necessarily linear or matrix code: If $\mathscr{C}$ is a code of length n (not necessarily linear) over F, then

$$\mathscr{C}^\perp = \{\mathbf{u} \in V(n,q) \mid \mathbf{u}\cdot\mathbf{v} = 0 \ \forall \mathbf{v} \in \mathscr{C}\}.$$

3. Every code word in $\mathscr{C}^\perp$ when $\mathscr{C}$ is linear is of the form $\mathbf{aH}$, where $\mathbf{a}$ is the vector related to a message sequence of length $n-k$.

Definition 5.6

A linear code $\mathscr{C}$ over F is called **self dual** if $\mathscr{C}^\perp = \mathscr{C}$.

Observe that the length of a self dual code is always even and the weight of every code word of a binary self dual code is also even.

Examples 5.3

Let $\mathscr{C}$ be a binary self dual code of length 4. Then its dimension is clearly 2. The vectors $(1 \quad 1 \quad 0 \quad 0)$ and $(1 \quad 0 \quad 1 \quad 0)$ are linearly independent over $\mathbb{B}$ and so generate a space of dimension 2. But

$$(1 \quad 1 \quad 0 \quad 0)(1 \quad 0 \quad 1 \quad 0)^t \neq 0$$

and so the space generated by these two vectors is not a self dual code. However, the code

$$\mathscr{C} = \{0000, 1100, 0011, 1111\}$$

generated by 1100 and 0011 is self dual. Following are the other self dual codes of length 4:

$$\{0000, 0101, 1010, 1111\} \qquad \{0000, 1001, 0110, 1111\}$$

Exercises 5.3

1. Construct a binary self dual code of length 6.
2. Construct a binary self dual code of length 8.
3. Prove that 1201 and 1012 generate a ternary self dual code of length 4. Also find all the code words of this code.
4. Prove that the weight of every code word of a ternary self dual is divisible by 3.
5. Prove that the (4, 7) binary Hamming code is not a polynomial code. Is this code equivalent to a polynomial code?

Next, we consider self dual codes over GF(q), q an odd prime. First we have a couple of number theoretic results which we again need in Chapter 8 when we study quadratic residue codes.

Definition 5.7

Let p be an odd prime. Recall that a positive integer a is called a **quadratic residue modulo p** if $x^2 \equiv a(\text{mod } p)$ for some integer x. If there is no such x, then a is called a **quadratic non-residue modulo p**. If b is another positive integer such that $b \equiv a(\text{mod } p)$, then b is a quadratic residue modulo p iff a is a quadratic residue modulo p. We may thus think of a as an element of the field $F = \text{GF}(p)$ rather than as an integer. We may similarly regard x as an element of F. Let λ denote a primitive element of F. Every non-zero element of F is then a power of λ and it follows that $a \in F$ is a residue mod p if $a = \lambda^{2k}$ for some k and a is a non-residue mod p if $a = \lambda^{2k+1}$ for some k. As a consequence we have the following proposition.

Proposition 5.6

If Q denotes the set of all quadratic residues modulo p and N the set of all quadratic non-residues modulo p, then:

(i) order of Q = order of $N = (p-1)/2$
(ii) $ab \in Q$ if both $a, b \in Q$ or $a, b \in N$
(iii) $ab \in N$ if one of a, b is in Q and the other is in N
(iv) $-1 \in Q$ if p is of the form $4k+1$ and $-1 \in N$ if p is of the form $4k-1$.

Proof
We need only to prove item (iv).
Let

$$\beta = \lambda^{(p-1)/2}$$

Then

$$\beta^2 = \lambda^{p-1} = 1$$

so that

$$(\beta - 1)(\beta + 1) = 0$$

The element λ being primitive, $\beta \neq 1$. Therefore, $\beta + 1 = 0$, i.e.

$$-1 = \beta = \lambda^{(p-1)/2}$$

which is an even power of λ if $p = 4k + 1$ while it is an odd power of λ if $p = 4k - 1$.

Proposition 5.7
If p is a prime of the form $4k + 1$ then there exist integers a and b such that $p = a^2 + b^2$.

Proof
Since $p \equiv +1 (\bmod 4)$, -1 is a quadratic residue mod p. Let s be an integer such that $s^2 \equiv -1 (\bmod p)$. Consider the set

$$S = \{(u, v) | 0 \leq u \leq \sqrt{p}, 0 \leq v \leq \sqrt{p}\}$$

of ordered pairs with u, v integers. This set contains $(1 + [\sqrt{p}])^2$ elements, where $[x]$ denotes the number of non-negative integers at most x. Now

$$(1 + [\sqrt{p}])^2 = 1 + 2[\sqrt{p}] + [\sqrt{p}]^2$$

Also

$$\sqrt{p} = [\sqrt{p}] + x \quad 0 < x < 1$$

and

$$\begin{aligned}(1 + [\sqrt{p}])^2 &= 1 + 2(\sqrt{p} - x) + p - 2\sqrt{p}x + x^2 \\ &= (1 - x)^2 + p + 2(1 - x)\sqrt{p} > p\end{aligned}$$

Therefore, $\{u - sv | (u, v) \in S\}$ has more than p numbers and, therefore, we have

$$u_2 - sv_2 \equiv u_1 - sv_1 (\bmod p)$$

for some $(u_1, v_1) \neq (u_2, v_2)$ in S.

Let $u_0 = u_2 - u_1$, $v_0 = v_2 - v_1$. Then $|u_0| < \sqrt{p}$, $|v_0| < \sqrt{p}$ and both $|u_0|$, $|v_0|$ cannot be simultaneously zero. Hence,

$$1 \leq u_0^2 + v_0^2 < 2p \tag{5.1}$$

Also

$$u_0^2 + v_0^2 \equiv s^2 v_0^2 + v_0^2 \pmod{p}$$
$$\equiv 0 \pmod{p} \tag{5.2}$$

It follows from (5.1) and (5.2) that $u_0^2 + v_0^2 = p$.

Proposition 5.8
Given any positive integer m and a prime p of the form $4k+1$, there always exists a self dual code of length $2m$ and dimension m over GF(p).

Proof
Let $a, b \in \mathrm{GF}(p)$ such that $a^2 + b^2 = p$. For any i, $1 \le i \le m$, let

$$e^i = e_1^i e_2^i \text{---} e_{2m}^i$$

be the word of length $2m$ with

$$e_{2i-1}^i = a$$
$$e_{2i}^i = b$$
$$e_j^i = 0 \quad \text{for every other } j$$

Then $e^1, e^2, \dots, e^m$ are linearly independent, and e^i, e^j for any i, j, $1 \le i, j \le m$, are orthogonal. Therefore, these generate a self dual code of dimension m and length $2m$ over GF(p).

Examples 5.4

Case (i)
Consider the ternary words

2 0 2 1 0 0 0 0   1 1 2 0 0 0 0 0
0 0 0 0 0 1 1 1   2 1 0 2 0 0 0 0

of length 8. These words are self-orthogonal and any two of these are orthogonal to each other. Therefore, these words generate a ternary self dual code of length 8 and dimension 4.

Case (ii)
The ternary words

1 1 2 0 0 0 0 0   0 0 0 0 1 1 2 0
2 1 0 2 0 0 0 0   0 0 0 0 2 1 0 2

of length 8 are self-orthogonal and any two of these are orthogonal to each other. These, thus, generate a ternary self dual code of length 8 and dimension 4.

Case (iii)
Observe that 112000, 211000, 000111 generate a ternary self dual code of length 6 and dimension 3.

Case (iv)
1211, 0123 generate a self dual code of length 4 and dimension 2 over GF(7).

Case (v)
12110000, 01230000, 00001211, 00000123 generate a self dual code of length 8 and dimension 4 over GF(7).

Exercises 5.4

1. Does there exist a self dual code of length 6 and dimension 3 over GF(7)?
2. Given an odd prime p, does there exist a self dual code of length
   (i) 8 and dimension 4, and
   (ii) 16 and dimension 8 over GF(p)?

(Hint: Every odd prime can be expressed as a sum of four squares.)

3. Given an odd prime p and a positive integer m, does there always exist a self dual code of length $4m$ and dimension $2m$ over GF(p)?
4. Find all possible self dual codes of length
   (i) 4;
   (ii) 6;
   over the field of four elements.
5. Does there exist a self dual code of length 6 over the field of 9 elements?
6. Describe (if possible) a self dual code of length 8 over GF(4).

5.3 WEIGHT DISTRIBUTION OF THE DUAL CODE OF A BINARY LINEAR CODE

In this section, we prove one of the most important results in algebraic coding theory. This is a result of F. J. MacWilliams (MacWilliams and Sloane, 1978) which says that the weight enumerator of the dual code $\mathscr{C}^{\perp}$ is completely determined once the weight enumerator of $\mathscr{C}$ is known.

Definition 5.8
Let $\mathscr{C}$ be an $[n, k, d]$ linear code over a finite field F and let $\mathscr{C}^{\perp}$ be its dual code. Recall that $\mathscr{C}^{\perp}$ is a linear $[n, n-k, -]$ code over F. Let A_i denote the number of code words in $\mathscr{C}$ which are of weight i. We call the polynomial

$$\sum_{i=0}^{n} A_i x^{n-i} y^i$$

the **weight enumerator** of $\mathscr{C}$ and denote it by $W_{\mathscr{C}}(x, y)$. This is a homogeneous polynomial of degree n in the variables x and y. Observe that we can rewrite this polynomial as

$$W_{\mathscr{C}}(x, y) = \sum_{u \in \mathscr{C}} x^{n - \text{wt}(u)} y^{\text{wt}(u)}$$

We denote by A'_i the number of code words of weight i in the dual code $\mathscr{C}^{\perp}$. Then we can similarly have the weight enumerator of the dual code $\mathscr{C}^{\perp}$ by

$$W_{\mathscr{C}^{\perp}}(x, y) = \sum_{i=0}^{n} A'_i x^{n-i} y^i$$

Examples 5.5

Case (i)
The weight enumerator of the code of Case (iv)(a) of Examples 5.1 is

$$W_{\mathscr{C}}(x, y) = x^4 + x^3 y + 3x^2 y^2 + 3xy^3$$

while that of its dual (refer to Case (i) of Examples 5.2) is

$$W_{\mathscr{C}^{\perp}} = x^4 + xy^3$$

Case (ii)
The weight enumerator of $[7, 4, 3]$ Hamming code is (refer to Case (ii) of Examples 5.2):

$$W_{\mathscr{C}}(x, y) = x^7 + 7x^4 y^3 + 7x^3 y^4 + y^7$$

while that of its dual is

$$W_{\mathscr{C}^{\perp}}(x, y) = x^7 + 7x^3 y^4$$

Case (iii)
The weight enumerator of $(4, 7)$ polynomial code $\mathscr{C}$ generated by the polynomial $1 + X + X^3$ (refer to Case (iii) of Examples 2.1) is given by

$$W_{\mathscr{C}}(x, y) = x^7 + 7x^4 y^3 + 7x^3 y^4 + y^7$$

which is the same as that of the $[7, 4, 3]$ Hamming code. The weight enumerator of its dual code is

$$W_{\mathscr{C}^{\perp}}(x, y) = x^7 + 7x^3 y^4$$

again the same (as it should be in view of MacWilliams's Identity to be proved later) as that of the dual of $[7, 4, 3]$ Hamming code.

Before going over to MacWilliams's result, we have a couple of auxiliary results that we need for its proof.

Lemma 5.1
If $\mathscr{C}$ is a binary linear code of length n and $\mathbf{v} \notin \mathscr{C}^{\perp}$, then

$$\sum_{u \in \mathscr{C}} (-1)^{\mathbf{u} \cdot \mathbf{v}} = 0$$

Proof
$\mathbf{v} \notin \mathscr{C}^{\perp}$ implies that there exists at least one $\mathbf{u} \in \mathscr{C}$ such that $\mathbf{u} \cdot \mathbf{v} \neq 0$. Let $\mathbf{u}^1, \mathbf{u}^2, \ldots, \mathbf{u}^r$ be all the elements of $\mathscr{C}$ such that $\mathbf{u}^i \cdot \mathbf{v} = 1$ for $1 \leq i \leq r$ and $\mathbf{w}^1, \mathbf{w}^2, \ldots, \mathbf{w}^s$ be all the elements of $\mathscr{C}$ such that $\mathbf{w}^j \cdot \mathbf{v} = 0$ for $1 \leq j \leq s$. Then $\mathbf{u}^1 + \mathbf{w}^1, \mathbf{u}^1 + \mathbf{w}^2, \ldots, \mathbf{u}^1 + \mathbf{w}^s$ are distinct elements of $\mathscr{C}$ and

$$(\mathbf{u}^1 + \mathbf{w}^j) \cdot \mathbf{v} = 1 \quad \text{for } 1 \leq j \leq s$$

This shows that $s \leq r$.

Again $\mathbf{u}^1 + \mathbf{u}^1, \mathbf{u}^1 + \mathbf{u}^2, \ldots, \mathbf{u}^1 + \mathbf{u}^r$ are distinct elements in $\mathscr{C}$ and

$$(\mathbf{u}^1 + \mathbf{u}^i) \cdot \mathbf{v} = 0 \quad \forall i, 1 \leq i \leq r$$

Hence $r \leq s$ and we then have $r = s$.

Then,

$$\begin{aligned} \sum_{\mathbf{u} \in \mathscr{C}} (-1)^{\mathbf{u} \cdot \mathbf{v}} &= \sum_{i=1}^{r} (-1)^{\mathbf{u}^i \cdot \mathbf{v}} + \sum_{j=1}^{r} (-1)^{\mathbf{w}^j \cdot \mathbf{v}} \\ &= \sum_{i=1}^{r} (-1) + \sum_{j=1}^{r} (-1)^0 \\ &= 0 \end{aligned}$$

Definition 5.9
Let f be a mapping defined on $V(n,q)$–the space of all vectors of length n over F. Suppose that f takes values in a set in which addition and subtraction are defined so that we can add and subtract the values $f(\mathbf{u})$. Then the **Hadamard transform** $\hat{f}$ of f is defined by

$$\hat{f}(\mathbf{u}) = \sum_{\mathbf{v} \in V(n,q)} (-1)^{\mathbf{u} \cdot \mathbf{v}} f(\mathbf{v}) \quad \mathbf{u} \in V(n,q)$$

Lemma 5.2
If $\mathscr{C}$ is an $[n,k,-]$ binary linear code, then

$$\sum_{\mathbf{u} \in \mathscr{C}^{\perp}} f(\mathbf{u}) = \frac{1}{|\mathscr{C}|} \sum_{\mathbf{u} \in \mathscr{C}} \hat{f}(\mathbf{u})$$

where $|\mathscr{C}|$ denotes the order of $\mathscr{C}$.

Proof

$$\sum_{\mathbf{u}\in\mathscr{C}} \hat{f}(\mathbf{u}) = \sum_{\mathbf{u}\in\mathscr{C}} \sum_{\mathbf{v}\in V(n,q)} (-1)^{\mathbf{u}\cdot\mathbf{v}} f(\mathbf{v})$$

$$= \sum_{\mathbf{v}\in V(n,q)} f(\mathbf{v}) \sum_{\mathbf{u}\in\mathscr{C}} (-1)^{\mathbf{u}\cdot\mathbf{v}}$$

$$= \sum_{\mathbf{v}\in\mathscr{C}^\perp} f(\mathbf{v}) \sum_{\mathbf{u}\in\mathscr{C}} (-1)^{\mathbf{u}\cdot\mathbf{v}} + \sum_{\mathbf{v}\notin\mathscr{C}^\perp} f(\mathbf{v}) \sum_{\mathbf{u}\in\mathscr{C}} (-1)^{\mathbf{u}\cdot\mathbf{v}}$$

$$= \sum_{\mathbf{v}\in\mathscr{C}^\perp} f(\mathbf{v}) \sum_{\mathbf{u}\in\mathscr{C}} (-1)^{\mathbf{u}\cdot\mathbf{v}} \qquad \text{(by Lemma 5.1)}$$

$$= |\mathscr{C}| \sum_{\mathbf{v}\in\mathscr{C}^\perp} f(\mathbf{v})$$

and the result follows.

Theorem 5.3

(MacWilliams's Identity for binary linear codes.) If $\mathscr{C}$ is an $[n,k,-]$ binary linear code with dual code $\mathscr{C}^\perp$, then

$$W_{\mathscr{C}^\perp}(x,y) = \frac{1}{|\mathscr{C}|} W_{\mathscr{C}}(x+y, x-y)$$

where $|\mathscr{C}| = 2^k$ is the number of code words in $\mathscr{C}$. Equivalently,

$$\sum_{j=0}^{n} A'_j x^{n-j} y^j = \frac{1}{|\mathscr{C}|} \sum_{i=0}^{n} A_i (x+y)^{n-i} (x-y)^i$$

or

$$\sum_{\mathbf{u}\in\mathscr{C}^\perp} x^{n-\mathrm{wt}(\mathbf{u})} y^{\mathrm{wt}(\mathbf{u})} = \frac{1}{|\mathscr{C}|} \sum_{\mathbf{u}\in\mathscr{C}} (x+y)^{n-\mathrm{wt}(\mathbf{u})} (x-y)^{\mathrm{wt}(\mathbf{u})}$$

Proof

Define a map $f: V(n,q) \to \mathbb{B}[x,y]$, where $\mathbb{B}[x,y]$ is the polynomial ring in the two commuting variables x and y, by

$$f(\mathbf{u}) = x^{n-\mathrm{wt}(\mathbf{u})} y^{\mathrm{wt}(\mathbf{u})} \quad \mathbf{u}\in V(n,q)$$

Of course, here $q = 2$. Let $\hat{f}$ denote the Hadamard transform of f. Then, we have

$$\hat{f}(\mathbf{u}) = \sum_{\mathbf{v}\in V(n,q)} (-1)^{\mathbf{u}\cdot\mathbf{v}} x^{n-\mathrm{wt}(\mathbf{v})} y^{\mathrm{wt}(\mathbf{v})}$$

Let

$$\mathbf{u} = (u_1 \quad u_2 \quad \cdots \quad u_n)$$

and

$$\mathbf{v} = (v_1 \quad v_2 \quad \cdots \quad v_n)$$

Then

$$\begin{aligned}
\hat{f}(\mathbf{u}) &= \sum_{\mathbf{v}\in V(n,q)} (-1)^{u_1v_1+\cdots+u_nv_n} \prod_{i=1}^{n} x^{1-v_i}y^{v_i} \\
&= \sum_{v_1=0}^{1}\sum_{v_2=0}^{1} \cdots \sum_{v_n=0}^{1}\prod_{i=1}^{n}(-1)^{u_iv_i}x^{1-v_i}y^{v_i} \\
&= \prod_{i=1}^{n}\sum_{w=0}^{1}(-1)^{u_iw}x^{1-w}y^{w}
\end{aligned}$$

If $u_i = 0$, then

$$\sum_{w=0}^{1}(-1)^{u_iw}x^{1-w}y^{w} = x + y$$

while if $u_i = 1$, then

$$\sum_{w=0}^{1}(-1)^{u_iw}x^{1-w}y^{w} = x - y$$

Therefore,

$$\hat{f}(\mathbf{u}) = (x+y)^{n-\mathrm{wt}(\mathbf{u})}(x-y)^{\mathrm{wt}(\mathbf{u})}$$

From Lemma 5.2, it follows that

$$\begin{aligned}
\sum_{\mathbf{u}\in\mathscr{C}^{\perp}} f(\mathbf{u}) &= \frac{1}{|\mathscr{C}|}\sum_{\mathbf{u}\in\mathscr{C}}\hat{f}(\mathbf{u}) \\
&= \frac{1}{|\mathscr{C}|}\sum_{\mathbf{u}\in\mathscr{C}}(x+y)^{n-\mathrm{wt}(\mathbf{u})}(x-y)^{\mathrm{wt}(\mathbf{u})}
\end{aligned}$$

or

$$\sum_{\mathbf{u}\in\mathscr{C}^{\perp}} x^{n-\mathrm{wt}(\mathbf{u})}y^{\mathrm{wt}(\mathbf{u})} = \frac{1}{|\mathscr{C}|}\sum_{\mathbf{u}\in\mathscr{C}}(x+y)^{n-\mathrm{wt}(\mathbf{u})}(x-y)^{\mathrm{wt}(\mathbf{u})}$$

or

$$W_{\mathscr{C}^{\perp}}(x,y) = \frac{1}{|\mathscr{C}|}W_{\mathscr{C}}(x+y,x-y)$$

Exercises 5.5

1. Determine the code words of the (4, 7) binary polynomial code generated by $1 + X^2 + X^3$. Also find the weight enumerator of its dual.
2. Determine the code words of the (3, 6) binary polynomial code generated by $1 + X + X^2$. Also find the weight enumerator of its dual.

3. Determine the code words of the (i) (2, 4); (ii) (2, 5) ternary code generated by the polynomial (a) X^2+X-1; (b) X^3+2X+1. Also find their duals. Write down the weight enumerators of the two codes and their duals.
4. Determine the (3, 5) ternary code generated by X^2+X-1. Also find its dual. Write down the weight enumerators of the code and its dual.
5. Find the weight enumerators of the duals of the codes of Case (iv) in Example 5.1 and question 3 of Exercise 5.3.

We have proved MacWilliams's Identity for binary linear codes. MacWilliams's Identity giving weight enumerator of the dual code $\mathscr{C}^{\perp}$ once the weight enumerator of $\mathscr{C}$ is known is available for linear codes over any finite field of q elements:

Theorem 5.4

If $\mathscr{C}$ is a linear code over a field F of q elements, then

$$W_{\mathscr{C}^{\perp}}(x,y)=\frac{1}{|\mathscr{C}|}W_{\mathscr{C}}(x+(q-1)y,x-y)$$

We do not go into the proof of this theorem.

5.4 NEW CODES OBTAINED FROM GIVEN CODES

Let $\mathscr{C}$ be a linear $[n,k,d]$ code over GF(q) with generator matrix **G** and parity check matrix **H**. There are several ways in which $\mathscr{C}$ can be modified to yield new codes. However, we here discuss only three such modifications.

5.4.1 Extending a code

Let $\hat{\mathscr{C}}$ be the set of words of length $n+1$ obtained from the words of $\mathscr{C}$ such that the first n symbols form a word in $\mathscr{C}$ and the $(n+1)$th symbol is the negative of the sum of the first n symbols. Then the sum of all the symbols of any word in $\hat{\mathscr{C}}$ is always zero. Clearly $\hat{\mathscr{C}}$ is again a linear code over GF(q) and is of length $n+1$. Given a basis $c(1),c(2),\ldots,c(k)$ of $\mathscr{C}$, we extend every word in the basis by adding an overall parity check. The resulting set of words is again linearly independent over GF(q). On the other hand, if

$$c'=c_1c_2\text{---}c_nc_{n+1}\in\hat{\mathscr{C}}$$

then $c=c_1c_2\text{---}c_n$ is in $\mathscr{C}$ and so is a linear combination of $c^{(1)},c^{(2)},\ldots,c^{(k)}$. Let

$$c=\alpha_1c^{(1)}+\alpha_2c^{(2)}+\cdots+\alpha_kc^{(k)}\quad \alpha_i\in\mathrm{GF}(q) \tag{5.3}$$

Then

$$c_i=\alpha_1c_i^{(1)}+\alpha_2c_i^{(2)}+\cdots+\alpha_kc_i^{(k)}\quad 1\le i\le n$$

Therefore

$$\sum_{i=1}^{n} c_i = \alpha_1 \sum_{i=1}^{n} c_i^{(1)} + \alpha_2 \sum_{i=1}^{n} c_i^{(2)} + \cdots + \alpha_k \sum_{i=1}^{n} c_i^{(k)}$$
$$= -\alpha_1 c_{n+1}^{(1)} - \alpha_2 c_{n+1}^{(2)} - \cdots - \alpha_k c_{n+1}^{(k)}$$

or

$$c_{n+1} = \alpha_1 c_{n+1}^{(1)} + \alpha_2 c_{n+2}^{(2)} + \cdots + \alpha_k c_{n+1}^{(k)} \tag{5.4}$$

Equations (5.3) and (5.4) together show that c' is a linear combination of the words obtained from $c^{(1)}, \ldots, c^{(k)}$ by adding an overall parity check. This proves that $\hat{\mathscr{C}}$ is of dimension k. It is clear that the minimum distance $\hat{d}$ of the code $\hat{\mathscr{C}}$ is d or $d+1$.

If $\mathscr{C}$ is a binary linear code, observe that the weight of every non-zero code word in $\hat{\mathscr{C}}$ is even. Moreover, if d is odd, then $\hat{d} = d+1$. However, in the non-binary case $\hat{d}$ may be d again.

For example, let $\mathscr{C}$ be the linear code over GF(3) generated by the matrix

$$\mathbf{G} = \begin{pmatrix} 1 & -1 & 0 & 1 \\ 0 & 1 & -1 & 1 \end{pmatrix}$$

The code words of $\mathscr{C}(\hat{\mathscr{C}})$ are:

Message words		Code words of $\mathscr{C}$				Code words of $\hat{\mathscr{C}}$				
0	0	0	0	0	0	0	0	0	0	0
1	0	1	−1	0	1	1	−1	0	1	−1
−1	0	−1	1	0	−1	−1	1	0	−1	1
0	1	0	1	−1	1	0	1	−1	1	−1
1	1	1	0	−1	−1	1	0	−1	−1	1
−1	1	−1	−1	−1	0	−1	−1	−1	0	0
0	−1	0	−1	1	−1	0	−1	1	−1	1
1	−1	1	1	1	0	1	1	1	0	0
−1	−1	−1	0	1	1	−1	0	1	1	−1

Thus $d = 3 = \hat{d}$.

Let $b = b_1 b_2 \cdots b_n$ be a code word in $\mathscr{C}$ and $c = c_1 \cdots c_n c_{n+1}$ be the corresponding code word in $\hat{\mathscr{C}}$. Then $c_1 + c_2 + \cdots + c_{n+1} = 0$ and, therefore, parity check matrix $\hat{\mathbf{H}}$ of $\hat{\mathscr{C}}$ must contain an all one word of length $n+1$. It is then clear that

$$\hat{\mathbf{H}} = \begin{pmatrix} 1 & 1 & \cdots & 1 \\ & & \mathbf{H} & \mathbf{0} \\ & & & \vdots \\ & & & 0 \end{pmatrix}$$

which is an $(n-k+1) \times (n+1)$ matrix.

Definition 5.10

The above process of obtaining the code $\hat{\mathscr{C}}$ from the given code $\mathscr{C}$ is called **extending a code** and also $\hat{\mathscr{C}}$ is called the **extended code of** $\mathscr{C}$.

Example 5.5

The (4, 7) binary Hamming code has a parity check matrix

$$\mathbf{H} = \begin{pmatrix} 0 & 0 & 0 & 1 & 1 & 1 & 1 \\ 0 & 1 & 1 & 0 & 0 & 1 & 1 \\ 1 & 0 & 1 & 0 & 1 & 0 & 1 \end{pmatrix}$$

and, therefore, a parity check matrix of extended Hamming code is

$$\hat{\mathbf{H}} = \begin{pmatrix} 1 & 1 & 1 & 1 & 1 & 1 & 1 & 1 \\ 0 & 0 & 0 & 1 & 1 & 1 & 1 & 0 \\ 0 & 1 & 1 & 0 & 0 & 1 & 1 & 0 \\ 1 & 0 & 1 & 0 & 1 & 0 & 1 & 0 \end{pmatrix}$$

The code words of the extended Hamming code are:

$$\begin{array}{cccccccc} 0&0&0&0&0&0&0&0 \\ 1&1&0&1&0&0&1&0 \\ 0&1&0&1&0&1&0&1 \\ 1&0&0&1&1&0&0&1 \\ 1&1&1&0&0&0&0&1 \\ 1&0&0&0&0&1&1&1 \\ 0&1&0&0&1&0&1&1 \\ 0&0&1&1&0&0&1&1 \end{array} \qquad \begin{array}{cccccccc} 1&1&0&0&1&1&0&0 \\ 1&0&1&1&0&1&0&0 \\ 0&1&1&1&1&0&0&0 \\ 0&1&0&1&1&1&1&1 \\ 0&1&1&0&0&1&1&0 \\ 1&0&1&0&1&0&1&0 \\ 0&0&1&0&1&1&0&1 \\ 1&1&1&1&1&1&1&1 \end{array}$$

Exercise 5.6

Find the extended code $\hat{\mathscr{C}}$ by adding an overall parity check where $\mathscr{C}$ is the code generated by

(a) $\begin{pmatrix} 1 & 1 & 0 & 0 & 0 \\ 0 & 0 & 1 & 1 & 1 \end{pmatrix}$ over $\mathbb{B}$

(b) $\begin{pmatrix} 1 & 2 & 1 & 0 & 0 \\ 0 & 0 & 1 & 2 & 1 \end{pmatrix}$ over GF(3)

5.4.2 Expurgating a code

Let $\mathscr{C}$ be a binary linear code of length n having code words of both even and odd weights. Then it is easy to see that exactly half the code words in $\mathscr{C}$ are of even weight and the other half are of odd weight. This follows from the observation that sum of two words both of odd weight or both of even weight is of even weight while the sum of two words one of which is of odd weight and the other of even weight is always of odd weight. Also it follows from this that $\mathscr{C}'$ the set of all even weight words is a subspace of $\mathscr{C}$ of dimension $k-1$ if $\mathscr{C}$ is of dimension k. The process of omitting code words is called **expurgation** or expurgating a code. The minimum distance d' of $\mathscr{C}'$ is always at least d while if d is odd strict inequality holds: $d' > d$.

Example 5.6

The $[7,4,3]$ binary Hamming code has code words of odd as well as even weights. Expurgating this code by throwing away odd weight code words gives a $[7,3,4]$ code. As Hamming code has minimum distance 3, the expurgated code has minimum distance 4.

Expurgation of codes is not available just for binary codes but may be defined for non-binary codes as well. If $\mathscr{C}$ is a linear $[n,k,d]$ code over $\mathrm{GF}(q)$ with parity check matrix $\mathbf{H}$, we may expurgate $\mathscr{C}$ by changing $\mathbf{H}$ to $\mathbf{H}_1$ by adding a row of all ones. If the row of all ones is linearly dependent on the rows of $\mathbf{H}$, then no code words are thrown away in this process. However, if the all ones row is linearly independent of the rows of $\mathbf{H}$, then precisely those code words of $\mathscr{C}$ will belong to the expurgated code $\mathscr{C}'$ the sum of all the entries of which is zero.

Exercises 5.7

1. Obtain the expurgated codes of the codes of Case (iv) of Examples 5.1 and Case (iii) of Examples 5.2 and also of their duals. Find also their extended codes.
2. Obtain the extended and the expurgated codes of the ternary self dual code of length 4 generated by 1201 and 1012.

5.4.3 Augmenting a code by adding new code words

The reverse process of expurgating is called **augmenting** a code. While by adding a row of all ones to parity check matrix removed certain words from $\mathscr{C}$, adding a row of all ones to the generator matrix $\mathbf{G}$ of $\mathscr{C}$ may result in adding new code words. If the row of all ones is a linear combination of the rows of $\mathbf{G}$, the augmented code of $\mathscr{C}$ is $\mathscr{C}$ itself. However, if the row of all ones is not a linear combination of the rows of $\mathbf{G}$, then the augmented code is the linear space generated by $\mathscr{C}$ and the all ones word. In this case, dimension of the augmented code becomes $k+1$ when k is the dimension of $\mathscr{C}$.

Exercises 5.8

1. What effect the addition of row of all ones to the parity check matrix of the [7, 4, 3] binary Hamming code have on the given code $\mathscr{C}$?
2. Obtain the augmented codes of the codes of Case (iv) of Examples 5.1, and Case (iii) of Examples 5.2 and their duals.

3. Obtain the augmented code of the ternary self dual code of length 4 generated by 1201 and 1012.
4. If $\mathscr{C}$ is a ternary code with minimum distance d, then the minimum distance $\hat{d}$ of the extended code $\hat{\mathscr{C}}$ is given by

$$\hat{d} = \begin{cases} d & \text{if } \mathscr{C} \text{ has a minimum distance word } c \text{ with} \\ & \text{wt}(c) \equiv 0 (\text{mod}\, 3) \\ d+1 & \text{if } \mathscr{C} \text{ has no minimum distance word } c \text{ with} \\ & \text{wt}(c) \equiv 0 (\text{mod}\, 3). \end{cases}$$

Can we have this or a similar observation about codes over GF(p) for any prime p?

6

Cyclic codes

6.1 CYCLIC CODES

Let $F = \mathrm{GF}(q)$ be a field of q elements and $F^{(n)} = V(n, q)$, as before, be the vector space of all vectors (or sequences) of length n over F. Then $V(n, q)$ is of dimension n over F. **We suppose that** $(n, q) = 1$.

Definition 6.1

A linear code $\mathscr{C}$ of length n over F is called **cyclic** if any cyclic shift of a code word is again a code word, i.e. if $(a_0, a_1, \ldots, a_{n-1})$ is in $\mathscr{C}$ then so is $(a_{n-1}, a_0, \ldots, a_{n-2})$.

Algebraic description of cyclic codes

There is a beautiful algebraic description of cyclic codes. To obtain this we define a map

$$\theta\colon V(n, q) \to F[X]/\langle X^n - 1\rangle$$

where $\langle X^n - 1\rangle$ denotes the ideal of the polynomial ring $F[X]$ generated by $X^n - 1$, by

$$\theta(a_0, a_1, \ldots, a_{n-1}) = a_0 + a_1X + \cdots + a_{n-1}X^{n-1} + \langle X^n - 1\rangle \qquad \forall a_i \in F, 0 \leq i \leq n-1$$

Observe that $F[X]/\langle X^n - 1\rangle$ is also a vector space over F and θ is a vector space isomorphism. Let $\mathscr{C}$ be a linear code of length n over F, i.e. $\mathscr{C}$ is a subspace of $V(n, q)$. Then $\theta(\mathscr{C})$ is a subspace of $F[X]/\langle X^n - 1\rangle$. Let $a = (a_0, a_1, \ldots, a_{n-1}) \in \mathscr{C}$. Then $(a_{n-1}, a_0, \ldots, a_{n-2}) \in \mathscr{C}$ iff

$$\begin{aligned} &a_{n-1} + a_0X + \cdots + a_{n-2}X^{n-1} + \langle X^n - 1\rangle \\ &= X(a_0 + a_1X + \cdots + a_{n-1}X^{n-1}) + \langle X^n - 1\rangle \end{aligned}$$

is in $\theta(\mathscr{C})$. From this it follows that $\mathscr{C}$ is a cyclic code iff $\theta(\mathscr{C})$ is an ideal in the quotient ring $F[X]/\langle X^n - 1\rangle$. Identifying the element $(a_0, a_1, \ldots, a_{n-1})$ in

$\mathscr{C}$ with the corresponding element

$$a_0 + a_1X + \cdots + a_{n-1}X^{n-1} + \langle X^n - 1\rangle$$

or with the polynomial $a_0 + a_1X + \cdots + a_{n-1}X^{n-1}$ of degree at most $n-1$, we may regard a cyclic code $\mathscr{C}$ of length n as an ideal of the quotient ring $F[X]/\langle X^n - 1\rangle$.

Theorem 6.1

Let $\mathscr{C}$ be a non-zero cyclic code of length n over F.

(a) There is a unique monic polynomial $g(X)$ of minimal degree in $\mathscr{C}$ which generates it.
(b) $g(X)$ is a factor of $X^n - 1$.
(c) Let $\deg g(X) = r$. Then the dimension of $\mathscr{C}$ is $n-r$ and any $a(X) \in \mathscr{C}$ has a unique representation of the form $a(X) = b(X)g(X)$, where $\deg b(X) < n-r$.
(d) If $g(X) = g_0 + g_1X + \cdots + g_rX^r$, then the $(n-r)\times n$ matrix

$$\mathbf{G} = \begin{pmatrix} g_0 & g_1 & \cdots & g_r & 0 & \cdots & 0 \\ 0 & g_0 & & g_{r-1} & g_r & \cdots & 0 \\ \vdots & \vdots & \ddots & \vdots & & \ddots & \\ 0 & 0 & & \cdots & g_0 & \cdots & g_r \end{pmatrix}$$

is a generator matrix of $\mathscr{C}$.

Proof

Let I denote the ideal $\langle X^n - 1\rangle$ of $F[X]$ generated by $X^n - 1$. Let $N = \{\deg a(X)/a(X) + I \in \mathscr{C}\}$. The set of non-negative integers being well ordered, N has a least element. Let $g(X)$ be a polynomial of minimal degree such that $g(X) + I \in \mathscr{C}$. F being a field, we can take $g(X)$ to be a monic polynomial. If $g'(X)$ is another monic polynomial of minimal degree such that $g'(X) + I \in \mathscr{C}$, then $g(X) - g'(X) + I \in \mathscr{C}$ and

$$\deg(g(X) - g'(X)) < \deg g(X)$$

Hence $g'(X) - g(X) = 0$ and $g(X)$ is the unique monic polynomial with $g(X) + I$ in $\mathscr{C}$. Let $a(X) + I$ be any element in $\mathscr{C}$. F being a field, $F[X]$ is a Euclidean domain. Therefore, there exist polynomials $b(X), r(X)$ in $F[X]$ such that

$$a(X) = b(X)g(X) + r(X)$$

where $r(X) = 0$ or $\deg r(X) < \deg g(X)$. If $r(X) \neq 0$, then

$$r(X) + I = a(X) - b(X)g(X) + I \in \mathscr{C}$$

giving a contradiction. Hence $r(X) = 0$ and

$$a(X) + I = (b(X) + I)(g(X) + I)$$

i.e. $g(X)$ generates $\mathscr{C}$.

Again, let

$$X^n - 1 = a(X)g(X) + r(X)$$

such that $\deg r(X) < \deg g(X)$ if $r(X) \neq 0$. This shows that

$$r(X) + I = -a(X)g(X) + I \in \mathscr{C}$$

and this gives a contradiction to the choice of $g(X)$. This proves part (b).

Observe that every element of $F[X]/I$ can be uniquely written as $a(X) + I$, where $a(X)$ is a polynomial of degree at most $n - 1$ and, so, that is in particular true for every element of $\mathscr{C}$. As such, the elements

$$g(X) + I, Xg(X) + I, \ldots, X^{n-r-1}g(X) + I$$

of $\mathscr{C}$ are linearly independent over F. Let $a(X)$ be a polynomial of degree at most $n - 1$ such that $a(X) + I \in \mathscr{C}$. Then

$$a(X) + I = b(X)g(X) + I$$

so that

$$a(X) = b(X)g(X) + (X^n - 1)c(X)$$

But $g(X) | X^n - 1$. Let $X^n - 1 = g(X)p(X)$. Then

$$a(X) = (b(X) + c(X)p(X))g(X) = d(X)g(X) \tag{6.1}$$

where $d(X) = b(X) + c(X)p(X)$. Also it follows from (6.1) and the degree considerations that $\deg d(X) < n - r$. Hence $a(X) + I$ is a linear combination of

$$g(X) + I, Xg(X) + I, \ldots, X^{n-r-1}g(X) + I$$

Thus it follows that

$$g(X) + I, Xg(X) + I, \ldots, X^{n-r-1}g(X) + I$$

is a basis of $\mathscr{C}$ over F, $\mathscr{C}$ is of dimension $n - r$, and every element of $\mathscr{C}$ can be uniquely written as $a(X)g(X) + I$, where $\deg a(X) < n - r$. Now $\mathscr{C}$ becomes a polynomial code and part (d) follows from Theorem 2.4.

Examples 6.1

Case (i)

We have seen earlier that over $\mathbb{B}$,

$$X^7 + 1 = (X + 1)(X^3 + X + 1)(X^3 + X^2 + 1)$$

Therefore the (4, 7) polynomial codes generated by $X^3 + X + 1$ and $X^3 + X^2 + 1$ are binary cyclic codes of length 7. (Refer to Examples 2.1 for the sets of code words.)

Case (ii)

We show that every polynomial code need not be a cyclic code. Consider the binary cyclic code of length 5 generated by $1 + X + X^3$. The code words of this code are:

$$
\begin{aligned}
1(1+X+X^3) &\longrightarrow 1\ \ 1\ \ 0\ \ 1\ \ 0\\
X(1+X+X^3) &\longrightarrow 0\ \ 1\ \ 1\ \ 0\ \ 1\\
(1+X)(1+X+X^3) &\longrightarrow 1\ \ 0\ \ 1\ \ 1\ \ 1\\
0(1+X+X^3) &\longrightarrow 0\ \ 0\ \ 0\ \ 0\ \ 0
\end{aligned}
$$

which is not a cyclic code.

Case (iii)

We next construct a binary cyclic code of length 15 and dimension 11.

The polynomial X^4+X^3+1 is irreducible over $\mathbb{B}$ and is a divisor of $X^{15}-1$. From this, we observe that

$$K=\mathbb{B}[X]/\langle X^4+X^3+1\rangle$$

is a field of order 16 and X^4+X^3+1 is the minimal polynomial of

$$\alpha = X+\langle X^4+X^3+1\rangle$$

As every non-zero element of the field K is a root of $X^{15}-1$, α is also a root of this polynomial and hence its minimal polynomial X^4+X^3+1 divides $X^{15}-1$. Therefore, the ideal

$$\langle X^4+X^3+1+\langle X^{15}-1\rangle\rangle$$

generated by X^4+X^3+1 is a cyclic code of length 15. The dimension of this code is $(15-4=)11$.

Exercise 6.1

1. Determine all the binary cyclic codes of length 9.
2. Determine all the ternary cyclic codes of length 8.
3. Determine all the binary cyclic codes of length 5.
4. Prove that the polynomial X^6+X^3+1 is irreducible over the field $\mathbb{B}$ of 2 elements. Use this to construct a binary cyclic code of length 9 and dimension 3.
5. Given a prime p and a positive integer n coprime to p. Does there always exist a cyclic code of length n over GF(p)?
6. Construct a cyclic code of length 4 and dimension 2 over the field GF(5) of 5 elements.
7. Let $F=\mathbb{B}[X]/\langle X^2+X+1\rangle=\{0,1,w,w^2\}$ with $1+w+w^2=0$, $w^3=1$, $2w=0$, be the field of four elements. Prove that the polynomials $1+wX+X^2$ and $1+w^2X+X^2$ are irreducible over F. Use these to construct a cyclic code of length 5 and dimension (i) 3 and (ii) 2 over the field F of 4 elements.
8. Using the irreducible polynomial X^2+X-1 over GF(3), construct a field F of 9 elements. Construct, if possible, a cyclic code of length 5 and dimension (i) 2 and (ii) 3 over F.

6.2 CHECK POLYNOMIAL

Let $\mathscr{C}$ be a cyclic code of length n over F with generator polynomial $g(X)$ of degree r. Let $h(X)$ be the polynomial of degree $n-r$ with

$$X^n - 1 = g(X)h(X)$$

Any code word $c(X)+I$ is of the form

$$c(X) + I = a(X)g(X) + I$$

where $I = \langle X^n - 1 \rangle$ and, therefore, $c(X)h(X) = 0$, i.e. $c(X)h(X)$ is zero in $F[X]/I$. For this reason $h(X)$ is called the **check polynomial** of the code $\mathscr{C}$. We have seen that $\mathscr{C}$ is a matrix code and so $\mathscr{C}$ must have some sort of a parity check matrix. Before we define this in the general case we consider an example.

Example 6.1
Let $\mathscr{C}$ be a binary cyclic code of length 7 defined by the polynomial

$$g(X) = X^3 + X + 1$$

Then

$$h(X) = (X+1)(X^3 + X^2 + 1) = X^4 + X^2 + X + 1$$

Let

$$c(X) = c_0 + c_1 X + \cdots + c_6 X^6$$

be a code word in $\mathscr{C}$. Then $c(X)h(X) = 0$ in $\mathbb{B}[X]/\langle X^7 - 1 \rangle$, i.e.

$$(c_0 + c_1 X + \cdots + c_6 X^6)(1 + X + X^2 + X^4) = 0$$

in $\mathbb{B}[X]/\langle X^7 - 1 \rangle$. This means that

$$\begin{aligned}(c_0 + c_6 + c_5 + c_3) &+ (c_1 + c_0 + c_6 + c_4)X + (c_2 + c_1 + c_0 + c_5)X^2 \\ &+ (c_3 + c_2 + c_1 + c_6)X^3 + (c_4 + c_3 + c_2 + c_0)X^4 + (c_5 + c_4 + c_3 + c_1)X^5 \\ &+ (c_6 + c_5 + c_4 + c_2)X^6 = 0\end{aligned}$$

So the parity check equations are

$$\begin{aligned} c_0 + c_6 + c_5 + c_3 &= 0 \\ c_1 + c_0 + c_6 + c_4 &= 0 \\ c_2 + c_1 + c_0 + c_5 &= 0 \\ c_3 + c_2 + c_1 + c_6 &= 0 \\ c_4 + c_3 + c_2 + c_0 &= 0 \\ c_5 + c_4 + c_3 + c_1 &= 0 \\ c_6 + c_5 + c_4 + c_2 &= 0 \end{aligned}$$

In the matrix form, these may be rewritten as

$$(c_0 \quad c_1 \quad \cdots \quad c_6)\begin{pmatrix} 0 & 0 & 1 & 0 & 1 & 1 & 1 \\ 0 & 1 & 0 & 1 & 1 & 1 & 0 \\ 1 & 0 & 1 & 1 & 1 & 0 & 0 \\ 0 & 1 & 1 & 1 & 0 & 0 & 1 \\ 1 & 1 & 1 & 0 & 0 & 1 & 0 \\ 1 & 1 & 0 & 0 & 1 & 0 & 1 \\ 1 & 0 & 0 & 1 & 0 & 1 & 1 \end{pmatrix} = 0$$

or $\mathbf{cH} = 0$, where $\mathbf{H}$ is the 7×7 matrix

$$\begin{pmatrix} 0 & 0 & h_4 & h_3 & h_2 & h_1 & h_0 \\ 0 & h_4 & h_3 & h_2 & h_1 & h_0 & 0 \\ h_4 & h_3 & h_2 & h_1 & h_0 & 0 & 0 \\ h_3 & h_2 & h_1 & h_0 & 0 & 0 & h_4 \\ h_2 & h_1 & h_0 & 0 & 0 & h_4 & h_3 \\ h_1 & h_0 & 0 & 0 & h_4 & h_3 & h_2 \\ h_0 & 0 & 0 & h_4 & h_3 & h_2 & h_1 \end{pmatrix}$$

Definition 6.1

Now let $\mathscr{C}$ be a cyclic code of length n over F with generator polynomial $g(X)$ of degree r. Let

$$h(X) = h_0 + h_1 X + \cdots + h_{n-r} X^{n-r}$$

be its check polynomial. Then a word

$$c(X) + I = c_0 + c_1 X + \cdots + c_{n-1} X^{n-1} + I$$

is in $\mathscr{C}$, iff $c(X)h(X) = 0$ in $F[X]/I$, where as before $I = \langle X^n - 1 \rangle$. This is equivalent to saying that

$$\sum_{i=0}^{n-1} c_i h_{j-i} = 0 \quad j = 0, 1, 2, \ldots, n-1$$

where the subscripts are taken modulo n and where it is also understood that $h_j = 0$ if $j > n - r$. These are the n check equations and, in the matrix form, may be rewritten as

$$\mathbf{cH} = (c_0 \quad c_1 \quad \cdots \quad c_{n-1})\mathbf{H} = 0$$

where $\mathbf{H}$ is the $n \times n$ matrix

$$\begin{pmatrix} 0 & 0 & \cdots & 0 & h_{n-r} & \cdots & h_1 & h_0 \\ 0 & & & h_{n-r} & h_{n-r-1} & \cdots & h_0 & 0 \\ \vdots & \vdots & \vdots & \cdot^{\cdot^{\cdot}} & \cdot^{\cdot^{\cdot}} & \cdot^{\cdot^{\cdot}} & & \vdots \\ h_0 & 0 & \cdots & \cdots & \cdots & \cdots & h_2 & h_1 \end{pmatrix}$$

the rows of which are defined inductively as follows:

The first row is taken as

$$0 \quad \cdots \quad 0 \quad h_{n-r} \quad \cdots \quad h_1 \quad h_0$$

and once the ith row is defined, the $(i+1)$th row is obtained by giving a cyclic shift to the ith. The matrix $\mathbf{H}$ thus obtained is called the **parity check matrix** of the cyclic code $\mathscr{C}$ (mark the difference from the usual definition of a parity check matrix!)

Thus $c(X)+I$ is in $\mathscr{C}$ iff $\mathbf{cH}=0$ or equivalently $\mathbf{H}^t\mathbf{c}^t=0$. From this, it follows that the dual code $\mathscr{C}^\perp$ contains the linear code generated by the rows of $\mathbf{H}^t=\mathbf{H}$, as $\mathbf{H}$ is clearly a symmetric matrix. The dual code $\mathscr{C}^\perp$ is of dimension $n-(n-r)=r$ and clearly the first r rows of $\mathbf{H}$ are linearly independent. Therefore, the linear code generated by $\mathbf{H}$ must be of dimension r, and hence, it must equal $\mathscr{C}^\perp$. Also, we could as well have taken the $r\times n$ matrix

$$\mathbf{H}_1=\begin{pmatrix} 0 & 0 & 0 & h_{n-r} & \cdots & h_1 & h_0 \\ 0 & & h_{n-r} & h_{n-r-1} & \cdots & h_0 & 0 \\ \vdots & \vdots & \cdot^{\cdot^{\cdot}} & \cdot^{\cdot^{\cdot}} & \cdot^{\cdot^{\cdot}} & \vdots & \vdots \\ h_{n-r} & \cdots & \cdots & \cdots & h_0 & 0 & 0 \end{pmatrix}$$

as the parity check matrix of $\mathscr{C}$. (Now we have the usual form of the parity check matrix!)

Since the rows of the matrix $\mathbf{H}$ are all the possible cyclic shifts of the vector

$$0 \quad \cdots \quad 0 \quad h_{n-r} \quad \cdots \quad h_1 \quad h_0$$

of length n, any linear combination over F of the rows of $\mathbf{H}$ will again be a linear combination of the rows of $\mathbf{H}$. Hence $\mathscr{C}^\perp$, the code generated by $\mathbf{H}$ (or $\mathbf{H}_1$) is again cyclic.

Reversing the order of the rows of $\mathbf{H}_1$, we may take the parity check matrix of $\mathscr{C}$ (and, so, also the generator matrix of $\mathscr{C}^\perp$) as

$$\mathbf{H}_2=\begin{pmatrix} h_{n-r} & \cdots & h_1 & h_0 & 0 & \cdots & \cdots & 0 \\ 0 & h_{n-r} & \cdots & h_2 & h_1 & h_0 & 0 & 0 \\ \vdots & \vdots & \cdot^{\cdot^{\cdot}} & \cdot^{\cdot^{\cdot}} & \cdot^{\cdot^{\cdot}} & \cdot^{\cdot^{\cdot}} & \vdots & \vdots \\ 0 & \cdots & \cdots & 0 & h_{n-r} & \cdots & h_1 & h_0 \end{pmatrix}$$

Thus it follows (see Theorem 2.4) that code $\mathscr{C}^\perp$ is the polynomial code generated by

$$\begin{aligned} k(X) &= h_{n-r}+h_{n-r-1}X+\cdots+h_0X^{n-r} \\ &= X^{n-r}(h_0+h_1X^{-1}+\cdots+h_{n-r}X^{-n+r}) \end{aligned}$$

Now

$$X^n-1=g(X)h(X)\Rightarrow X^{-n}-1=g(X^{-1})h(x^{-1})$$

or

$$1-X^n=X^rg(X^{-1})k(X)$$

showing that $k(X)\mid X^n-1$ as it should.

Theorem 6.2

Let $\mathscr{C}$ be a cyclic code of length n over F with generator polynomial $g(X)$ of degree r and $h(X)$ as its check polynomial. Then:

(i) the dual code $\mathscr{C}^{\perp}$ is also cyclic with $k(X) = X^{n-r}h(X^{-1})$ as a generator polynomial; and
(ii) the dual code $\mathscr{C}^{\perp}$ is equivalent to the code generated by $h(X)$.

Proof

We only have to prove (ii).

Consider the permutation matrix $\mathbf{P} = (p_{ij})$ where

$$p_{ij} = \begin{cases} 1 & \text{if } i + j = n + 1 \\ 0 & \text{otherwise} \end{cases}$$

Let $\mathbf{c}_1, \ldots, \mathbf{c}_n$ denote the columns of the generator matrix $\mathbf{H}_1$ of $\mathscr{C}^{\perp}$. Then

$$\begin{aligned} \mathbf{H}_1\mathbf{P} &= (\mathbf{c}_1 \quad \cdots \quad \mathbf{c}_n)P \\ &= \left(\sum_i \mathbf{c}_i p_{i1} \quad \sum \mathbf{c}_i p_{i2} \quad \cdots \quad \sum \mathbf{c}_i p_{in}\right) \\ &= (\mathbf{c}_n \quad \mathbf{c}_{n-1} \quad \cdots \quad \mathbf{c}_1) \\ &= \begin{pmatrix} h_0 & h_1 & \cdots & h_{n-r} & 0 & \cdots & \cdots & 0 \\ 0 & h_0 & \cdots & h_{n-r-1} & h_{n-r} & 0 & \cdots & 0 \\ \vdots & \vdots & \iddots & \iddots & \iddots & \iddots & \iddots & \vdots \\ 0 & \cdots & \cdots & 0 & h_0 & h_1 & \cdots & h_{n-r} \end{pmatrix} \end{aligned}$$

The linear code generated by $\mathbf{H}_1\mathbf{P}$ is thus the code $\langle h(X) + I \rangle$ generated by $h(X)$. But the code generated by $\mathbf{H}_1\mathbf{P}$ is equivalent to $\mathscr{C}^{\perp}$. Hence the result.

Exercise 6.2

1. Is a code equivalent to a cyclic code cyclic?
2. Determine the check polynomials and also parity check matrices of the cyclic codes constructed in Exercise 6.1.
3. Determine the duals of the codes constructed in Exercise 6.1.

6.3 BCH AND HAMMING CODES AS CYCLIC CODES

Let $\mathscr{C}$ be a cyclic code of length n over F (i.e. an ideal in $F[X]/I, I = \langle X^n - 1 \rangle$) with generator matrix $g(X)$ of degree r. Let $\alpha_1, \alpha_2, \ldots, \alpha_r$ be the roots of $g(X)$ in a suitable extension field of F. Then

$$g(X) = (X - \alpha_1)\cdots(X - \alpha_r)$$

Observe that $g(X)$ divides a polynomial $a(X)$ iff $\alpha_1, \ldots, \alpha_r$ are among the roots of $a(X)$. Therefore $a(X) + I$ is in $\mathscr{C}$ iff $\alpha_1, \ldots, \alpha_r$ are among the roots of $a(X)$.

Given a positive integer m, we defined a binary $(2^m - m - 1, 2^m - 1)$ Hamming code by taking $\mathbf{H}^t$ as the parity check matrix where $\mathbf{H}$ is the $m \times (2^m - 1)$ matrix the ith row of which, $1 \le i \le 2^m - 1$, is the binary representation of the number i and by insisting that in a code word $b_1 b_2 \cdots b_n$ $(n = 2^m - 1), b_1, b_2, b_{2^2}, \ldots, b_{2^{m-1}}$ are the check symbols. If we do not insist on the condition about the position of the check symbols, we may define the **binary Hamming code** as the code with parity check matrix $\mathbf{H}^t$.

With the parity check matrix given, it is not easy to find the code words. For finding these, we observe that $\mathbf{H}^t$ contains m columns, a suitable permutation of which forms the identity matrix $\mathbf{I}_m$ of order m. Let σ be a permutation of the columns of $\mathbf{H}^t$ which, when applied, transforms $\mathbf{H}^t$ into $(\mathbf{A} \quad \mathbf{I}_m) = \mathbf{H}_1$. The corresponding generating matrix then becomes

$$\mathbf{G}_1 = (\mathbf{I}_{n-m} \quad \mathbf{A}^t)$$

An application of the permutation σ^{-1} to the columns of $\mathbf{G}_1$ gives a generator matrix $\mathbf{G}$ of the Hamming code. The code words of the Hamming code are then given by $a\mathbf{G}$, $a \in V(n-m, 2)$.

We illustrate this procedure for the case $m = 3$. Here,

$$\mathbf{H}^t = \begin{pmatrix} 0 & 0 & 0 & 1 & 1 & 1 & 1 \\ 0 & 1 & 1 & 0 & 0 & 1 & 1 \\ 1 & 0 & 1 & 0 & 1 & 0 & 1 \end{pmatrix}$$

Applying the permutation

$$\sigma = \begin{pmatrix} 1 & 2 & 4 & 3 & 5 & 6 & 7 \\ 7 & 6 & 5 & 1 & 2 & 3 & 4 \end{pmatrix}$$

to the columns of $\mathbf{H}^t$ gives

$$\mathbf{H}_1 = \begin{pmatrix} 0 & 1 & 1 & 1 & 1 & 0 & 0 \\ 1 & 0 & 1 & 1 & 0 & 1 & 0 \\ 1 & 1 & 0 & 1 & 0 & 0 & 1 \end{pmatrix}$$

The corresponding generator matrix is

$$\mathbf{G}_1 = \begin{pmatrix} 1 & 0 & 0 & 0 & 0 & 1 & 1 \\ 0 & 1 & 0 & 0 & 1 & 0 & 1 \\ 0 & 0 & 1 & 0 & 1 & 1 & 0 \\ 0 & 0 & 0 & 1 & 1 & 1 & 1 \end{pmatrix}$$

Applying σ^{-1} to the columns of $\mathbf{G}_1$ gives the generator matrix of the code corresponding to $\mathbf{H}^t$ as

$$\mathbf{G} = \begin{pmatrix} 1 & 1 & 1 & 0 & 0 & 0 & 0 \\ 1 & 0 & 0 & 1 & 1 & 0 & 0 \\ 0 & 1 & 0 & 1 & 0 & 1 & 0 \\ 1 & 1 & 0 & 1 & 0 & 0 & 1 \end{pmatrix}$$

For a message word $a = a_1a_2a_3a_4$, we have

$$\mathbf{aG} = (a_1 + a_2 + a_4 \quad a_1 + a_3 + a_4 \quad a_1 \quad a_2 + a_3 + a_4 \quad a_2 \quad a_3 \quad a_4)$$

Observe that the message symbols occupy the 3rd, 5th, 6th and 7th positions while the 1st, 2nd and 4th positions are occupied by check symbols as they do in the case of Hamming codes defined originally.

Going back to the case of arbitrary m, we may define

$$\boldsymbol{\sigma} = \begin{pmatrix} 1 & 2 & 2^2 & \cdots & 2^{m-1} & 3 & 5 & 6 & 7 & 9\cdots \\ n & n-1 & n-2 & \cdots & n-m+1 & 1 & 2 & 3 & 4 & 5\cdots \end{pmatrix}$$

Applying $\boldsymbol{\sigma}$ to the columns of $\mathbf{H}^{\mathrm{t}}$ gives

$$\mathbf{H}_1 = (\mathbf{A} \quad \mathbf{I}_m)$$

so that the corresponding generating matrix is

$$\mathbf{G}_1 = (\mathbf{I}_{n-m} \quad \mathbf{A}^{\mathrm{t}})$$

Now applying $\boldsymbol{\sigma}^{-1}$ to the columns of $\mathbf{G}_1$ gives the generator matrix $\mathbf{G}$ of the code corresponding to the parity check matrix $\mathbf{H}^{\mathrm{t}}$. With this generating matrix $\mathbf{G}$, we find that the code word in the Hamming code corresponding to the message word a is the same as the code word corresponding to a for the Hamming code originally defined. Thus, the two definitions of the Hamming code give the *same* code.

Let $\mathscr{C}$ be a Hamming code of length $n = 2^r - 1$ so that $\mathscr{C}$ is a code with a parity check matrix $\mathbf{H}$ of order $r \times n$ in which the columns are the binary representations of the numbers $1, 2, \ldots, n$. Then no two columns of $\mathbf{H}$ are identical and so the code is single error correcting (Theorem 1.5). But then the minimum distance of the code is at least 3 (Theorem 1.2). Thus we have an alternative proof of Theorem 3.1.

Examples 6.1

Case (i)

As a first illustration of the use of the above discussion we obtain all the code words of binary Hamming code of length $7 = 2^3 - 1$. A parity check matrix of this code is

$$\mathbf{H} = \begin{pmatrix} 0 & 0 & 0 & 1 & 1 & 1 & 1 \\ 0 & 1 & 1 & 0 & 0 & 1 & 1 \\ 1 & 0 & 1 & 0 & 1 & 0 & 1 \end{pmatrix}$$

Applying the permutation

$$\boldsymbol{\sigma} = \begin{pmatrix} 1 & 2 & 3 & 4 & 5 & 6 & 7 \\ 7 & 6 & 1 & 5 & 2 & 3 & 4 \end{pmatrix}$$

to the columns of **H** gives a parity check matrix

$$\mathbf{H}_1 = \begin{pmatrix} 0 & 1 & 1 & 1 & 1 & 0 & 0 \\ 1 & 0 & 1 & 1 & 0 & 1 & 0 \\ 1 & 1 & 0 & 1 & 0 & 0 & 1 \end{pmatrix}$$

The corresponding generator matrix is

$$\mathbf{G}_1 = \begin{pmatrix} 1 & 0 & 0 & 0 & 0 & 1 & 1 \\ 0 & 1 & 0 & 0 & 1 & 0 & 1 \\ 0 & 0 & 1 & 0 & 1 & 1 & 0 \\ 0 & 0 & 0 & 1 & 1 & 1 & 1 \end{pmatrix}$$

Now apply the permutation $\boldsymbol{\sigma}^{-1}$ to the columns of $\mathbf{G}_1$ to obtain the generator **G** matrix corresponding to the parity check matrix **H**:

$$\mathbf{G} = \begin{pmatrix} 1 & 1 & 1 & 0 & 0 & 0 & 0 \\ 1 & 0 & 0 & 1 & 1 & 0 & 0 \\ 0 & 1 & 0 & 1 & 0 & 1 & 0 \\ 1 & 1 & 0 & 1 & 0 & 0 & 1 \end{pmatrix}$$

The code words of this Hamming code are:

```
0 0 0 0 0 0 0   1 0 0 0 0 1 1   0 1 0 1 1 1 1
1 1 0 1 0 0 1   0 1 0 0 1 0 1   0 1 1 0 0 1 1
0 1 0 1 0 1 0   0 0 1 1 0 0 1   1 0 1 0 1 0 1
1 0 0 1 1 0 0   1 1 0 0 1 1 0   0 0 1 0 1 1 0
1 1 1 0 0 0 0   1 0 1 1 0 1 0   1 1 1 1 1 1 1
                0 1 1 1 1 0 0
```

Case (ii)

Our second illustration results in a generator matrix of binary Hamming code of length $15 = 2^4 - 1$. A parity check matrix of this code is

$$\mathbf{H} = \begin{pmatrix} 0 & 0 & 0 & 0 & 0 & 0 & 0 & 1 & 1 & 1 & 1 & 1 & 1 & 1 & 1 \\ 0 & 0 & 0 & 1 & 1 & 1 & 1 & 0 & 0 & 0 & 0 & 1 & 1 & 1 & 1 \\ 0 & 1 & 1 & 0 & 0 & 1 & 1 & 0 & 0 & 1 & 1 & 0 & 0 & 1 & 1 \\ 1 & 0 & 1 & 0 & 1 & 0 & 1 & 0 & 1 & 0 & 1 & 0 & 1 & 0 & 1 \end{pmatrix}$$

To obtain a canonical form of parity check matrix, we consider a permutation

$$\boldsymbol{\sigma} = \begin{pmatrix} 1 & 2 & 3 & 4 & 5 & 6 & 7 & 8 & 9 & 10 & 11 & 12 & 13 & 14 & 15 \\ 15 & 14 & 1 & 13 & 2 & 3 & 4 & 12 & 5 & 6 & 7 & 8 & 9 & 10 & 11 \end{pmatrix}$$

Then

$$\mathbf{H}_1 = \boldsymbol{\sigma}(\mathbf{H}) = \begin{pmatrix} 0 & 0 & 0 & 0 & 1 & 1 & 1 & 1 & 1 & 1 & 1 & 1 & 0 & 0 & 0 \\ 0 & 1 & 1 & 1 & 0 & 0 & 0 & 1 & 1 & 1 & 1 & 0 & 1 & 0 & 0 \\ 1 & 0 & 1 & 1 & 0 & 1 & 1 & 0 & 0 & 1 & 1 & 0 & 0 & 1 & 0 \\ 1 & 1 & 0 & 1 & 1 & 0 & 1 & 0 & 1 & 0 & 1 & 0 & 0 & 0 & 1 \end{pmatrix}$$

Corresponding to $\mathbf{H}_1$ the generator matrix is $\mathbf{G}_1 = (\mathbf{I}_{11} \quad \mathbf{A}^t)$, where

$$\mathbf{A} = \begin{pmatrix} 0 & 0 & 0 & 0 & 1 & 1 & 1 & 1 & 1 & 1 & 1 \\ 0 & 1 & 1 & 1 & 0 & 0 & 0 & 1 & 1 & 1 & 1 \\ 1 & 0 & 1 & 1 & 0 & 1 & 1 & 0 & 0 & 1 & 1 \\ 1 & 1 & 0 & 1 & 1 & 0 & 1 & 0 & 1 & 0 & 1 \end{pmatrix}$$

Applying the permutation $\boldsymbol{\sigma}^{-1}$ to the columns of $\mathbf{G}_1$, we obtain the generator matrix of the Hamming code as

$$\mathbf{G} = \begin{pmatrix} 1 & 1 & 1 & 0 & 0 & 0 & 0 & 0 & 0 & 0 & 0 & 0 & 0 & 0 & 0 \\ 1 & 0 & 0 & 1 & 1 & 0 & 0 & 0 & 0 & 0 & 0 & 0 & 0 & 0 & 0 \\ 0 & 1 & 0 & 1 & 0 & 1 & 0 & 0 & 0 & 0 & 0 & 0 & 0 & 0 & 0 \\ 0 & 1 & 0 & 1 & 0 & 0 & 1 & 0 & 0 & 0 & 0 & 0 & 0 & 0 & 0 \\ 1 & 0 & 0 & 0 & 0 & 0 & 0 & 1 & 1 & 0 & 0 & 0 & 0 & 0 & 0 \\ 0 & 1 & 0 & 0 & 0 & 0 & 0 & 1 & 0 & 1 & 0 & 0 & 0 & 0 & 0 \\ 1 & 1 & 0 & 0 & 0 & 0 & 0 & 1 & 0 & 0 & 1 & 0 & 0 & 0 & 0 \\ 0 & 0 & 0 & 1 & 0 & 0 & 0 & 1 & 0 & 0 & 0 & 1 & 0 & 0 & 0 \\ 1 & 0 & 0 & 1 & 0 & 0 & 0 & 1 & 0 & 0 & 0 & 0 & 1 & 0 & 0 \\ 0 & 1 & 0 & 1 & 0 & 0 & 0 & 1 & 0 & 0 & 0 & 0 & 0 & 1 & 0 \\ 1 & 1 & 0 & 1 & 0 & 0 & 0 & 1 & 0 & 0 & 0 & 0 & 0 & 0 & 1 \end{pmatrix}$$

Theorem 6.3

The binary cyclic code of length $n = 2^m - 1$ for which the generator is the minimal polynomial of a primitive element of $\mathrm{GF}(2^m)$ is equivalent to the $(n - m, n)$ Hamming code.

Proof

Let α be a primitive element of $\mathrm{GF}(2^m)$. Let $m(X)$ be the minimal polynomial of α over $\mathbb{B}$. Since for $\beta \in \mathrm{GF}(2^m)$, β and β^2 have the same minimal polynomial, $\alpha, \alpha^2, \ldots, \alpha^{2^{m-1}}$ are distinct roots of $m(X)$. Since the degree $[\mathrm{GF}(2^m){:}\mathbb{B}] = m$, the degree of the minimal polynomial of any element of $\mathrm{GF}(2^m)$ over $\mathbb{B}$ is at most m. Hence

$$m(X) = (X - \alpha)(X - \alpha^2)\cdots(X - \alpha^{2^{m-1}})$$

The elements $1, \alpha, \alpha^2, \ldots, \alpha^{m-1}$ form a basis of $\mathrm{GF}(2^m)$ over $\mathbb{B}$ and, therefore, every element of $GF(2^m)$ can be uniquely written as

$$\sum_{i=0}^{m-1} e_i\alpha^i \quad e_i \in \mathbb{B}$$

For $0 \leq j \leq 2^m - 2$, let

$$\alpha^j = \sum_{i=0}^{m-1} e_{ij}\alpha^i$$

and let $\mathbf{H}$ be the $m \times n$ matrix, the $(j+1)$th column of which is

$$(e_{0j} \quad e_{1j} \quad \cdots \quad e_{m-1,j})^{\mathrm{t}}$$

Every row vector

$$(e_{0j} \quad e_{1j} \quad \cdots \quad e_{m-1,j})$$

gives the binary representation of one and only one positive integer at most n.

Now $a(X) + \langle X^n - 1 \rangle$ belongs to the cyclic code generated by $m(X)$ iff $a(\alpha) = 0$. But this is so iff $\mathbf{Ha}^{\mathrm{t}} = 0$, where $a = (a_0 a_1 \text{---} a_{n-1})$ with

$$a(X) = a_0 + a_1 X + \cdots + a_{n-1} X^{n-1}$$

Therefore, the cyclic code generated by $m(X)$ is the same as the code given by the parity check matrix $\mathbf{H}$. But $\mathbf{H}$ is obtained by permutating the binary representations of the numbers $1, 2, \ldots, n$. This completes the proof.

Remark 6.1

Observe that, in the above proof, we have also shown that a binary Hamming code is a BCH code (up to equivalence).

We have seen earlier that every non-zero element of $GF(2^m)$ is a root of the polynomial $X^n - 1$ with $n = 2^m - 1$. Therefore, the minimal polynomial of every element β of $GF(2^m)$ divides $X^n - 1$. Also the minimal polynomials of two elements are either identical or relatively coprime. Hence, if α is a primitive element of $GF(2^m)$ and $d \geq 2$ is a positive integer then

$$g(X) = \mathrm{LCM}\,\{m_1(X), \ldots, m_{d-1}(X)\}$$

where $m_i(X)$ denotes the minimal polynomial of α^i, divides $X^n - 1$. It then follows that the polynomial code of length n generated by $g(X)$ is the same as the cyclic code with generator $g(X)$. Thus, every binary BCH code is a cyclic code.

6.4 NON-BINARY HAMMING CODES

We have so far restricted ourselves only to binary Hamming codes. However, Hamming codes may be defined over any finite field $GF(q)$.

Definition 6.2

A Hamming code of length $n = (q^m - 1)/(q - 1)$ over GF(q) is defined to be the code given by an $m \times n$ parity check matrix **H**, the columns of which are all non-zero m-tuples over GF(q) with the first non-zero entry in each column equal to 1.

There are m columns in the parity check matrix **H** a suitable permutation of which form identity matrix of order m and it follows that the Hamming code given by **H** is a vector space of dimension $n - m$ over GF(q).

As examples we construct two Hamming codes over GF(3).

Examples 6.2

Case (i) – Hamming code of length 4 over GF(3)
As $4 = (3^2 - 1)/(3 - 1)$, the parity check matrix is a 2×4 matrix given by

$$\mathbf{H} = \begin{pmatrix} 0 & 1 & 1 & 1 \\ 1 & 0 & 1 & 2 \end{pmatrix}$$

Applying the permutation

$$\boldsymbol{\sigma} = \begin{pmatrix} 1 & 2 & 3 & 4 \\ 4 & 3 & 1 & 2 \end{pmatrix}$$

to the columns of **H**, gives

$$\mathbf{H}_1 = \begin{pmatrix} 1 & 1 & 1 & 0 \\ 1 & 2 & 0 & 1 \end{pmatrix}$$

The corresponding generator matrix is then given by

$$\mathbf{G}_1 = \begin{pmatrix} 1 & 0 & 1 & 1 \\ 0 & 1 & 1 & 2 \end{pmatrix}$$

Applying the permutation $\boldsymbol{\sigma}^{-1}$ to the columns of $\mathbf{G}_1$ gives the generator matrix corresponding to **H** as

$$\mathbf{G} = \begin{pmatrix} 1 & 1 & 1 & 0 \\ 2 & 1 & 0 & 1 \end{pmatrix}$$

All the code words of the (2, 4) ternary Hamming code are then given by:

Message word	Code word
0 0	0 0 0 0
0 1	2 1 0 1
0 2	1 2 0 2
1 0	1 1 1 0
1 1	0 2 1 1
1 2	2 0 1 2
2 0	2 2 2 0
2 1	1 0 2 1
2 2	0 1 2 2

The minimum distance of the code is 3.

Case (ii) – Hamming code of length 13 over GF(3)
As $13 = (3^3 - 1)/(3 - 1)$, the parity check matrix is a 3×13 matrix and is given by

$$\mathbf{H} = \begin{pmatrix} 0 & 0 & 0 & 0 & 1 & 1 & 1 & 1 & 1 & 1 & 1 & 1 & 1 \\ 0 & 1 & 1 & 1 & 0 & 0 & 0 & 1 & 1 & 1 & 2 & 2 & 2 \\ 1 & 0 & 1 & 2 & 0 & 1 & 2 & 0 & 1 & 2 & 0 & 1 & 2 \end{pmatrix}$$

Applying the permutation

$$\boldsymbol{\sigma} = \begin{pmatrix} 1 & 2 & 3 & 4 & 5 & 6 & 7 & 8 & 9 & 10 & 11 & 12 & 13 \\ 13 & 12 & 1 & 2 & 11 & 3 & 4 & 5 & 6 & 7 & 8 & 9 & 10 \end{pmatrix}$$

to the columns of $\mathbf{H}$, gives

$$\mathbf{H}_1 = \begin{pmatrix} 0 & 0 & 1 & 1 & 1 & 1 & 1 & 1 & 1 & 1 & 1 & 0 & 0 \\ 1 & 1 & 0 & 0 & 1 & 1 & 1 & 2 & 2 & 2 & 0 & 1 & 0 \\ 1 & 2 & 1 & 2 & 0 & 1 & 2 & 0 & 1 & 2 & 0 & 0 & 1 \end{pmatrix}$$

The corresponding generator matrix is

$$\mathbf{G}_1 = \begin{pmatrix} 1 & 0 & 0 & 0 & 0 & 0 & 0 & 0 & 0 & 0 & 0 & 1 & 1 \\ 0 & 1 & 0 & 0 & 0 & 0 & 0 & 0 & 0 & 0 & 0 & 1 & 2 \\ 0 & 0 & 1 & 0 & 0 & 0 & 0 & 0 & 0 & 0 & 1 & 0 & 1 \\ 0 & 0 & 0 & 1 & 0 & 0 & 0 & 0 & 0 & 0 & 1 & 0 & 2 \\ 0 & 0 & 0 & 0 & 1 & 0 & 0 & 0 & 0 & 0 & 1 & 1 & 0 \\ 0 & 0 & 0 & 0 & 0 & 1 & 0 & 0 & 0 & 0 & 1 & 1 & 1 \\ 0 & 0 & 0 & 0 & 0 & 0 & 1 & 0 & 0 & 0 & 1 & 1 & 2 \\ 0 & 0 & 0 & 0 & 0 & 0 & 0 & 1 & 0 & 0 & 1 & 2 & 0 \\ 0 & 0 & 0 & 0 & 0 & 0 & 0 & 0 & 1 & 0 & 1 & 2 & 1 \\ 0 & 0 & 0 & 0 & 0 & 0 & 0 & 0 & 0 & 1 & 1 & 2 & 2 \end{pmatrix}$$

Applying the permutation $\boldsymbol{\sigma}^{-1}$ to the columns of $\mathbf{G}_1$ gives

$$\mathbf{G} = \begin{pmatrix} 1 & 1 & 1 & 0 & 0 & 0 & 0 & 0 & 0 & 0 & 0 & 0 & 0 \\ 2 & 1 & 0 & 1 & 0 & 0 & 0 & 0 & 0 & 0 & 0 & 0 & 0 \\ 1 & 0 & 0 & 0 & 1 & 1 & 0 & 0 & 0 & 0 & 0 & 0 & 0 \\ 2 & 0 & 0 & 0 & 1 & 0 & 1 & 0 & 0 & 0 & 0 & 0 & 0 \\ 0 & 1 & 0 & 0 & 1 & 0 & 0 & 1 & 0 & 0 & 0 & 0 & 0 \\ 1 & 1 & 0 & 0 & 1 & 0 & 0 & 0 & 1 & 0 & 0 & 0 & 0 \\ 2 & 1 & 0 & 0 & 1 & 0 & 0 & 0 & 0 & 1 & 0 & 0 & 0 \\ 0 & 2 & 0 & 0 & 1 & 0 & 0 & 0 & 0 & 0 & 1 & 0 & 0 \\ 1 & 2 & 0 & 0 & 1 & 0 & 0 & 0 & 0 & 0 & 0 & 1 & 0 \\ 2 & 2 & 0 & 0 & 1 & 0 & 0 & 0 & 0 & 0 & 0 & 0 & 1 \end{pmatrix}$$

which is a generator matrix of the (3, 13) ternary Hamming code. Corresponding to message word $a_1a_2\text{---}a_{10} \in V(10, q)$ is code word

$$a_1 + 2a_2 + a_3 + 2a_4 + a_6 + 2a_7 + a_9 + 2a_{10}, a_1 + a_2 + a_5 + a_6 + a_7 + 2a_8 + 2a_9 + 2a_{10}, a_1, a_2, a_3 + a_4 + a_5 + a_6 + a_7 + a_8 + a_9 + a_{10}, a_3, a_4, a_5, a_6, a_7, a_8, a_9, a_{10}$$

Exercise 6.3

1. Find a parity check matrix of a Hamming code of length 6 over GF(5).
2. Find a parity check matrix and the corresponding generator matrix of a Hamming code of length 5 over GF(4).
3. Find a parity check matrix of a Hamming code of length 21 over GF(4).
4. Find a parity check matrix and the corresponding generator matrix of a Hamming code of length
   (i) 8 over GF(7);
   (ii) 12 over GF(11);
   (iii) 14 over GF(13).
5. Find a parity check matrix of a ternary Hamming code of length 4.

While working with **non-binary codes**, the syndrome decoding procedure with a parity check matrix **H** needs to be modified as follows. Let $r = r_1\text{---}r_n$ be the word received and $\mathbf{s} = \mathbf{Hr}^t$ be the vector associated with its syndrome.

(i) If **s** equals a constant multiple of a unique column of **H**, say the ith, i.e.

$$\mathbf{s} = \lambda \mathbf{H}_i \quad 0 \neq \lambda \in \mathrm{GF}(q)$$

where $\mathbf{H}_1, \mathbf{H}_2, \ldots, \mathbf{H}_n$ are the columns of **H**, we assume that an error in transmission occurred in the ith position and take

$$c = r_i\text{---}r_{i-1}(r_i - \lambda)r_{i+1}\text{---}r_n$$

as the code word transmitted.

(ii) If **s** is not a multiple of any column of **H** then at least two errors occurred in transmission.

(iii) If **s** equals a multiple of $\mathbf{H}_i$ and also of $\mathbf{H}_j$ with $i \neq j$, there is the case of decoding failure.

Proceeding as in the binary case, we can prove the following proposition.

Proposition 6.1
Let $\mathscr{C}$ be a linear code over GF(q) with an $(n - m) \times n$ parity check matrix **H**. The code is capable of correcting all single errors iff every two columns of **H** are linearly independent.

As the first non-zero entry in every column of the parity check matrix **H** of the Hamming code over GF(q), $q \neq 2$, is 1, it follows that no column of **H** is a scalar multiple of any other column. As such we have the following corollary.

Corollary
Hamming codes over GF(q) are single error correcting.

As a more general result than Proposition 6.1 we have the following proposition.

Proposition 6.2
The minimum distance of a code over GF(q) with parity check matrix $\mathbf{H}$ is at least $k+1$ iff every set of k columns of $\mathbf{H}$ is linearly independent.

Definition 6.3
Let F be a field of order q and for a positive integer n, let $F^{(n)} = V(n, q)$ denote, as before, the space of all n-tuples of length n over F. Let $\rho > 0$ and $\mathbf{x} \in V(n, q)$. Then the sphere in $V(n, q)$ of radius ρ with centre at the point $\mathbf{x}$ is defined by

$$S_\rho(\mathbf{x}) = \{\mathbf{y} \in V(n, q)/d(\mathbf{x}, \mathbf{y}) \leq \rho\}$$

Observe that the sphere $S_1(\mathbf{x})$:

(i) in $\mathbb{B}^n$ contains exactly $n+1$ elements;
(ii) in $V(n, 3)$ contains exactly $2n+1$ elements; and
(iii) in $V(n, q)$ contains exactly $n(q-1)+1$ elements.

Definition 6.4
An e-error-correcting code $\mathscr{C}$ of length n over GF(q) is called **perfect** if

$$\bigcup_{\mathbf{x} \in \mathscr{C}} S_e(X) = V(n, q)$$

Proposition 6.3
Hamming codes are single error correcting perfect codes.

Proof
We know that Hamming codes are single error correcting. So, we only need to prove that these are perfect. Let $\mathscr{C}$ be a Hamming code of length $n = (q^r - 1)/(q-1)$ over GF(q) so that a parity check matrix of $\mathscr{C}$ is an $r \times n$ matrix over GF(q). Therefore, the dimension of $\mathscr{C}$ over GF(q) is $n - r$ and order of $\mathscr{C}$ is q^{n-r}. The minimum distance of the code being at least 3, every sphere $S_1(\mathbf{x})$, $\mathbf{x} \in \mathscr{C}$, contains exactly one code word inside it, namely, the vector associated with the word x itself. Also, for $\mathbf{x}, \mathbf{y} \in \mathscr{C}, \mathbf{x} \neq \mathbf{y}$, $S_1(\mathbf{x})$ and $S_1(\mathbf{y})$ are disjoint. Therefore,

$$\mathrm{O}\left(\bigcup_{\mathbf{x} \in \mathscr{C}} S_1(\mathbf{x})\right) = (n(q-1)+1)\mathrm{O}(\mathscr{C}) = (n(q-1)+1)q^{n-r} = q^n$$
$$= \mathrm{O}(V(n, q))$$

Hence

$$\bigcup_{\mathbf{x}\in\mathscr{C}} S_1(\mathbf{x}) = V(n,q)$$

Having defined spheres, we now obtain the **sphere-packing** or **Hamming bound**.

Theorem 6.4
A k-error-correcting binary code of length n containing M code words must satisfy:

$$M\left\{1+\binom{n}{1}+\binom{n}{2}+\cdots+\binom{n}{k}\right\}\leq 2^n$$

Proof
Since the code is k error correcting, distance between any two code words is at least $2k+1$. Therefore, the spheres of radius k around the code words are disjoint. But every sphere around a code word contains the code word and vectors which are at a distance $1 \leq d \leq k$ from this code word. The number of vectors which are at distance d from the code word is $\binom{n}{d}$. Therefore, the number of all vectors of length n contained within the above spheres is

$$M\left\{1+\binom{n}{1}+\binom{n}{2}+\cdots+\binom{n}{k}\right\}$$

Since the total number of vectors of length n is 2^n, the result follows. ∎

As an application of Proposition 6.2 we obtain the following theorem.

Theorem 6.5 (The Gilbert–Varshamov bound)
There exists a binary linear code of length n, with at most r parity checks and minimum distance at least d, provided

$$1+\binom{n-1}{1}+\cdots+\binom{n-1}{d-2}<2^r$$

Proof
We know that if $\mathscr{C}$ is a code with parity check matrix $\mathbf{H}$ such that all $d-1$ columns of $\mathbf{H}$ are linearly independent, then the minimum distance of the code is at least d. Furthermore, if $\mathbf{H}$ is an $r \times n$ matrix, then the number of parity checks is r. Therefore, we need to construct an $r \times n$ matrix $\mathbf{H}$ in which all $d-1$ columns are linearly independent.

The first column can be chosen to be any non-zero r-tuple. Suppose that we have chosen i columns so that no $d-1$ columns out of these are linearly dependent. There are at most

$$\binom{i}{1}+\binom{i}{2}+\cdots+\binom{i}{d-2}$$

distinct linear combinations of these i columns taken at most $d-2$ at a time. If this number is less than 2^r-1, we can certainly find an r-tuple which does not equal any of these linear combinations and, thus, can add a column such that any $d-1$ columns of the new $r \times (i+1)$ array are linearly independent. We can go on doing this as long as

$$1+\binom{i}{1}+\cdots+\binom{i}{d-2}<2^r$$

Therefore, if

$$1+\binom{n}{1}+\cdots+\binom{n}{d-2}<2^r$$

we can certainly find an $r \times n$ matrix in which all $d-1$ columns are linearly independent.

Theorem 6.6 (The Gilbert–Varshamov bound – non-binary case)
There exists a linear code over a field of q elements, having length n, at most r parity checks, and minimum distance at least d, provided

$$\sum_{i=0}^{d-2}(q-1)^i\binom{n-1}{i}<q^r$$

Proof
As in the binary case, we have to construct an $r \times n$ matrix $\mathbf{H}$ in which all $d-1$ columns are linearly independent.

The first column may be chosen to be any non-zero r-tuple. Suppose that we have chosen i columns so that no $d-1$ columns out of these are linearly dependent. Out of the i columns, k columns can be chosen in $\binom{i}{k}$ ways. Also there are $(q-1)^k$ linear combinations of any k chosen columns. Therefore, there are at most

$$(q-1)\binom{i}{1}+(q-1)^2\binom{i}{2}+\cdots+(q-1)^{d-2}\binom{i}{d-2}$$

distinct linear combinations of these i columns taken at most $d-2$ at a time. The rest of the argument is the same as in the binary case.

Next, we give a bound on the minimum distance of any code whether linear or not. A code of length n over a set (or field) of q elements is called non-linear if the set of all code words is not a vector space. First, we have a simple number theoretic lemma.

Lemma 6.1
If $a_1, a_2, \ldots, a_k$ are non-negative numbers with

$$a_1 + a_2 + \cdots + a_k = N$$

then

$$\sum_{i=1}^{k} a_i^2$$

is the least where

$$a_1 = a_2 = \cdots = a_k = N/k$$

Proof
We prove the lemma by induction on k. For $k = 2$,

$$\begin{aligned} a_1^2 + a_2^2 &= a_1^2 + (N - a_1)^2 \\ &= 2\left\{\frac{N^2}{4} + \left(a_1 - \frac{N}{2}\right)^2\right\} \end{aligned}$$

which is the least when $a_1 - N/2 = 0$, i.e. $a_1 = N/2$. But then $a_2 = N/2$ also. Suppose that the result holds for k numbers. Consider non-negative numbers $a_1, a_2, \ldots, a_{k+1}$ with

$$a_1 + a_2 + \cdots + a_{k+1} = N$$

Then

$$\sum_{i=1}^{k} a_i = N - a_{k+1}$$

Now

$$\begin{aligned} \sum_{i=1}^{k+1} a_i^2 &= \sum_{i=1}^{k} a_i^2 + a_{k+1}^2 \\ &\geq k\left(\frac{N - a_{k+1}}{k}\right)^2 + a_{k+1}^2 \qquad \text{(by induction hypothesis)} \\ &= \frac{1}{k}\{N^2 - 2Na_{k+1} + (k+1)a_{k+1}^2\} \\ &= \frac{k+1}{k}\left\{\frac{N^2}{k+1} + \left(a_{k+1} - \frac{N}{k+1}\right)^2 - \left(\frac{N}{k+1}\right)^2\right\} \end{aligned}$$

and this is the least when

$$a_{k+1} = \frac{N}{k+1}$$

Already for

$$\sum_{i=1}^{k} a_i^2$$

minimum, we have by the induction hypothesis that

$$a_1 = a_2 = \cdots = a_k = \frac{N - a_{k+1}}{k}$$

But

$$\frac{N - a_{k+1}}{k} = \frac{N - N/(k+1)}{k} = \frac{N}{k+1}$$

Therefore,

$$\sum_{i=1}^{k+1} a_i^2$$

is minimum only when

$$a_1 = a_2 = \cdots = a_{k+1} = N/(k+1)$$

Theorem 6.7 (Plotkin bound)
If $\mathscr{C}$ is a block code of length n, order N and minimum distance d over an alphabet set of order q, then

$$d \le \frac{nN(q-1)}{(N-1)q}$$

Proof
Consider the number

$$A = \sum_{u,v \in \mathscr{C}} d(u,v)$$

where as usual $d(u,v)$ denotes the distance between u and v. If $u \neq v$, then $d(u,v) \geq d$ and is 0 otherwise. We can choose $u \neq v$ in $N(N-1)$ ways. Therefore $A \geq N(N-1)d$. We next obtain an upper bound for the number A. We consider the first entry of all the code words in $\mathscr{C}$. Write $0, 1, 2, \ldots, q-1$ as the elements of the set over which the code $\mathscr{C}$ is given. Among the first entries of all the code words of $\mathscr{C}$, let b_i be each equal to i, $0 \leq i \leq q-1$. Then

$$\sum_{i=0}^{q-1} b_i = N$$

If u, v are two words in $\mathscr{C}$ with the first entry i, then the first entries of u, v contribute 0 to the sum A. Consider now u with the first entry i. There are $N - b_i$ words in $\mathscr{C}$ with first entry $\neq i$. Therefore u with each of these $N - b_i$

words contributes $N - b_i$ to the sum A. There being b_i such code words the total contribution of the first entries of these words taken with the rest is $b_i(N - b_i)$. Considering all other j's, we find that the contribution to A because of the first entries of all words of $\mathscr{C}$ is B (say) where

$$B = \sum_{i=0}^{q-1} b_i(N - b_i)$$

Now

$$\begin{aligned} B &= \sum_{i=0}^{q-1} b_i(N - b_i) \\ &= N \sum_{i=0}^{q-1} b_i - \sum_{i=0}^{q-1} b_i^2 \\ &= N^2 - \sum_{i=0}^{q-1} b_i^2 \end{aligned}$$

B takes the largest value when

$$\sum_{i=0}^{q-1} b_i^2$$

is the least which in turn happens when

$$b_1 = b_2 = \cdots = b_{q-1} = \frac{N}{q}$$

Hence

$$B \leq N^2 - q\left(\frac{N}{q}\right)^2 = \frac{N^2(q-1)}{q}$$

Since ith entries of all the words of $\mathscr{C}$ contribute the same number B to the sum A for all i, $1 \leq i \leq n$,

$$A = nB \leq \frac{nN^2(q-1)}{q}$$

Using the lower bound obtained earlier for A, gives

$$N(N-1)d \leq \frac{nN^2(q-1)}{q}$$

or

$$d \leq \frac{nN(q-1)}{(N-1)q}$$

Remark 6.2

The bound for d is attained iff each symbol i occurs exactly N/q times in jth entries of all the code words of $\mathscr{C}$ for $1 \leq j \leq n$. This means that N/q must be an integer, i.e. q must divide N. Observe that for linear codes, this is no restriction for if $\mathscr{C}$ is of dimension k over GF(q), then $N = q^k$.

6.5 IDEMPOTENTS

Theorem 6.8

Let $\mathscr{C}$ be a cyclic code of length n over F and I be the ideal of $F[X]$ generated by $X^n - 1$. Then there exists a unique element $c(X) + I \in \mathscr{C}$ such that:

(i) $c(X) + I = c^2(X) + I$
(ii) $c(X) + I$ generates $\mathscr{C}$
(iii) $\forall f(X) + I$ in $\mathscr{C}$

$$c(X)f(X) + I = f(X) + I$$

i.e. $c(X) + I$ is an identity for $\mathscr{C}$.

Proof

Let $g(X) \in F[X]$ be a generator of $\mathscr{C}$ and $h(X)$ be its check polynomial. Then

$$g(X)h(X) = X^n - 1$$

Since $(n, q) = 1$ where $q = \mathrm{O}(F)$, $X^n - 1$ does not have multiple zeros. Therefore $(g(X), h(X)) = 1$ and there exist elements $a(X), b(X)$ in the Euclidean ring $F[X]$ such that

$$g(X)a(X) + h(X)b(X) = 1 \tag{6.2}$$

Take $c(X) = g(X)a(X)$. Multiplying both sides of the relation (6.2) by $c(X)$ and working in the quotient ring $F[X]/I$, gives

$$c^2(X) + g(X)h(X)a(X)b(X) + I = c(X) + I$$

or

$$c^2(X) + I = c(X) + I$$

This proves part (i).

Clearly

$$\langle c(X) + I \rangle \subseteq \langle g(X) + I \rangle$$

Again, multiplying both sides of the relation (6.2) by $g(X)$ and going to the quotient ring $F[X]/I$, gives

$$g(X)c(X) + g(X)h(X)b(X) + I = g(X) + I$$

or

$$g(X)c(X) + I = g(X) + I \tag{6.3}$$

This shows that

$$\langle g(X)+I\rangle \subseteq \langle c(X)+I\rangle$$

and hence

$$\langle g(X)+I\rangle = \langle c(X)+I\rangle$$

which proves part (ii). The relation (iii) also follows from (6.3).

Let $d(X)\in F[X]$ be another polynomial with the properties (i), (ii) and (iii). Since $c(X)$ and $d(X)$ both satisfy (iii), we have

$$c(X)d(X)+I=c(X)+I=d(X)+I$$

Definition 6.5

The unique element $c(X)+I\in\mathscr{C}$ with the properties (i), (ii) and (iii) of the theorem is called the **idempotent of** $\mathscr{C}$.

Remarks 6.3

(i) In a ring R, an element e is called an idempotent if $e^2=e$. In general, a ring R may have many idempotents. For example, if R is a ring with identity and e is an idempotent, then $1-e$ is another idempotent in R. Thus, we are in the above taking a very special idempotent in the ideal – namely the one that generates $\mathscr{C}$.

(ii) From the definition of $c(X)$, $g(X)|c(X)$. Also $g(X)|(X^n-1)$. Therefore $g(X)|d(x)$, where

$$d(x)=\text{g.c.d.}(c(X),X^n-1)$$

Let $d(X)=g(X)\lambda(X)$. Then $\lambda(X)|\text{g.c.d.}(h(X),c(X))$. But it follows from (6.2) that

$$\text{g.c.d.}(c(X),h(X))=1$$

Therefore $\lambda(X)=1$ and $g(X)=\text{g.c.d.}(c(X),X^n-1)$.

(iii) Let β be a root of X^n-1 such that $c(\beta)=0$. Then β is a common root of $c(X)$ and X^n-1 and hence is a root of $g(X)$. As $g(X)|c(X)$, every root of $g(X)$ is a root of $c(X)$. Hence if α is a primitive nth root of unity in a suitable extension of F, then $c(\alpha^i)=0$ iff $g(\alpha^i)=0$.

From this it follows that the polynomial $c(X)$ such that $c(X)+I$ is the unique idempotent in $\mathscr{C}$ that generates it is a power of $g(X)$ multiplied by a power of X.

Examples 6.3

Case (i)

Let $\mathscr{C}$ be the binary cyclic code of length 7 generated by $g(X)=X^3+X^2+1$. Then

$$h(X)=(X+1)(X^3+X+1)=X^4+X^3+X^2+1$$

Observe that

$$X^3g(X)+(X^2+1)h(X)$$
$$=X^6+X^5+X^3+X^6+X^5+X^4+X^2+X^4+X^3+X^2+1=1$$

Therefore

$$c(X)+\langle X^7-1\rangle = X^3g(X)+I = X^6+X^5+X^3+I$$

is the unique idempotent in $\mathscr{C}$ that generates it.

Case (ii)
Let $\mathscr{C}$ be the binary cyclic code of length 15 generated by $g(X)=X^4+X+1$. Then

$$h(X)=\frac{(X^{15}+1)}{g(X)}$$
$$=(X+1)(X^4+X^3+1)(X^6+X^4+X^3+X^2+1)$$
$$=X^{11}+X^8+X^7+X^5+X^3+X^2+X+1$$

Observe that

$$g(X)^3=(X^4+X+1)^3$$
$$=X^{12}+X^9+X^8+X^6+X^4+X^3+X^2+X+1$$

Therefore,

$$g(X)^3+Xh(X)=1$$

and hence

$$c(X)=g(X)^3=X^{12}+X^9+X^8+X^6+X^4+X^3+X^2+X+1$$

and $c(X)+\langle X^{15}+1\rangle$ is the unique idempotent in $\mathscr{C}$ that generates it.

6.6 SOME SOLVED EXAMPLES AND AN INVARIANCE PROPERTY

Examples 6.4

Case (i)
A $(3,9)$ binary linear code V is defined by $(a_1,a_2,\ldots,a_9)\in V$ iff $a_1=a_2=a_3$, $a_4=a_5=a_6$ and $a_7=a_8=a_9$. Show that V is equivalent to a cyclic code and determine the generator.

Solution

$$X^9-1=(X^3-1)(X^6+X^3+1)$$
$$=(X-1)(X^2+X+1)(X^6+X^3+1)$$

Let V_1 be the cyclic code of length 9 generated by $X^6 + X^3 + 1$. The code word corresponding to the message word $a_1a_2a_3$ in this code is

$$(a_1 + a_2X + a_3X^2)(1 + X^3 + X^6) + \langle X^9 - 1 \rangle$$
$$= a_1 + a_2X + a_3X^2 + a_1X^3 + a_2X^4 + a_3X^5 + a_1X^6 + a_2X^7 + a_3X^8 + \langle X^9 - 1 \rangle$$

or the word

$$a_1a_2a_3a_1a_2a_3a_1a_2a_3 = c_1c_2\text{---}c_9$$

(say). Consider the permutation $\boldsymbol{\sigma}$ of the set $\{1, 2, \ldots, 9\}$ defined by

$$\boldsymbol{\sigma} = \begin{pmatrix} 1 & 2 & 3 & 4 & 5 & 6 & 7 & 8 & 9 \\ 1 & 4 & 7 & 2 & 5 & 8 & 3 & 6 & 9 \end{pmatrix}$$

Then

$$\boldsymbol{\sigma}(c) = c_1c_4c_7c_2c_5c_8c_3c_6c_9 = a_1a_1a_1a_2a_2a_2a_3a_3a_3$$

Hence the code V_1 is equivalent to the code V. Generator matrix of the code V_1 is

$$\mathbf{G}_1 = \begin{pmatrix} 1 & 0 & 0 & 1 & 0 & 0 & 1 & 0 & 0 \\ 0 & 1 & 0 & 0 & 1 & 0 & 0 & 1 & 0 \\ 0 & 0 & 1 & 0 & 0 & 1 & 0 & 0 & 1 \end{pmatrix}$$

The permutation matrix $\mathbf{P}$ corresponding to $\boldsymbol{\sigma}$ is

$$\mathbf{P} = \begin{pmatrix} 1 & 0 & 0 & 0 & 0 & 0 & 0 & 0 & 0 \\ 0 & 0 & 0 & 1 & 0 & 0 & 0 & 0 & 0 \\ 0 & 0 & 0 & 0 & 0 & 0 & 1 & 0 & 0 \\ 0 & 1 & 0 & 0 & 0 & 0 & 0 & 0 & 0 \\ 0 & 0 & 0 & 0 & 1 & 0 & 0 & 0 & 0 \\ 0 & 0 & 0 & 0 & 0 & 0 & 0 & 1 & 0 \\ 0 & 0 & 1 & 0 & 0 & 0 & 0 & 0 & 0 \\ 0 & 0 & 0 & 0 & 0 & 1 & 0 & 0 & 0 \\ 0 & 0 & 0 & 0 & 0 & 0 & 0 & 0 & 1 \end{pmatrix}$$

Therefore, the generator matrix of V is

$$\mathbf{G} = \mathbf{G}_1\mathbf{P} = \begin{pmatrix} 1 & 1 & 1 & 0 & 0 & 0 & 0 & 0 & 0 \\ 0 & 0 & 0 & 1 & 1 & 1 & 0 & 0 & 0 \\ 0 & 0 & 0 & 0 & 0 & 0 & 1 & 1 & 1 \end{pmatrix}$$

Case (ii)

A binary cyclic code of length 63 is generated by $X^5 + X^4 + 1$. Find the minimum distance of this code.

Solution
The polynomial $X^5 + X^4 + 1$ being a generator, there is a code word of weight 3. Therefore the minimum distance of the code is at most 3. The relation

$$a(X)(X^5 + X^4 + 1) \equiv X^i \ (\text{mod}\ X^{63} - 1)$$

is not possible for any $i \geq 0$ as

$$X^5 + X^4 + 1 = (X^2 + X + 1)(X^3 + X + 1)$$

is a divisor of $X^{63} - 1$.
Hence, there is no code word of weight 1.

$$X^3 + 1 | X^{21} + 1$$

and also

$$X^7 + 1 | X^{21} + 1$$

Therefore, $X^5 + X^4 + 1 | X^{21} + 1$ over $\mathbb{B}$. Taking

$$a(X) = \frac{(X^{21} + 1)}{(X^5 + X^4 + 1)}$$

we find that

$$a(X)(X^5 + X^4 + 1) \equiv X^{21} + 1 (\text{mod}\ X^{63} - 1)$$

Hence there are code words of weight 2 and the minimum distance of the code is 2.

Case (iii)
Consider the binary cyclic code $\mathscr{C}$ of length 7 generated by $1 + X^2 + X^3$. Any word of length 7 which is obtained from a code word in $\mathscr{C}$ by changing 0s into 1s and 1s into 0s is a linear combination over $\mathbb{B}$ of the polynomials $X + X^4 + X^5 + X^6, 1 + X^2 + X^5 + X^6, 1 + X + X^3 + X^6$ and $1 + X + X^2 + X^4$. Now

$$\begin{aligned}
X + X^4 + X^5 + X^6 &= X + X^3 + X^4 + X^3 + X^5 + X^6 \\
&= (X + X^3)(1 + X^2 + X^3) \\
1 + X^2 + X^5 + X^6 &= 1 + X^2 + X^3 + X^3 + X^5 + X^6 \\
&= (1 + X^3)(1 + X^2 + X^3) \\
1 + X + X^3 + X^6 &= 1 + X^2 + X^3 + X + X^2 + X^6 \\
&= 1 + X^2 + X^3 + X(1 + X^2 + X^3) + X^2 + X^3 + X^4 + X^6 \\
&= (1 + X^2 + X^3)(1 + X) + X^2(1 + X^2 + X^3) \\
&\quad + X^3(1 + X^2 + X^3) \\
&= (1 + X + X^2 + X^3)(1 + X^2 + X^3)
\end{aligned}$$

and

$$1 + X + X^2 + X^4 = 1 + X^2 + X^3 + X + X^3 + X^4$$
$$= (1 + X)(1 + X^2 + X^3)$$

Thus each one of these is in $\mathscr{C}$ and, hence, $\mathscr{C}$ is invariant under the operation of changing 0s into 1s and 1s into 0s in the code words.

Case (iv)

A similar argument can be used to prove that the binary cyclic code of length 7 generated by $1 + X + X^3$ is invariant under the operation of changing 0s into 1s and 1s into 0s in the code words.

We can make a general observation about the invariance of codes as mentioned in Cases (iii) and (iv) of Examples 6.4 above.

Definition 6.6

Let $a = a_0a_1 \cdots a_{n-1}$ be a binary word of length n. Call the word obtained from a by changing 0s into 1s and 1s into 0s the **complement** of a and denote it by a'. Let the process of interchanging 0s and 1s be called **complementation**.

Let $\mathscr{C}$ be a binary linear code of length n and dimension k. Let $e^1, e^2, \ldots, e^k$ be a basis of $\mathscr{C}$. Then the set $\mathscr{C}' = \{a' | a \in \mathscr{C}\}$ is again a linear code generated by the complements of the basis elements. If $\mathscr{C}' = \mathscr{C}$, then we say that $\mathscr{C}$ is **invariant under complementation**.

Theorem 6.9

Let $\mathscr{C}$ be a binary cyclic code of length n with generator polynomial

$$g(X) = 1 + g_1X + \cdots + g_{r-1}X^{r-1} + X^r$$

Then the code $\mathscr{C}$ is invariant under complementation iff $1 + X + X^2 + \cdots + X^{n-1}$ is divisible by $g(X)$. Equivalently, a binary cyclic code is invariant under complementation iff it contains the all 1 word.

Proof

Every code polynomial in $\mathscr{C}$ is a linear combination of the polynomials $g(X)$, $Xg(X), \ldots, X^{n-r-1}g(X)$. Let $k(X)$ denote the complement

$$(g_1 + 1)X + \cdots + (g_{r-1} + 1)X^{r-1} + X^{r+1} + \cdots + X^{n-1}$$

of the generator polynomial $g(X)$.

Then

$$k(X) = g(X) + (1 + X + \cdots + X^{n-1})$$

Modulo $X^n - 1$

$$X^i k(X) = X^i g(X) + (1 + X + \cdots + X^{n-1}) \; \forall i \geq 1$$

Therefore, the complement of any code polynomial is a linear combination of the polynomials

$$X^i g(X) + (1 + X + \cdots + X^{n-1}) \quad 0 \le i \le n - r - 1$$

Any such linear combination is divisible by $g(X)$ iff

$$g(X) | (1 + X + \cdots + X^{n-1})$$

and the result follows.

Corollary

A binary cyclic code of odd length n with generator polynomial $g(X)$ is invariant under interchange of 0s and 1s iff $1 + X$ does not divide $g(X)$.

We know that

$$X^{15} + 1 = (X + 1)(X^4 + X^3 + 1)(X^4 + X + 1)(X^2 + X + 1) \times (X^4 + X^3 + X^2 + X + 1)$$

Also

$$X^{15} + 1 = (X + 1)(1 + X + \cdots + X^{14})$$

Therefore the binary cyclic codes of length 15 generated by

(i) $X^4 + X^3 + 1$
(ii) $X^4 + X + 1$
(iii) $X^4 + X^3 + X^2 + X + 1$
(iv) $(X^2 + X + 1)(X^4 + X + 1)$
(v) $(X^2 + X + 1)(X^4 + X^3 + 1)$

are invariant under interchange of 0s and 1s.

Exercise 6.5

1. A (4, 12) binary linear code $\mathscr{C}$ is defined by $(a_1, a_2, \ldots, a_{12}) \in \mathscr{C}$ iff $a_1 = a_2 = a_3 = a_4$, $a_5 = a_6 = a_7 = a_8$, $a_9 = a_{10} = a_{11} = a_{12}$. Is the code $\mathscr{C}$ equivalent to a cyclic code?
2. A (5, 15) binary linear code $\mathscr{C}$ is defined by $(a_1, a_2, \ldots, a_{15}) \in \mathscr{C}$ iff $a_1 = a_2 = \cdots = a_5$, $a_6 = a_7 = \cdots = a_{10}$, $a_{11} = a_{12} = \cdots = a_{15}$. Is $\mathscr{C}$ equivalent to a cyclic code?
3. Determine m and k if the (m, mk) binary linear code $\mathscr{C}$ defined by $(a_1, \ldots, a_{mk}) \in \mathscr{C}$ iff $a_1 = \cdots = a_m$, $a_{m+1} = \cdots = a_{2m}, \ldots, a_{m(k-1)+1} = \cdots = a_{mk}$ is equivalent to a cyclic code.

6.7 CYCLIC CODES AND GROUP ALGEBRAS

Let A be a group and F a field. Let FA denote the set of all finite formal sums of the form

$$\alpha_1 a_1 + \alpha_2 a_2 + \cdots + \alpha_n a_n$$

where $\alpha_i \in F$, $a_i \in A$. Two elements $\sum \alpha_i a_i$, $\sum \beta_i a_i$ in FA are equal iff $\alpha_i = \beta_i \forall i$. For two elements $\sum \alpha_i a_i$, $\sum \beta_i a_i$, define

$$\sum \alpha_i a_i + \sum \beta_i a_i = \sum (\alpha_i + \beta_i) a_i$$

In FA, we define multiplication distributively using the group multiplication in A. Explicitly

$$(\sum \alpha_i a_i)(\sum \beta_j b_j) = \sum (\alpha_i \beta_j) a_i b_j$$

With the addition and multiplication defined above, FA becomes an algebra over F. Also, it is a free F-module with the elements of A as a basis.

Theorem 6.10

If $A = \langle a \rangle$ is a finite cyclic group of order n, then FA and $F[X]/I$, I the ideal $\langle X^n - 1 \rangle$ of $F[X]$ generated by $X^n - 1$, are isomorphic F-algebras.

Proof

Define a map $\theta: F[X] \to FA$ by

$$\theta(\sum \alpha_i X^i) = \sum \alpha_i a^i \quad \alpha_i \in F$$

which is clearly an onto F-algebra homomorphism. The homomorphism θ maps the ideal I of $F[X]$ onto 0 and so θ induces an epimorphism $\theta: F[X]/I \to FA$,

$$\theta(\sum \alpha_i X^i + I) = \sum \alpha_i a^i \quad \alpha_i \in F$$

Any element of $F[X]/I$ is of the form $\sum \alpha_i X^i + I$, where $\alpha_i = 0$ for $i \geq n$. Therefore, if

$$\theta\left(\sum_{i=0}^{n-1} \alpha_i X^i + I\right) = \sum_{i=0}^{n-1} \alpha_i a^i = 0$$

then $\alpha_i = 0$ $\forall i$, $0 \leq i \leq n-1$. Thus θ is a monomorphism and, hence, an isomorphism.

Remark 6.3

The map $\theta: F[X]/I \to FA$ being an algebra isomorphism, the ideals of $F[X]/I$ get mapped onto ideals of FA. Thus a cyclic code of length n over F may be regarded as an ideal of the group algebra FA of a cyclic group A of order n over F and conversely.

Generalizing this way of looking at cyclic codes, Berman (1967) has defined and extensively studied Abelian codes of length n as ideals of the group algebra FA where A is an Abelian group of order n. However, we do not pursue the subject matter regarding Abelian codes any further except for the following:

Recall that an Abelian group C is called the direct sum of its subgroups A and B if every element of C can be uniquely written as ab, $a \in A$, $b \in B$.

Exercise 6.6

1. Determine all non-cyclic binary Abelian codes of length 9. (Hint: Compute all the ideals in the group algebra FA, where A is direct sum of two cyclic groups of order 3 each.)
2. Determine all non-cyclic ternary Abelian codes of length 4. (As in 1 above, here we need all ideals of the group algebra FA when F is the field of 3 elements and A is direct sum of two cyclic groups each of order 2.)

6.8 SELF DUAL BINARY CYCLIC CODES

Let $\mathscr{C}$ be a binary cyclic code of length n with generator polynomial $g(X)$. Let $h(X)$ be its check polynomial. Then the dual $\mathscr{C}^{\perp}$ is generated by

$$k(X) = X^{n-r}h(X^{-1})$$

The relation $X^n + 1 = g(X)h(X)$ shows that

$$\begin{aligned} X^n + 1 &= X^n g(X^{-1})h(X^{-1}) \\ &= X^r g(X^{-1}) \times X^{n-r}h(X^{-1}) \\ &= X^r g(X^{-1})k(X) \end{aligned}$$

Thus

$$k(X) = -\frac{X^n + 1}{X^r g(X^{-1})}$$

It then follows that $\mathscr{C}$ is self dual iff

$$g(X) = -\frac{X^n + 1}{X^r g(X^{-1})}$$

or iff

$$X^r g(X)g(X^{-1}) = X^n + 1$$

We then have the following theorem.

Theorem 6.11
A binary cyclic code of length n generated by $g(X)$ of degree r is self dual iff

$$X^n + 1 = X^r g(X)g(X^{-1})$$

In a similar fashion we can prove that a cyclic code of length n generated by $g(X)$ of degree r over GF(q) is self dual iff

$$X^r g(X)g(X^{-1}) = -(X^n - 1)$$

Examples 6.5

The binary polynomial $X^{14}+1$ factors as

$$X^{14}+1=(X^7+1)^2=(X+1)^2(X^3+X^2+1)^2(X^3+X+1)^2$$

The polynomial

$$g(X)=(X+1)(X^3+X+1)^2$$

is such that

$$X^7g(X)g(X^{-1})=(X+1)(X^3+X+1)^2(X+1)(X^3+X^2+1)^2=X^{14}+1$$

Hence the binary cyclic code of length 14 generated by $g(X)$ is self dual.

The binary polynomial $X^{30}+1$ factors as

$$X^{30}+1=(X^{15}+1)^2=(X^5+1)^2(X^{10}+X^5+1)^2$$

and the factor

$$g(X)=(X^5+1)(X^{10}+X^5+1)$$

is such that

$$X^{15}g(X)g(X^{-1})=X^{30}+1$$

Hence $g(X)$ generates a binary self dual code of length 30.

The binary polynomial $X^{42}+1$ factors over $\mathbb{B}$ as

$$\begin{aligned}X^{42}+1&=(X^6+1)(X^{18}+X^6+1)(X^{18}+X^{12}+1)\\&=(X+1)^2(X^2+X+1)^2(X^{18}+X^6+1)(X^{18}+X^{12}+1)\end{aligned}$$

Consider the binary code of length 42 generated by

$$g(X)=(X+1)(X^2+X+1)(X^{18}+X^6+1)$$

Clearly

$$X^{21}g(X)g(X^{-1})=X^{42}+1$$

Therefore the code is self dual.

Exercise 6.7

1. Find the minimum distance of the binary self dual code of length (i) 14; (ii) 30; (iii) 42, constructed above.
2. Construct some cyclic self dual codes over GF(4).
3. Construct some cyclic self dual codes over GF(8).

We now show that no cyclic self dual codes over an odd order field exist.

Let $F=\mathrm{GF}(q)$ where $q=p^r$ and p is an odd prime. If possible, let $\mathscr{C}$ be a cyclic self dual code of length n over F generated by $g(X)$ (say). Let $h(X)$ be its check polynomial. Then

$$g(X)=aX^{n-m}h(X^{-1})$$

where $0 \neq a \in F$, and $m = \deg g(X)$ $(= n/2)$. The relation $X^n - 1 = g(X)h(X)$ then implies that

$$X^n - 1 = -aX^m g(X)g(X^{-1}) \tag{6.4}$$

If n and p are coprime, $X^n - 1$ has no repeated roots in any extension field of F. But (6.4) shows that 1 is a root of $X^n - 1$ repeated at least twice – a contradiction.

If n, p are not coprime, let $n = p^t u$, where u and p and coprime. Then

$$X^n - 1 = (X^u - 1)^{p^t}$$

and $X^u - 1$ has no repeated roots. Therefore, every root of $X^n - 1$ and, in particular, 1 is repeated exactly p^t times. However, (6.4) shows that 1 is a root repeated an even number of times – again a contradiction. Hence, a cyclic self dual code over an odd order field does not exist.

7

Factorization of polynomials

7.1 FACTORS OF $X^n - 1$

In the construction of finite fields, we were required to find certain irreducible factors of $X^n - 1$ where $n = p^m - 1$, p a prime. Again, while studying cyclic codes we encountered the same problem. Here we study the problem of factorization of $X^n - 1$ as a product of irreducible polynomials. However, first we consider a couple of results about finite fields.

Theorem 7.1

(i) $\mathrm{GF}(p^r)$, p a prime contains a subfield of order p^s iff $s|r$.
(ii) If $\mathrm{GF}(p^s)$ is a subfield of $\mathrm{GF}(p^r)$ and $\beta \in \mathrm{GF}(p^r)$, then

$$\beta \in \mathrm{GF}(p^s) \quad \text{iff } \beta^{p^s} = \beta$$

Proof

Part (i)
Suppose that $s|r$ and let k be a positive integer such that $r = ks$. Let α be a primitive element of $\mathrm{GF}(p^r)$. Then

$$\mathrm{GF}(p^r) = \{0, 1, \alpha, \ldots, \alpha^{p^r - 1}\}$$

Let

$$F = \{\beta \in \mathrm{GF}(p^r) | \beta^{p^s} = \beta\}$$

Since $\mathrm{GF}(p^r)$ is of characteristic p, if the elements $\beta, \gamma \in F$ then so do the elements $\beta \pm \gamma$ and $\beta\gamma^{-1}$, $\gamma \neq 0$. Thus F is a subfield of $\mathrm{GF}(p^r)$. As the polynomial

$$X^{p^s} - X$$

has p^s roots in its splitting field, $O(F) \le p^s$. Take

$$\lambda = \alpha^{1 + p^s + \dots + p^{(k-1)s}} = \alpha^{(p^{ks} - 1)/(p^s - 1)}$$

The elements

$$1, \lambda, \lambda^2, \dots, \lambda^{p^s - 2}$$

are all distinct and

$$\lambda^{ip^s} = \lambda^i \quad \text{for } 0 \le i < p^s - 1$$

Thus $O(F) \ge p^s$ and, therefore, F is a subfield of $GF(p^r)$ of order p^s.

Part (ii)
That

$$\beta^{p^s} = \beta \quad \text{if } \beta \in GF(p^s)$$

is clear and the converse follows from the construction of F in the above proof.

Lemma 7.1
Let n, r, s be positive integers with $n \ge 2$. Then $n^s - 1 | n^r - 1$ iff $s | r$.

Proof
Write $r = sa + b$ where $0 \le b < s$. Then

$$\frac{n^r - 1}{n^s - 1} = \frac{n^{sa+b} - n^b + n^b - 1}{n^s - 1}$$

$$= n^b \frac{n^{sa} - 1}{n^s - 1} + \frac{n^b - 1}{n^s - 1}$$

Since $(n^s - 1) | (n^{sa} - 1)$, it follows that $(n^s - 1) | (n^r - 1)$ iff $b = 0$.

Theorem 7.2
If p is a prime

$$X^{p^m} - X$$

is a product of all monic polynomials, irreducible over $GF(p)$, whose degree divides m.

Proof
Let $f(X)$ be an irreducible monic polynomial over $GF(p)$ of degree d, where $d | m$. The case $f(X) = X$ is trivial, so assume that $f(X) \ne X$. Use $f(X)$ to construct a field F of order p^d. Then $f(X)$ is the minimal polynomial of one of the elements of F and so

$$f(X) | X^{p^d - 1} - 1$$

Now

$$d|m \Rightarrow (p^d - 1)|(p^m - 1)$$
$$\Rightarrow X^{p^d - 1} - 1 | X^{p^m - 1} - 1$$

Hence

$$f(X)|X^{p^m} - X$$

Conversely, let

$$f(X) \text{ be a divisor of } X^{p^m} - X$$

irreducible and of degree d. We have to prove that $d|m$. We can again assume that $f(X) \neq X$ so that

$$f(X)|(X^{p^m - 1} - 1)$$

Use $f(X)$ to construct a field F of order p^d. Let $\alpha \in F$ be a root of $f(X)$ and let β be a primitive element of F, say

$$\beta = a_0 + a_1\alpha + \cdots + a_{d-1}\alpha^{d-1} \quad a_i \in \mathrm{GF}(p)$$

Now

$$f(\alpha) = 0 \Rightarrow \alpha^{p^m} = \alpha$$

Also

$$a_i^p = a_i \forall i, 0 \leq i \leq d - 1$$

It then follows that

$$\beta^{p^m} = \beta$$

and so

$$\beta^{p^m - 1} = 1$$

But the multiplicative order of β is $p^d - 1$. Therefore $(p^d - 1)|(p^m - 1)$ and so $d|m$.

Finally, the formal derivative of

$$X^{p^m} - X$$

being

$$p^m X^{p^m - 1} - 1 = -1$$

the polynomial

$$X^{p^m} - X$$

does not have repeated roots and so no repeated irreducible factors. ■

As applications of this theorem, we have the following for polynomials over $\mathbb{B}$.

$$X^2 + X = X(X+1)$$

$$X^4 + X = X(X+1)(X^2+X+1)$$

$$X^8 + X = X(X+1)(X^3+X+1)(X^3+X^2+1) \qquad \text{(Proposition 4.4)}$$

$$X^{16} + X = X(X+1)(X^2+X+1)(X^4+X+1)(X^4+X^3+1) \times (X^4+X^3+X^2+X+1) \qquad \text{(Examples 4.3)}$$

Over the field of 3 elements, we have the following decomposition as product of irreducible polynomials

$$X^9 - X = X^{3^2} - X = X(X+1)(X+2)(X^2+1)(X^2+X+2)(X^2+2X+2)$$

$$X^{27} - X = X^{3^3} - X = X(X+1)(X+2)(X^3+2X+2)(X^3+2X+1)(X^3+X^2+2) \times (X^3+2X^2+1)(X^3+X^2+X+2)(X^3+X^2+2X+1) \times (X^3+2X^2+X+1)(X^3+2X^2+2X+2) \qquad \text{(Proposition 4.4)}$$

$$X^3 - X = X(X+1)(X+2)$$

7.2 FACTORIZATION THROUGH CYCLOTOMIC COSETS

Definition 7.1

In this section, p is a prime, n is a positive integer not divisible by p and q is a power of p. To obtain factorization of $X^n - 1$ over GF(q), we first define cyclotomic classes and partition the set $S = \{0, 1, 2, \ldots, n-1\}$ of integers into cyclotomic classes or cosets modulo n over GF(q). Since g.c.d.$(n, q) = 1$, there exists a smallest positive integer m such that $q^m \equiv 1 \pmod{n}$ or $q^m - 1$ is divisible by n (by Euler–Fermat theorem and also $m = \phi(n)$ – the Euler totient function). This m is called the **multiplicative order** of q modulo n. In S define a relation '$\sim$' as follows:

For $a, b \in S$, say that $a \sim b$ if $a \equiv bq^i \pmod{n}$ for some i. This relation is an equivalence relation and partitions the set S into equivalence classes. Each such equivalence class is called a **cyclotomic class** or **coset mod** n over GF(q). Observe that the cyclotomic class containing $s \in S$ is

$$C_s = \{s, qs, \ldots, q^{m_s - 1}s\}$$

where m_s is the smallest positive integer with

$$sq^{m_s} \equiv s \pmod{n}$$

Then $m_1 = m$ – the multiplicative order of q modulo n.

Examples 7.1

(i) For $n = 7$, $q = 2$,

$$C_0 = \{0\} \qquad C_1 = \{1, 2, 4\} \qquad C_3 = \{3, 6, 5\}$$

(ii) For $n = 11$, $q = 2$,

$$C_0 = \{0\} \qquad C_1 = \{1, 2, 4, 8, 5, 10, 9, 7, 3, 6\}$$

(iii) For $n = 15$, $q = 2$,

$$C_0 = \{0\} \qquad C_1 = \{1, 2, 4, 8\} \qquad C_3 = \{3, 6, 12, 9\}$$

$$C_5 = \{5, 10\} \qquad C_7 = \{7, 14, 13, 11\}$$

Lemma 7.2

If s is relatively prime to n, then C_s has m elements.

Proof

Since $q^m \equiv 1 \pmod{n}$, $sq^m \equiv s \pmod{n}$. Also

$$sq^{m_s} \equiv s \pmod{n}$$

Therefore

$$sq^{m_s}(q^{m - m_s} - 1) \equiv 0 \pmod{n}$$

But g.c.d.$(s, n) = 1$ and, so

$$q^{m - m_s} - 1 \equiv 0 \pmod{n}$$

But m being the smallest positive integer with this property, $m = m_s$. ■

From now on we assume that $q = p$ itself. Let $\mathrm{GF}(p^m)$ be an extension of $\mathrm{GF}(p)$ and α be a primitive nth root of unity. Then $\alpha \in \mathrm{GF}(p^m)$.

Theorem 7.3

If C_s is the cyclotomic coset mod n over $\mathrm{GF}(p)$ containing the integer s, then

$$\prod_{i \in C_s} (X - \alpha^i)$$

is the minimal polynomial of α^s over $\mathrm{GF}(p)$.

Proof

Let $M_s(X)$ denote the minimal polynomial of α^s over $\mathrm{GF}(p)$. Since the elements β and β^p of $\mathrm{GF}(p^m)$ have the same minimal polynomial over $\mathrm{GF}(p)$ (Proposition 4.2)

$$\prod_{i \in C_s} (X - \alpha^i)$$

is a factor of $M_s(X)$ over $\mathrm{GF}(p^m)$. In particular

$$\deg \prod_{i \in C_s} (X - \alpha^i) \leq \deg M_s(X) \tag{7.1}$$

Let

$$C_s(\alpha) = \{\alpha^s, \alpha^{ps}, \ldots, \alpha^{sp^{m_s - 1}}\}$$

Since

$$sp^{m_s} \equiv s (\bmod n)$$

then

$$\alpha^{sp^{m_s}} = \alpha^s$$

Therefore, with $A^p = \{a^p | a \in A\}$ for any subset A of $\mathrm{GF}(p^m)$, we find that

$$C_s(\alpha)^p = C_s(\alpha)$$

It follows that if

$$\sigma_1, \sigma_2, \ldots, \sigma_{m_s}$$

denote the symmetric polynomials in the elements of $C_s(\alpha)$, then

$$\sigma_i^p = \sigma_i \quad \forall 1 \leq i \leq m_s$$

and, hence, $\sigma_i \in \mathrm{GF}(p)$ for $1 \leq i \leq m_s$ (by Theorem 7.1 Part (ii)). Therefore

$$\prod_{i \in C_s} (X - \alpha^i)$$

is a polynomial over $\mathrm{GF}(p)$ having α^s as a root. But then

$$M_s(X) \mid \prod_{i \in C_s} (X - \alpha^i)$$

over $\mathrm{GF}(p)$. Both the polynomials being monic, it follows from this and (7.1) that

$$M_s(X) = \prod_{i \in C_s} (X - \alpha^i)$$

Corollary

$$X^n - 1 = \prod_s M_s(X)$$

where s runs over a set of representatives of cyclotomic cosets modulo n over $\mathrm{GF}(p)$.

Examples 7.2

Case (i)

We use this theorem first to obtain factorization of X^9+1 over $\mathbb{B}$. Here the least m such that $9|2^m-1$ is 6. So, we have to construct a field of order 64. The polynomial X^6+X+1 is irreducible over $\mathbb{B}$ and so

$$F=\mathbb{B}[X]/\langle X^6+X+1\rangle$$

is a field of order 64. Let

$$\alpha=X+\langle X^6+X+1\rangle$$

It is a primitive element of F and so $\beta=\alpha^7=\alpha^2+\alpha$ is a primitive 9th root of unity. The cyclotomic classes modulo 9 over $\mathbb{B}$ are:

$$C_0=\{0\} \qquad C_1=\{1,2,4,8,7,5\} \qquad C_3=\{3,6\}$$

Then, $M_0(X)=X+1$ and $M_3(X)=X^2+X+1$ – this being the only irreducible polynomial of degree 2 over $\mathbb{B}$. Also then, it follows that

$$\beta^6+\beta^3+1=0$$

The polynomial X^6+X^3+1 being irreducible, it follows that

$$M_1(X)=X^6+X^3+1$$

Hence

$$X^9+1=(X+1)(X^2+X+1)(X^6+X^3+1)$$

Case (ii)

Now we factorize $X^{13}-1$ over GF(3). The order of 3 modulo 13 is 3. So, we have to construct a field of order 27. The polynomial X^3+2X+1 is irreducible over F and so

$$F_1=F[X]/\langle X^3+2X+1\rangle$$

is a field of order 27. The element

$$\alpha=X+\langle X^3+2X+1\rangle$$

of F_1 is primitive and so $\beta=\alpha^2$ is a primitive 13th root of unity. The cyclotomic classes modulo 13 over F are:

$$C_0=\{0\} \qquad C_1=\{1,3,9\} \qquad C_2=\{2,6,5\}$$
$$C_4=\{4,12,10\} \qquad C_7=\{7,8,11\}$$

so we have to find the minimal polynomials of β, β^2, β^4 and β^7.

The minimal polynomial of α is X^3+2X+1. Therefore,

$$\alpha^3=\alpha+2$$

Now

$$\beta^2 = \alpha^4 = \alpha^2 + 2\alpha$$
$$\beta^3 = \alpha^4 + 2\alpha^3$$
$$= \alpha^2 + 2\alpha + 2\alpha + 1$$
$$= \alpha^2 + \alpha + 1$$

Clearly, $\beta^3 + \beta^2 + \beta = 1$ and as $X^3 + X^2 + X + 2$ is irreducible over F, it is the minimal polynomial of β. Now,

$$M_2(X) = X^3 - X^2(\beta^2 + \beta^5 + \beta^6) + X(\beta^7 + \beta^8 + \beta^{11}) - 1$$
$$M_7(X) = X^3 - X^2(\beta^7 + \beta^8 + \beta^{11}) + X(\beta^2 + \beta^5 + \beta^6) - 1$$
$$M_4(X) = X^3 - X^2(\beta^4 + \beta^{10} + \beta^{12}) + X(\beta + \beta^3 + \beta^9) - 1$$

Now

$$\beta^3 = -\beta^2 - \beta + 1$$

and so

$$\beta^6 = \beta^4 + \beta^2 + 1 - \beta^3 + \beta^2 + \beta$$
$$= \beta^4 - \beta^3 - \beta^2 + \beta + 1$$

and

$$\beta^2 + \beta^5 + \beta^6 = \beta^2 - \beta^4 - \beta^3 + \beta^2 + \beta^4 - \beta^3 - \beta^2 + \beta + 1$$
$$= \beta^3 + \beta^2 + \beta + 1$$
$$= 2$$
$$\beta^7 = \beta^5 - \beta^4 - \beta^3 + \beta^2 + \beta$$
$$\beta^9 = -\beta^6 - \beta^3 + 1$$

and so

$$\beta^{11} = -\beta^8 - \beta^5 + \beta^2$$

Therefore

$$\beta^7 + \beta^8 + \beta^{11} = -\beta^4 - \beta^3 - \beta^2 + \beta$$
$$= -\beta(\beta^3 + \beta^2 + \beta) + \beta = 0$$

Again

$$\beta^4 = -\beta^3 - \beta^2 + \beta$$
$$\beta^{12} = -\beta^9 - \beta^6 + \beta^3 = -\beta^3 - 1$$
$$\beta^{10} = -\beta^7 - \beta^4 + \beta$$

and so

$$\begin{aligned}\beta^4+\beta^{10}+\beta^{12} &= -\beta^7+\beta-\beta^3-1\\ &= -\beta^5+\beta^4-\beta^2-1\\ &= -\beta^4+\beta^3+\beta^2-1\\ &= -\beta^3-\beta^2-\beta-1=1\end{aligned}$$

Also

$$\begin{aligned}\beta^3+\beta^9 &= -\beta^6+1\\ &= -\beta^4+\beta^3+\beta^2-\beta\\ &= -\beta^3-\beta^2-2\beta\\ &= -1-\beta\end{aligned}$$

or

$$\beta+\beta^3+\beta^9=2$$

Thus

$$M_2(X)=X^3+X^2+2 \qquad M_7(X)=X^3+2X+2$$
$$M_4(X)=X^3+2X^2-X+2$$

and

$$\begin{aligned}X^{13}-1 = {} &(X-1)(X^3+X^2+X+2)(X^3+X^2+2)(X^3+2X+2)\\ &\times(X^3+2X^2+2X+2)\end{aligned}$$

Exercise 7.1

1. Prove that over any field $X^s-1|X^r-1$ iff $s|r$.
2. Show that g.c.d.$(X^s-1, X^r-1)=X^d-1$, where $d=$ g.c.d.(r,s).
3. Let q be a prime power, n a positive integer relatively coprime to q and m be the multiplicative order of q mod n. Prove that X^n-1 has all its roots in an extension field GF(q^m) of GF(q).
4. Let α be a primitive nth root of unity in an extension GF(2^m) of $\mathbb{B}$ and let

$$f(X)=\prod_{i\in K}(X+\alpha^i)$$

   where K is a subset of $\{0,1,2,\ldots,n-1\}$. Show that the coefficients of $f(X)$ are in $\mathbb{B}$ iff $k\in K$ implies $2k\in K$ modulo n.
5. Find the irreducible factors of X^n+1 over $\mathbb{B}$ for $n\le 15$.
6. Find the irreducible factors of X^n-1 over GF(3) for $n\le 8$.
7. Write down some symmetric polynomials in commuting variables
   (i) X, Y;
   (ii) X, Y, Z.

8. Let α be a primitive nth root of unity in an extension $GF(p^m)$ of $GF(p)$ and let

$$f(X) = \prod_{i \in K} (X - \alpha^i)$$

where K is a subset of $\{0,1,2,\ldots,n-1\}$. The coefficients of $f(X)$ are in $GF(p)$ iff $k \in K$ implies $pk \in K$ modulo n. Comment!

7.3 BERLEKAMP'S ALGORITHM FOR FACTORIZATION OF POLYNOMIALS

We have earlier considered two simple methods for factorization of a polynomial $(x^n - 1)$ over $GF(q)$. We here give an algorithm due to Berlekamp (1968) for factorization of an arbitrary polynomial.

Let $F = GF(q)$ be the field of q elements and

$$f(x) = \sum_{i=0}^{m} a_i x^i$$

be a monic polynomial of degree m over F. Let $\mathbf{Q} = (Q_{ij})$ be the square matrix of order m over F in which the ith row is represented by $x^{q(i-1)}$ reduced modulo $f(x)$. For example, if

$$f(x) = x^5 + x^3 + 1$$

and $q = 3$, then the third row of the $\mathbf{Q}$-matrix is $(x^6 \equiv -x - x^4 (\mathrm{mod}\, f(x)))$

$$0 \quad -1 \quad 0 \quad 0 \quad -1$$

Lemma 7.2

Given any polynomial

$$g(x) = \sum_{i=0}^{m-1} g_i x^i$$

over F of degree less than m,

$$g(x)^q - g(x) \equiv 0 (\mathrm{mod}\, f(x))$$

iff the row vector $(g_0 \quad g_1 \quad \cdots \quad g_m)$ is in the null space of $\mathbf{Q} - \mathbf{I}$, where $\mathbf{I}$ is the identity matrix of order m.

Proof

As $q\beta = 0$ for every $\beta \in F$,

$$\begin{aligned} g(x)^q &= g(x^q) \\ &= \sum_{i=0}^{m-1} g_i x^{iq} \\ &\equiv \sum_{i=0}^{m-1} g_i \left(\sum_{k=0}^{m-1} Q_{i+1,k+1} x^k \right) \qquad (\mathrm{mod}\, f(x)) \\ &\equiv \sum_{k=0}^{m-1} \left(\sum_{i=0}^{m-1} g_i Q_{i+1,k+1} \right) x^k \end{aligned}$$

Observe that

$$\sum_{i=0}^{m-1} g_i Q_{i+1,k+1}$$

is the $(k+1)$th entry of the product

$$(g_0 \quad g_1 \quad \cdots \quad g_{m-1})\mathbf{Q}$$

Also g_k is the $(k+1)$th entry of the product

$$(g_0 \quad g_1 \quad \cdots \quad g_{m-1})\mathbf{I}$$

Therefore,

$$\begin{aligned} g(x)^q - g(x) &\equiv \sum_{k=0}^{m-1} \left(\left(\sum_{i=0}^{m-1} g_i Q_{i+1,k+1} \right) - g_k \right) x^k \\ &= 0 \end{aligned}$$

iff

$$\left(\sum_{i=0}^{m-1} g_i Q_{i+1,k+1} \right) - g_k = 0 \quad \forall k, 0 \leq k \leq m-1$$

or equivalently

$$(g_0 \quad g_1 \quad \cdots \quad g_{m-1})(\mathbf{Q} - \mathbf{I}) = 0$$

Theorem 7.4

$$f(x) = \prod_{s \in F} [\text{g.c.d.}(f(x), g(x) - s)]$$

where

$$g(x) = \sum_{i=0}^{m-1} g_i x^i$$

is such that

$$(g_0 \quad g_1 \quad \cdots \quad g_{m-1})(\mathbf{Q} - \mathbf{I}) = 0$$

Proof
By the above lemma

$$f(x) | g(x)^q - g(x)$$

But

$$g(x)^q - g(x) = \prod_{s \in F} (g(x) - s)$$

since for any y

$$y^q - y = \prod_{s \in F} (y - s)$$

Therefore

$$f(x) \mid \prod_{s \in F} (g(x) - s)$$

and hence

$$f(x) = \text{g.c.d.}(f(x), \prod_{s \in F} (g(x) - s))$$

Also

$$\text{g.c.d.}\left(f(x), \prod_{s \in S} (g(x) - s) \right) \Bigg| \prod_{s \in F} (f(x), g(x) - s)$$

or

$$f(x) \Bigg| \prod_{s \in F} \text{g.c.d.}(f(x), g(x) - s) \tag{7.2}$$

On the other hand

$$\text{g.c.d.}(f(x), g(x) - s) \mid f(x)$$

and for $s \neq t$ in F, $g(x) - s$ and $g(x) - t$ are relatively coprime. Therefore, g.c.d.$(f(x), g(x) - s)$ and g.c.d.$(f(x), g(x) - t))$ are coprime, and so

$$\prod_{s \in F} \text{g.c.d.}(f(x), g(x) - s) \mid f(x) \tag{7.3}$$

The polynomial $f(x)$ being monic, it follows from (7.2) and (7.3) that

$$f(x) = \prod_{s \in F} \text{g.c.d.}(f(x), g(x) - s)$$

Examples 7.3

Case (i)

Consider first the polynomial

$$f(x) = x^5 + x^3 + 1$$

over GF(3). The successive powers of x needed are:

$$x^0 = 1$$

$$x^3$$

$$x^6 = -x - x^4$$

$$x^9 = x^2 - x^4 + x^5 = -1 + x^2 - x^3 - x^4$$

$$x^{12} = 1 + x + x^2 + x^4$$

Therefore

$$\mathbf{Q}-\mathbf{I}=\begin{pmatrix} 0 & 0 & 0 & 0 & 0 \\ 0 & -1 & 0 & 1 & 0 \\ 0 & -1 & -1 & 0 & -1 \\ -1 & 0 & 1 & 1 & -1 \\ 1 & 1 & 1 & 0 & 0 \end{pmatrix}$$

Let $(g_0 \quad g_1 \quad \cdots \quad g_4)$ be in the null space of $\mathbf{Q}-\mathbf{I}$. Then

$$-g_3+g_4=0$$

$$-g_1-g_2+g_4=0$$

$$-g_2+g_3+g_4=0$$

$$g_1+g_3=0$$

and

$$-g_2-g_3=0$$

Thus

$$g_1=g_2=-g_3=-g_4$$

and

$$g(x)=g_0+g_1(x+x^2-x^3-x^4) \tag{7.4}$$

Let $g_1=g_0=-1$ and take $s=0$. We then need to find the HCF of $x^4+x^3-x^2-x-1$ and x^5+x^3+1.

$$\begin{array}{l|l} x^4+x^3-x^2-x-1 & x^5 \qquad +x^3 \qquad\qquad +1 \\ & \underline{x^5+x^4-x^3-x^2-x} \\ & \quad -x^4-x^3+x^2+x+1 \end{array}$$

Therefore, the HCF is $x^4+x^3-x^2-x-1$ and we have

$$x^5+x^3+1=(x-1)(x^4+x^3-x^2-x-1)$$

By multiplying $g(x)$ as in (7.4) by $-g_1^{-1}$, we could have taken

$$g(x)=x^4+x^3-x^2-x+g_0'$$

The above HCF obtained corresponds to taking $s=-g_0'-1$. It is clear that by taking a different value of g_0', we find that $g(x)$ is coprime to x^5+x^3+1.

Since none of the elements of GF(3) is a root of

$$h(x)=x^4+x^3-x^2-x-1$$

this polynomial does not have a linear factor. The only monic irreducible polynomials of degree 2 over GF(3) are

$$x^2+x-1 \qquad x^2+1 \qquad x^2-x-1$$

and none of these is a factor of $h(x)$. Hence $h(x)$ is an irreducible polynomial over GF(3). Hence

$$x^5 + x^3 + 1 = (x - 1)h(x)$$

is a factorization of $f(x)$ as a product of irreducible polynomials.

Case (ii)

Next, consider the binary polynomial

$$f(x) = 1 + x + x^3 + x^7 + x^8$$

For writing the **Q**-matrix, the powers of x needed are

$$x^0 = 1 \qquad x^2 \qquad x^4 \qquad x^6$$

$$x^8 = 1 + x + x^3 + x^7$$

$$x^{10} = 1 + x^4 + x^5$$

$$x^{12} = x^2 + x^6 + x^7$$

$$x^{14} = x + x^2$$

Therefore,

$$\mathbf{Q} - \mathbf{I} = \begin{pmatrix} 0 & 0 & 0 & 0 & 0 & 0 & 0 & 0 \\ 0 & 1 & 1 & 0 & 0 & 0 & 0 & 0 \\ 0 & 0 & 1 & 0 & 1 & 0 & 0 & 0 \\ 0 & 0 & 0 & 1 & 0 & 0 & 1 & 0 \\ 1 & 1 & 0 & 1 & 1 & 0 & 0 & 1 \\ 1 & 0 & 0 & 0 & 1 & 0 & 0 & 0 \\ 0 & 0 & 1 & 0 & 0 & 0 & 0 & 1 \\ 0 & 1 & 1 & 0 & 0 & 0 & 0 & 1 \end{pmatrix}$$

Let $\mathbf{g} = (g_0 \quad g_1 \quad \cdots \quad g_7)$ be in the null space of $\mathbf{Q} - \mathbf{I}$. Then

$$g_4 + g_5 = 0 \qquad g_1 + g_4 + g_7 = 0 \qquad g_1 + g_2 + g_6 + g_7 = 0$$

$$g_3 + g_4 = 0 \qquad g_2 + g_4 + g_5 = 0 \qquad g_3 = 0 \qquad g_4 + g_6 + g_7 = 0$$

These equations yield

$$g_1 = g_2 = g_3 = g_4 = g_5 = g_6 = g_7 = 0$$

Therefore, $g(x) = g_0$ is a constant and the algorithm does not yield any factors of $f(x)$. Neither 0 nor 1 being a root of $f(x)$, $f(x)$ has no linear factors. As is easily seen, the only irreducible polynomial $x^2 + x + 1$ of degree 2 is not a divisor of $f(x)$. Also $x^3 + x + 1$ and $x^3 + x^2 + 1$ are the only irreducible polynomials of degree 3 and neither of these divides $f(x)$. None of the three irreducible polynomials

$$x^4 + x^3 + 1 \qquad x^4 + x + 1 \qquad x^4 + x^3 + x^2 + x + 1$$

of degree 4 is a divisor of $f(x)$ and it follows that $f(x)$ is an irreducible polynomial.

The above conclusion could easily have been drawn from the following general theorem (the proof of which we omit for the time being).

Theorem 7.5
The number of distinct irreducible factors of $f(x)$ is equal to the dimension of the null space of $\mathbf{Q}-\mathbf{I}$.

Let us now consider one more example:

Examples 7.3 *contd*

Case (iii)
Consider the binary polynomial

$$f(x)=x^7+x^5+x^4+x^2+x+1$$

The relevant powers of x are

$$\begin{aligned}
x^0 &= 1 \\
& x^2, x^4, x^6 \\
x^8 &= x+x^2+x^3+x^5+x^6 \\
x^{10} &= x^3+x^4+x^5+x^7+x^8 \\
&= x^3+x^4+x^5+1+x+x^2+x^4+x^5+x+x^2+x^3+x^5+x^6 \\
&= 1+x^5+x^6 \\
x^{12} &= x^2+x^7+x^8 \\
&= x^2+1+x+x^2+x^4+x^5+x+x^2+x^3+x^5+x^6 \\
&= 1+x^2+x^3+x^4+x^6
\end{aligned}$$

Therefore,

$$\mathbf{Q}-\mathbf{I}=\begin{pmatrix}
0 & 0 & 0 & 0 & 0 & 0 & 0 \\
0 & 1 & 1 & 0 & 0 & 0 & 0 \\
0 & 0 & 1 & 0 & 1 & 0 & 0 \\
0 & 0 & 0 & 1 & 0 & 0 & 1 \\
0 & 1 & 1 & 1 & 1 & 1 & 1 \\
1 & 0 & 0 & 0 & 0 & 0 & 1 \\
1 & 0 & 1 & 1 & 1 & 0 & 0
\end{pmatrix}$$

If $(g_0 \quad g_1 \quad \cdots \quad g_6)$ is in the null space of $\mathbf{Q}-\mathbf{I}$, then

$$\begin{array}{llll}
g_5+g_6=0 & g_1+g_4=0 & g_1+g_2+g_4+g_6=0 & \\
g_3+g_4+g_6=0 & g_2+g_4+g_6=0 & g_4=0 & g_3+g_4+g_5=0
\end{array}$$

These equations lead to

$$0 = g_4 = g_1 \qquad g_2 = g_3 = g_5 = g_6$$

Therefore

$$g(x) = x^6 + x^5 + x^3 + x^2 + g_0$$

We find the HCF of $g(x)$ with $g_0 = 1$ and $f(x)$:

$$\begin{array}{l|l} x^6 + x^5 + x^3 + x^2 + 1 & x^7 \quad + x^5 + x^4 \quad + x^2 + x + 1 \\ & \underline{x^7 + x^6 \quad + x^4 + x^3 \quad + x} \\ & \quad x^6 + x^5 \quad + x^3 + x^2 \quad + 1 \end{array}$$

Thus HCF is $x^6 + x^5 + x^3 + x^2 + 1$ and

$$x^7 + x^5 + x^4 + x^2 + x + 1 = (x + 1)(x^6 + x^5 + x^3 + x^2 + 1) \tag{7.5}$$

Observe that the dimension of the null space of $\mathbf{Q} - \mathbf{I}$ is 2. Therefore, $f(x)$ has two distinct irreducible factors. Therefore, either $h(x) = x^6 + x^5 + x^3 + x^2 + 1$ is irreducible or it is square of a binary cubic irreducible polynomial or it is cube of a binary quadratic irreducible polynomial. However,

$$\begin{aligned} (x^2 + x + 1)^3 &= x^6 + x^2(x + 1)(x^2 + x + 1) + (x + 1)^3 \\ &= x^6 + (x^3 + x^2)(x^2 + x + 1) + x^3 + x(x + 1) + 1 \\ &= x^6 + x^5 + x^4 + x^3 + x^4 + x^3 + x^2 + x^3 + x^2 + x + 1 \\ &= x^6 + x^5 + x^3 + x + 1 \neq h(x) \end{aligned}$$

$$(x^3 + x + 1)^2 = x^6 + x^2 + 1 \neq h(x)$$

$$(x^3 + x^2 + 1)^2 = x^6 + x^4 + 1 \neq h(x)$$

Hence $h(x)$ is irreducible and (7.5) expresses $f(x)$ as a product of prime factors.

The Chinese remainder theorem for integers states that given primes $p_1, p_2, \ldots, p_k$ and integers $a_1, a_2, \ldots, a_k$, the simultaneous congruences

$$x \equiv a_i (\text{mod } p_i^{e_i})$$

e_i, $1 \leq i \leq k$, positive integers, have a unique solution

$$\text{mod} \prod_i p_i^{e_i}$$

The main tool in the proof of this theorem is Euclid's division algorithm which holds for integers. If F is a field, the polynomial ring $F[x]$ is a Euclidean domain and so Euclid's division algorithm is also true for the ring $F[x]$.

Theorem 7.6 – Chinese remainder theorem for polynomials
Given irreducible polynomials $f_1(x), f_2(x), \ldots, f_k(x)$ and arbitrary polynomials $a_1(x), a_2(x), \ldots, a_k(x)$ over a field F, the simultaneous congruences

$$h(x) \equiv a_i(x) (\text{mod } f_i(x)^{e_i})$$

where $e_1, e_2, \ldots, e_k$ are positive integers, have a unique solution for

$$h(x) \bmod \prod_i f_i(x)^{e_i}$$

Proof

The ring $F[x]$ being a Euclidean domain, any two elements in $F[x]$ have a greatest common divisor and a g.c.d. of $f(x), g(x) \in F[x]$ can be expressed in the form

$$a(x)f(x) + b(x)g(x)$$

for some $a(x), b(x)$ in $F[x]$. Since the polynomials

$$\prod_{j \neq i} f_j(x)^{e_j} \quad \text{and} \quad f_i(x)^{e_i}$$

are relatively coprime, it follows from the above observation that there exist polynomials $b_i(x) \in F[x]$ such that

$$b_i(x) \prod_{j \neq i} f_j(x)^{e_j} \equiv 1 (\bmod f_i(x)^{e_i})$$

Then

$$a_i(x) b_i(x) \prod_{j \neq i} f_j(x)^{e_j} \equiv a_i(x) (\bmod f_i(x)^{e_i})$$

and

$$a_i(x) b_i(x) \prod_{j \neq i} f_j(x)^{e_j} \equiv 0 (\bmod f_l(x)^{e_l}) \quad \text{for } l \neq i$$

Set

$$h(x) = \sum_i a_i(x) b_i(x) \prod_{j \neq i} f_j(x)^{e_j}$$

Then

$$h(x) \equiv a_i(x) (\bmod f_i(x)^{e_i}) \quad \forall i, 1 \leq i \leq k$$

For proving uniqueness, suppose that $H(x)$ is also a solution of the simultaneous congruences, i.e.

$$H(x) \equiv a_i(x) (\bmod f_i(x)^{e_i}) \quad \forall i, 1 \leq i \leq k$$

Then

$$H(x) - h(x) \equiv 0 (\bmod f_i(x)^{e_i}) \quad \forall i, 1 \leq i \leq k$$

Since no two of the irreducible polynomials $f_1(x), f_2(x), \ldots, f_k(x)$ are equal, it follows that

$$H(x) - h(x) \equiv 0(\text{mod} \prod f_i(x)^{e_i})$$

or

$$H(x) \equiv h(x)(\text{mod} \prod f_i(x)^{e_i})$$

Proof of Theorem 7.5
Suppose that

$$f(x) = \prod_{i=1}^{n} p_i(x)^{e_i}$$

where each $p_i(x)$ is an irreducible polynomial over $F = \text{GF}(q)$, each e_i is a positive integer and no two of $p_1(x), p_2(x), \ldots, p_n(x)$ are equal. A polynomial $g(x) \in F[x]$ is in the null space of $\mathbf{Q} - \mathbf{I}$ iff

$$f(x) | g(x)^q - g(x) = \prod_{a_i \in F} (g(x) - a_i)$$

But this is so iff each

$$p_i(x)^{e_i} | g(x) - a_i$$

for some $a_i \in F$. On the other hand, given any elements $a_1, a_2, \ldots, a_n \in F$, it follows from Theorem 7.6 that there exists a unique $g(x)(\text{mod} f(x))$ in $F[x]$ with

$$g(x) \equiv a_i(\text{mod} p_i(x)^{e_i}) \quad 1 \leq i \leq n$$

As each a_i has q choices, we can choose $a_1, \ldots, a_n$ in q^n ways and hence there are q^n elements $g(x)$ in $F[x]$ with

$$g(x)^q - g(x) \equiv 0(\text{mod} f(x))$$

Thus the null space of $\mathbf{Q} - \mathbf{I}$ has q^n elements and, therefore, its dimension over F is n – the number of distinct irreducible factors of $f(x)$.

7.4 BERLEKAMP'S ALGORITHM – A SPECIAL CASE

As mentioned earlier, we are most of the time concerned with factorization of the polynomial $x^n - 1$. For such a polynomial, Berlekamp's algorithm takes a much simpler form as, in that case, we do not need to find the $\mathbf{Q}$-matrix. In this case, as we can assume that g.c.d.$(n, q) = 1$, the $\mathbf{Q}$-matrix $\mathbf{Q} = (\mathbf{Q}_{ij})$ has exactly one non-zero entry in every row and since for a given j there is exactly one i with

$$q(i-1) \equiv j(\text{mod} n)$$

there is exactly one non-zero entry in every column also. Moreover, this non-zero entry is always 1. In fact

$$Q_{i+1,j+1} = 1 \quad \text{iff } qi \equiv j \pmod n$$

From this it follows that if $(g_0 \quad g_1 \quad \cdots \quad g_{n-1})$ is in the null space of $\mathbf{Q} - \mathbf{I}$ and $qi \equiv j \pmod n$, then $g_i = g_j$. As a consequence we have

$$g_j = g_i \quad \forall j \in C_i$$

where C_i is the cyclotomic class modulo n relative to q containing i. Set

$$\bar{C}_i = \sum_{j \in C_i} x^j$$

Then the result of Theorem 7.4 takes the form

$$x^n - 1 = \prod_{s \in F} [\text{g.c.d.}(x^n - 1, g(x) - s)]$$

where $g(x)$ is a linear combination of $\bar{C}_i$ over F.

Examples 7.4

Case (i)

First, let $p = 23$, and consider the cyclotomic cosets modulo p relative to 2:

$$C_0 = \{0\}$$
$$C_1 = \{1, 2, 4, 8, 16, 9, 18, 13, 3, 6, 12\}$$
$$C_5 = \{5, 10, 20, 17, 11, 22, 21, 19, 15, 7, 14\}$$

so $x^{23} - 1$ factors as a product of $x - 1$ and two irreducible polynomials of degree 11 each (Corollary of Theorem 7.3). To achieve this factorization we use the algorithm of Berlekamp. The irreducible factors are among the common divisors of $x^{23} - 1$ and

$$1 + a(x + x^2 + x^3 + x^4 + x^6 + x^8 + x^9 + x^{12} + x^{13} + x^{16} + x^{18})$$
$$+ b(x^5 + x^7 + x^{10} + x^{11} + x^{14} + x^{15} + x^{17} + x^{19} + x^{20} + x^{21} + x^{22})$$

where $a, b \in \mathbb{B}$. We use the Euclidean algorithm to find the HCFs.

$$x^{18} + x^{16} + x^{13} + x^{12} + x^9 + x^8 + x^6 + x^4 + x^3 + x^2 + x + 1 \overline{\smash{)}x^{23} + 1}$$

The 1st remainder is:

$$x^{21} + x^{18} + x^{17} + x^{14} + x^{13} + x^{11} + x^9 + x^8 + x^7 + x^6 + x^5 + 1$$

The 2nd remainder is:

$$x^{19} + x^{18} + x^{17} + x^{16} + x^{15} + x^{14} + x^{13} + x^{12} + x^8 + x^4 + x^3 + 1$$

The 3rd remainder is:

$$x^{18} + x^{16} + x^{15} + x^{12} + x^{10} + x^9 + x^8 + x^7 + x^5 + x^2 + x + 1$$

The 4th remainder is:

$$x^{15} + x^{13} + x^{10} + x^7 + x^6 + x^5 + x^4 + x^3$$

$$x^{15} + x^{13} + x^{10} + x^7 + x^6 + x^5 + x^4 + x^3 \overline{\big)x^{18} + x^{16} + x^{13} + x^{12} + x^9 + x^8 + x^6 + x^4 + x^3 + x^2 + x + 1}$$

The 1st remainder is:

$$x^{12} + x^{10} + x^7 + x^4 + x^3 + x^2 + x + 1$$

Thus the HCF is:

$$\begin{aligned} & x^{12} + x^{10} + x^7 + x^4 + x^3 + x^2 + x + 1 \\ &= (x + 1)\{x^{10}(x + 1) + x^4(x^2 + x + 1) + x^2 + 1\} \\ &= (x + 1)(x^{11} + x^{10} + x^6 + x^5 + x^4 + x^2 + 1) \end{aligned}$$

To find the other irreducible factor of $x^{23} + 1$ of degree 11, we divide $x^{23} + 1$ by the HCF:

$$\begin{array}{r} x^{11} + x^9 + x^7 + x^6 + x^5 + x + 1 \\ x^{12} + x^{10} + x^7 + x^4 + x^3 + x^2 + x + 1 \overline{\big) x^{23} + 1 \qquad\qquad\qquad\qquad\quad} \end{array}$$

The 1st remainder is:

$$x^{21} + x^{18} + x^{15} + x^{14} + x^{13} + x^{12} + x^{11} + 1$$

The 2nd remainder is:

$$x^{19} + x^{18} + x^{16} + x^{15} + x^{14} + x^{10} + x^9 + 1$$

The 3rd remainder is:

$$x^{18} + x^{17} + x^{16} + x^{15} + x^{11} + x^8 + x^7 + 1$$

The 4th remainder is:

$$x^{17} + x^{15} + x^{13} + x^{11} + x^{10} + x^9 + x^6 + 1$$

The 5th remainder is:

$$x^{13} + x^{12} + x^{11} + x^{10} + x^8 + x^7 + x^5 + 1$$

The 6th remainder is:

$$x^{12} + x^{10} + x^7 + x^4 + x^3 + x^2 + x + 1$$

The 7th remainder is 0

Hence,

$$\begin{aligned} x^{23}+1 &= (x+1)(x^{11}+x^{10}+x^6+x^5+x^4+x^2+1) \\ &\quad \times (x^{11}+x^9+x^7+x^6+x^5+x+1) \end{aligned}$$

Case (ii)

Next we consider the factorization of $x^{11}-1$ over GF(3). The cyclotomic cosets modulo 11 are:

$$C_0=\{0\} \qquad C_1=\{1,3,9,5,4\} \qquad C_2=\{2,6,7,10,8\}$$

Therefore, $x^{11}-1$ factors as a product of $x-1$ and two irreducible factors of degree 5 each (Corollary of Theorem 7.3). The factors of $x^{11}-1$ are among the HCF of $x^{11}-1$ with

$$a+b(x+x^3+x^4+x^5+x^9)+c(x^2+x^6+x^7+x^8+x^{10})$$

where a, b, $c \in F_3 = \mathrm{GF}(3)$. We apply Euclid's algorithm to the case $a=1=b, c=0$.

$$\begin{array}{r|l} x^9+x^5+x^4+x^3+x+1 & x^{11} \qquad\qquad\qquad\qquad -1 \\ & x^{11}+x^7+x^6+x^5+x^3+x^2 \\ \hline & -x^7-x^6-x^5-x^3-x^2-1 \end{array}$$

$$\begin{array}{r|l} -x^7-x^6-x^5-x^3-x^2-1 & x^9 \qquad\qquad +x^5+x^4+x^3 \qquad +x+1 \\ & x^9+x^8+x^7 \quad +x^5+x^4 \qquad +x^2 \\ \hline & -x^8-x^7 \qquad\qquad +x^3-x^2+x+1 \\ & -x^8-x^7-x^6 \quad -x^4-x^3 \qquad -x \\ \hline & x^6 \quad +x^4-x^3-x^2-x+1 \\ \hline \end{array}$$

$$\begin{array}{r|l} x^6+x^4-x^3-x^2-x+1 & +x^7+x^6+x^5 \qquad +x^3+x^2 \qquad +1 \\ & x^7 \quad +x^5-x^4-x^3-x^2+x \\ \hline & x^6 \quad +x^4-x^3-x^2-x+1 \\ \hline \end{array}$$

Thus one of the factors of $x^{11}-1$ is $x^6+x^4-x^3-x^2-x+1$. We find the other factor by actual division.

$$
\begin{array}{rl}
 & x^5 \qquad -x^3+x^2-x-1 \\
x^6+x^4-x^3-x^2-x+1 & \overline{)\,x^{11} \qquad\qquad\qquad\qquad\qquad\qquad -1} \\
 & x^{11}+x^9-x^8-x^7-x^6-x^5 \\
 \hline
 & -x^9+x^8+x^7+x^6-x^5 \qquad\qquad -1 \\
 & -x^9 \qquad -x^7+x^6+x^5+x^4-x^3 \\
 \hline
 & x^8-x^7 \qquad +x^5-x^4+x^3 \qquad -1 \\
 & x^8 \qquad +x^6-x^5-x^4-x^3+x^2 \\
 \hline
 & -x^7-x^6-x^5 \qquad -x^3-x^2 \qquad -1 \\
 & -x^7 \qquad -x^5+x^4+x^3+x^2-x \\
 \hline
 & -x^6 \qquad -x^4+x^3+x^2+x-1 \\
 & -x^6 \qquad -x^4+x^3+x^2+x-1 \\
 \hline
 & 0 \\
 \hline
\end{array}
$$

Hence

$$
\begin{aligned}
x^{11}-1 &= (x^6+x^4-x^3-x^2-x+1)(x^5-x^3+x^2-x-1) \\
&= (x^6-x^5+x^5-x^4-x^4+x^3+x^3-x^2-x+1) \\
&\quad \times (x^5-x^3+x^2-x-1) \\
&= (x-1)(x^5+x^4-x^3+x^2-1)(x^5-x^3+x^2-x-1)
\end{aligned}
$$

As the factorization of $x^{11}-1$ has to have two irreducible factors of degree 5 each, both the above polynomials of degree 5 are indeed irreducible.

Case (iii)

We next consider the factorization of $x^{13}-1$ over GF(3). Here the cyclotomic classes modulo 13 relative to 3 are:

$$
C_0=\{0\} \qquad C_1=\{1,3,9\} \qquad C_2=\{2,6,5\} \qquad C_4=\{4,12,10\}
$$
$$
C_7=\{7,8,11\}
$$

Therefore $x^{13}-1$ is a product of $x-1$ and four irreducible polynomials of degree 3 each (by Corollary to Theorem 7.3). The factors are given by the HCF of

$$
x^{12}+x^{11}+\cdots+x+1
$$

and

$$a + b(x + x^3 + x^9) + c(x^2 + x^5 + x^6) + d(x^4 + x^{10} + x^{12}) + e(x^7 + x^8 + x^{11})$$

where $a, b, c, d, e \in \mathrm{GF}(3)$. We find these HCFs by Euclid's algorithm:

$$
\begin{array}{r|l}
x^9+x^3+x-1 & \overline{x^{12}+x^{11}+x^{10}+x^9+x^8+x^7+x^6+x^5+x^4+x^3+x^2+x+1} \\
 & x^{12} \qquad\qquad\qquad\qquad +x^6 \quad\; +x^4-x^3 \\
\hline
 & x^{11}+x^{10}+x^9+x^8+x^7 \qquad +x^5 \qquad -x^3+x^2+x+1 \\
 & x^{11} \qquad\qquad\qquad\qquad\quad +x^5 \qquad +x^3-x^2 \\
\hline
 & x^{10}+x^9+x^8+x^7 \qquad\qquad\qquad +x^3-x^2+x+1 \\
 & x^{10} \qquad\qquad\qquad\qquad +x^4 \qquad +x^2-x \\
\hline
 & x^9+x^8+x^7 \qquad\qquad -x^4+x^3+x^2-x+1 \\
 & x^9 \qquad\qquad\qquad\qquad\quad +x^3 \qquad +x-1 \\
\hline
 & x^8+x^7 \qquad\qquad -x^4 \qquad +x^2+x-1
\end{array}
$$

$$
\begin{array}{r|l}
x^8+x^7-x^4+x^2+x-1 & \overline{x^9 \qquad\qquad\qquad\qquad +x^3 \qquad +x-1} \\
 & x^9+x^8 \qquad -x^5 \qquad +x^3+x^2-x \\
\hline
 & -x^8 \qquad +x^5 \qquad\qquad -x^2-x-1 \\
 & -x^8-x^7 \qquad +x^4 \qquad -x^2-x+1 \\
\hline
 & x^7+x^5-x^4 \qquad\qquad\qquad +1
\end{array}
$$

$$
\begin{array}{r|l}
x^7+x^5-x^4+1 & \overline{x^8+x^7 \qquad\qquad -x^4+x^2+x-1} \\
 & x^8 \qquad +x^6-x^5 \qquad\qquad +x \\
\hline
 & x^7-x^6+x^5-x^4+x^2 \qquad -1 \\
 & x^7 \qquad +x^5-x^4 \qquad\qquad +1 \\
\hline
 & -x^6 \qquad\qquad +x^2 \qquad +1
\end{array}
$$

$$
\begin{array}{r|l}
-x^6+x^2+1 & \overline{x^7+x^5-x^4 \qquad\qquad +1} \\
 & x^7 \qquad\qquad -x^3-x \\
\hline
 & x^5-x^4+x^3+x+1
\end{array}
$$

$$
\begin{array}{rl}
x^5 - x^4 + x^3 + x + 1\,\big) & \overline{-x^6 \qquad\qquad\qquad + x^2 \quad + 1} \\
 & -x^6 + x^5 - x^4 \qquad - x^2 - x \\ \cline{2-2}
 & \qquad -x^5 + x^4 \qquad - x^2 + x + 1 \\
 & \qquad -x^5 + x^4 - x^3 \qquad - x - 1 \\ \cline{2-2}
 & \qquad\qquad\qquad x^3 - x^2 - x - 1 \\ \cline{2-2}
\end{array}
$$

$$
\begin{array}{rl}
x^3 - x^2 - x - 1\,\big) & \overline{x^5 - x^4 + x^3 \qquad + x + 1} \\
 & x^5 - x^4 - x^3 - x^2 \\ \cline{2-2}
 & \qquad -x^3 + x^2 + x + 1 \\
 & \qquad -x^3 + x^2 + x + 1 \\ \cline{2-2}
 & \qquad\qquad\qquad\qquad 0
\end{array}
$$

$$
\begin{array}{rl}
x^{12} + x^{10} + x^4 - 1\,\big) & \overline{x^{12} + x^{11} + x^{10} + x^9 + x^8 + x^7 + x^6 + x^5 + x^4 + x^3 + x^2 + x + 1} \\
 & x^{12} \qquad + x^{10} \qquad\qquad\qquad\qquad\qquad + x^4 \qquad\qquad\qquad - 1 \\ \cline{2-2}
 & \qquad x^{11} \qquad + x^9 + x^8 + x^7 + x^6 + x^5 \qquad + x^3 + x^2 + x - 1
\end{array}
$$

$$
\begin{array}{rl}
x^{11} + x^9 + x^8 + x^7 + x^6 + x^5 + x^3 + x^2 + x - 1\,\big) & \overline{x^{12} + x^{10} \qquad\qquad\qquad\qquad + x^4 \qquad\qquad - 1} \\
 & x^{12} + x^{10} + x^9 + x^8 + x^7 + x^6 + x^4 + x^3 + x^2 - x \\ \cline{2-2}
 & \qquad -x^9 - x^8 - x^7 - x^6 \qquad - x^3 - x^2 + x - 1
\end{array}
$$

$$
\begin{array}{rl}
-x^9 - x^8 - x^7 - x^6 - x^3 - x^2 + x - 1\,\big) & \overline{x^{11} \qquad + x^9 + x^8 + x^7 + x^6 + x^5 \qquad + x^3 + x^2 + x - 1} \\
 & x^{11} + x^{10} + x^9 + x^8 \qquad\qquad + x^5 + x^4 - x^3 + x^2 \\ \cline{2-2}
 & \quad -x^{10} \qquad + x^7 + x^6 \qquad - x^4 - x^3 \qquad + x - 1 \\
 & \quad -x^{10} - x^9 - x^8 - x^7 \qquad - x^4 - x^3 + x^2 - x \\ \cline{2-2}
 & \qquad\quad x^9 + x^8 - x^7 + x^6 \qquad\qquad - x^2 - x - 1 \\
 & \qquad\quad x^9 + x^8 + x^7 + x^6 \qquad\qquad + x^3 + x^2 - x + 1 \\ \cline{2-2}
 & \qquad\qquad\qquad x^7 \qquad\qquad - x^3 + x^2 \qquad + 1
\end{array}
$$

$$
\begin{array}{r l}
x^7 - x^3 + x^2 + 1 \big) & \overline{-x^9 - x^8 - x^7 - x^6 \qquad - x^3 - x^2 + x - 1} \\
& -x^9 \qquad + x^5 - x^4 \qquad - x^2 \\ \cline{2-2}
& -x^8 - x^7 - x^6 - x^5 + x^4 - x^3 \qquad + x - 1 \\
& -x^8 \qquad + x^4 - x^3 \qquad - x \\ \cline{2-2}
& -x^7 - x^6 - x^5 \qquad - x - 1 \\
& -x^7 \qquad + x^3 - x^2 \qquad - 1 \\ \cline{2-2}
& -x^6 - x^5 \qquad - x^3 + x^2 - x
\end{array}
$$

$$
\begin{array}{r l}
-x^6 - x^5 - x^3 + x^2 - x \big) & \overline{x^7 \qquad - x^3 + x^2 \qquad + 1} \\
& x^7 + x^6 \qquad + x^4 - x^3 + x^2 \\ \cline{2-2}
& -x^6 \qquad - x^4 \qquad + 1 \\
& -x^6 - x^5 \qquad - x^3 + x^2 - x \\ \cline{2-2}
& x^5 - x^4 + x^3 - x^2 + x + 1
\end{array}
$$

$$
\begin{array}{r l}
x^5 - x^4 + x^3 - x^2 + x + 1 \big) & \overline{-x^6 - x^5 \qquad - x^3 + x^2 - x} \\
& -x^6 + x^5 - x^4 + x^3 - x^2 - x \\ \cline{2-2}
& x^5 + x^4 + x^3 - x^2 \\
& x^5 - x^4 + x^3 - x^2 + x + 1 \\ \cline{2-2}
& -x^4 \qquad - x \qquad - 1
\end{array}
$$

$$
\begin{array}{r l}
-x^4 - x - 1 \big) & \overline{x^5 - x^4 + x^3 - x^2 + x + 1} \\
& x^5 \qquad + x^2 + x \\ \cline{2-2}
& -x^4 + x^3 + x^2 + 1 \\
& -x^4 \qquad - 1 - x \\ \cline{2-2}
& x^3 + x^2 + x - 1
\end{array}
$$

$$
\begin{array}{r l}
-x^3 + x^2 + x - 1 \big) & \overline{-x^4 - x - 1} \\
& -x^4 + x - x^3 - x^2 \\ \cline{2-2}
& x^3 + x^2 + x - 1 \\
& x^3 + x^2 + x - 1 \\ \cline{2-2}
& 0
\end{array}
$$

The two factors of $x^{13}-1$ we have obtained give the factor

$$\begin{aligned}(x^3+x^2+x-1)(x^3-x^2-x-1) &= (x^3-1)^2-(x^2+x)^2\\ &= x^6+x^3+1-x^4-x^2+x^3\\ &= x^6-x^4-x^3-x^2+1\end{aligned}$$

We divide $x^{12}+x^{11}+\cdots+x+1$ by this factor:

$$\begin{array}{r}
x^6+x^5-x^4-x^2+x+1\\
\hline
x^6-x^4-x^3-x^2+1\,\big)\,x^{12}+x^{11}+x^{10}+x^9+x^8+x^7+x^6+x^5+x^4+x^3+x^2+x+1\\
x^{12}\qquad -x^{10}-x^9-x^8\qquad +x^6\qquad\qquad\qquad\qquad\qquad\qquad\\
\hline
x^{11}-x^{10}-x^9-x^8+x^7\qquad +x^5+x^4+x^3+x^2+x+1\\
x^{11}\qquad -x^9-x^8-x^7\qquad +x^5\qquad\qquad\qquad\qquad\qquad\\
\hline
-x^{10}\qquad\qquad -x^7\qquad\qquad +x^4+x^3+x^2+x+1\\
-x^{10}\qquad +x^8+x^7+x^6\qquad -x^4\qquad\qquad\qquad\qquad\\
\hline
-x^8+x^7-x^6\qquad -x^4+x^3+x^2+x+1\\
\hline
-x^8+x^7-x^6\qquad\qquad +x^3+x^2+x+1-x^4\\
-x^8\qquad +x^6+x^5+x^4\qquad -x^2\qquad\qquad\qquad\\
\hline
x^7+x^6-x^5+x^4+x^3-x^2+x+1\\
x^7\qquad -x^5-x^4-x^3\qquad +x\qquad\\
\hline
x^6\qquad -x^4-x^3-x^2\qquad +1\\
x^6\qquad -x^4-x^3-x^2\qquad +1\\
\hline
0
\end{array}$$

Next consider the case $c=1$, $a=-1$:

$$\begin{array}{r}
x^6+x^5+x^2-1\,\big)\,x^6+x^5-x^4-x^2+x+1\\
x^6+x^5\qquad +x^2\qquad -1\\
\hline
-x^4+x^2+x-1
\end{array}$$

$$
\begin{array}{r|l}
-x^4+x^2+x-1 & x^6+x^5+x^2-1 \\
 & x^6-x^4-x^3+x^2 \\ \hline
 & x^5+x^4+x^3-1 \\
 & x^5-x^3-x^2+x \\ \hline
 & x^4-x^3+x^2-x-1 \\
 & x^4-x^2-x+1 \\ \hline
 & -x^3-x^2+1
\end{array}
$$

$$
\begin{array}{r|l}
-x^3-x^2+1 & -x^4+x^2+x-1 \\
 & -x^4-x^3+x \\ \hline
 & x^3+x^2-1 \\
 & x^3+x^2-1 \\ \hline
 & 0 \\ \hline
\end{array}
$$

Thus the 3rd irreducible factor of $x^{13}-1$ is x^3+x^2-1. To find the last irreducible factor of $x^{13}-1$, we divide $x^6+x^5-x^4-x^2+x+1$ by x^3+x^2-1:

$$
\begin{array}{r|l}
 & x^3-x-1 \\ \hline
x^3+x^2-1 & x^6+x^5-x^4-x^2+x+1 \\
 & x^6+x^5-x^3 \\ \hline
 & -x^4+x^3-x^2+x+1 \\
 & -x^4-x^3+x \\ \hline
 & -x^3-x^2+1 \\
 & -x^3-x^2+1 \\ \hline
 & 0 \\ \hline
\end{array}
$$

Hence

$$x^{13}-1=(x-1)(x^3-x^2-x-1)(x^3+x^2+x-1)(x^3+x^2-1)(x^3-x-1)$$

(Also refer to Case (ii) in Examples 7.2 for an alternative method of factorization of this polynomial.)

We end this section and also the chapter with the following examples which we shall return to in the chapter on quadratic residue codes.

Case (iv)
We consider the polynomial $x^{11}-1$ over the field GF(5) of 5 elements. The cyclotomic cosets relative to 5 modulo 11 are:

$$C_0=\{0\} \qquad C_1=\{1,5,3,4,9\} \qquad C_2=\{2,10,6,8,7\}$$

Therefore $x^{11}-1$ has two irreducible factors of degree 5 each. To find these, we find the HCF of

$$x^9+x^5+x^4+x^3+x+1 \quad \text{and} \quad \sum_{0\le i\le 10} x^i$$

$$x^9+x^5+x^4+x^3+x-1\,\big)\,x^{10}+x^9+x^8+x^7+x^6+x^5+x^4+x^3+x^2+x+1$$
$$x^{10}+x^6+x^5+x^4+x^2-x$$
$$\overline{x^9+x^8+x^7+x^3+2x+1}$$
$$x^9+x^5+x^4+x^3+x-1$$
$$\overline{x^8+x^7-x^5-x^4+x+2}$$

$$x^8+x^7-x^5-x^4+x+2\,\big)\,x^9+x^5+x^4+x^3+x-1$$
$$x^9+x^8-x^6-x^5+x^2+2x$$
$$\overline{-x^8+x^6+2x^5+x^4+x^3-x^2-x-1}$$
$$-x^8-x^7+x^5+x^4-x-2$$
$$\overline{x^7+x^6+x^5+x^3-x^2+1}$$

$$x^7+x^6+x^5+x^3-x^2+1\,\big)\,x^8+x^7-x^5-x^4+x+2$$
$$x^8+x^7+x^6+x^4-x^3+x$$
$$\overline{-x^6-x^5-2x^4+x^3+2}$$

$$-x^6-x^5-2x^4+x^3+2\,\big)\,x^7+x^6+x^5+x^3-x^2+1$$
$$+x^7+x^6+2x^5-x^4-2x$$
$$\overline{-x^5+x^4+x^3-x^2+2x+1}$$

$$\begin{array}{r|l} -x^5+x^4+x^3-x^2+2x+1 & -x^6-x^5-2x^4+x^3+2 \\ & -x^6+x^5+x^4-x^3+2x^2+x \\ \hline & -2x^5-3x^4+2x^3-2x^2-x+2 \\ & -2x^5+2x^4+2x^3-2x^2+4x+2 \\ \hline & 0 \end{array}$$

The HCF $f(x)=x^5-x^4-x^3+x^2-2x-1$ is one of the irreducible factors of degree 5 of $x^{11}-1$. The process of division gives the other factor as

$$g(x)=x^5+2x^4-x^3+x^2+x-1$$

Hence

$$x^{11}-1=(x-1)f(x)g(x)$$

Case (v)

We now consider the polynomial $x^{37}-1$ over the field GF(3) of 3 elements.

The cyclotomic cosets relative to 3 modulo 37 are:

$$C_0=\{0\}$$

$$C_1=\{1,3,9,27,7,21,26,4,12,36,34,28,10,30,16,11,33,25\}$$

$$C_2=\{2, 6, 18, 17, 14, 5, 15, 8, 24, 35, 31, 19, 20, 23, 32, 29, 13, 22\}$$

Thus $x^{37}-1$ factors into three irreducible factors one of which is $x-1$ and the other two are of degree 18 each. These factors are obtained as common factors of $x^{37}-1$ with $g(x)-s$ for some $s\in\mathrm{GF}(3)$ where

$$g(x)=\sum_{k\in C_1} x^k \quad \text{or} \quad g(x)=\sum_{k\in C_2} x^k$$

Using Euclid's division algorithm, we obtain these common factors and find that

$$x^{37}-1=(x-1)f(x)g(x)$$

where

$$\begin{aligned} f(x)=x^{18}+x^{17}-x^{16}-x^{15}+x^{14}-x^{13}-x^{12}-x^{10}-x^9-x^8-x^6 \\ -x^5+x^4-x^3-x^2+x+1 \end{aligned}$$

$$g(x)=x^{18}-x^{16}-x^{14}-x^{13}+x^{11}+x^7-x^5-x^4-x^2+1$$

Case (vi)

Again consider the polynomial $x^{61}-1$ over GF(3). The cyclotomic cosets relative to 3 modulo 61 are:

$$C_0=\{0\}$$
$$C_1=\{1,3,9,27,20,60,58,52,34,41\}$$
$$C_2=\{2,6,18,54,40,59,55,43,7,21\}$$
$$C_4=\{4,12,36,47,19,57,49,25,14,42\}$$
$$C_5=\{5,15,45,13,39,56,46,16,48,22\}$$
$$C_8=\{8,24,11,33,38,53,37,50,28,23\}$$
$$C_{10}=\{10,30,29,26,17,51,31,32,35,44\}$$

It follows that $x^{61}-1$ factors as a product of 7 irreducible polynomials over GF(3) one of which is $x-1$ and every other is of degree 10 each.

We use the algorithm of Berlekamp to find the factorization of $x^{61}-1$. Calculating the HCF of

$$1+\sum_{i\in C_8} x^i \quad \text{and} \quad x^{61}-1$$

gives a factor

$$f(x)=x^{20}-x^{19}-x^{18}+x^{17}-x^{16}+x^{14}-x^{13}-x^{12}+x^{11}$$
$$+x^9-x^8-x^7+x^6-x^4+x^3-x^2-x+1$$

of $x^{61}-1$. Then finding the HCF of

$$\frac{(x^{61}-1)}{f(x)} \quad \text{and} \quad 1+\sum_{i\in C_{10}} x^i$$

gives another factor of $x^{61}-1$ as:

$$g(x)=x^{20}+x^{18}+x^{16}+x^{15}+x^{14}-x^{11}-x^{10}-x^9$$
$$+x^6+x^5+x^4+x^2+1$$

Then the third factor of degree 20 is obtained by actual process of division as

$$h(x)=x^{20}-x^{19}+x^{17}-x^{15}-x^{14}-x^{13}+x^{12}-x^{11}-x^{10}-x^9+x^8$$
$$-x^7-x^6-x^5+x^3-x+1$$

We next need to factorize $f(x)$, $g(x)$ and $h(x)$.

To factorize $f(x)$, we have to find a square matrix of order 20 in which the ith row is represented by $x^{3(i-1)}$ reduced modulo $f(x)$. Let this matrix be called

$\mathbf{Q}_1$. Then

$$\mathbf{Q}_1-\mathbf{I}=\begin{pmatrix}
0&0&0&0&0&0&0&0&0&0&0&0&0&0&0&0&0&0&0&0\\
0&-1&0&1&0&0&0&0&0&0&0&0&0&0&0&0&0&0&0&0\\
0&0&-1&0&0&0&1&0&0&0&0&0&0&0&0&0&0&0&0&0\\
0&0&0&-1&0&0&0&0&0&1&0&0&0&0&0&0&0&0&0&0\\
0&0&0&0&-1&0&0&0&0&0&0&0&1&0&0&0&0&0&0&0\\
0&0&0&0&0&-1&0&0&0&0&0&0&0&0&0&1&0&0&0&0\\
0&0&0&0&0&0&-1&0&0&0&0&0&0&0&0&0&0&0&1&0\\
-1&0&-1&0&0&1&-1&-1&-1&0&-1&-1&0&-1&0&-1&1&0&0&-1\\
-1&-1&1&-1&1&-1&1&-1&1&-1&0&-1&-1&0&-1&0&1&1&0&1\\
1&0&-1&1&1&1&-1&1&1&-1&-1&1&-1&-1&1&-1&0&-1&-1&0\\
-1&-1&0&-1&-1&1&0&0&1&0&1&0&-1&-1&-1&0&0&-1&1&0\\
0&-1&1&0&1&1&-1&1&-1&1&-1&0&-1&0&1&1&-1&0&-1&1\\
-1&1&0&0&1&0&0&1&1&-1&1&0&1&1&1&0&0&0&0&1\\
-1&-1&1&-1&-1&0&1&0&1&0&-1&1&1&-1&0&-1&1&-1&1&1\\
1&0&0&0&0&-1&1&-1&0&-1&-1&1&0&-1&1&0&1&0&0&1\\
1&1&-1&1&-1&0&-1&1&1&1&-1&0&0&1&1&-1&0&1&-1&0\\
0&1&-1&0&-1&1&1&0&-1&1&-1&1&-1&1&-1&1&0&0&1&-1\\
1&-1&0&0&-1&0&0&1&1&1&-1&1&0&-1&1&1&1&-1&-1&1\\
-1&1&0&1&1&-1&0&0&0&0&-1&-1&-1&-1&1&1&-1&0&-1&1\\
1&1&-1&-1&0&0&0&-1&1&0&0&0&1&1&-1&-1&0&0&0&0
\end{pmatrix}$$

If $(g_1 \quad \cdots \quad g_{20})$ is a row vector over GF(3) which is in the null space of $\mathbf{Q}_1-\mathbf{I}$, we obtain 20 equations in the g_i which finally lead to the relations

$$\begin{aligned}
&g_3=g_4=g_5=g_6=g_7=g_9=g_{10}=g_{13}=g_{14}=g_{17}=0\\
&g_2=-g_8=-g_{11}=g_{12}=g_{15}=g_{16}=-g_{18}=g_{19}=g_{20}
\end{aligned}\qquad(7.6)$$

Calculating the HCF of $f(x)$ and

$$x^{19}+x^{18}-x^{17}+x^{15}+x^{14}+x^{11}-x^{10}-x^7+x$$

gives one irreducible factor of $f(x)$ as

$$f_1(x)=x^{10}+x^9-x^8-x^7-x^6-x^4-x^3-x^2+x+1$$

and by the actual process of division the other factor is

$$f_2(x)=x^{10}+x^9-x^8+x^7-x^6-x^5-x^4+x^3-x^2+x+1$$

If $\mathbf{Q}_2$ denotes the $\mathbf{Q}$-matrix for the factorization of $h(x)$, then

$$\mathbf{Q}_2-\mathbf{I}=\begin{pmatrix}
0&0&0&0&0&0&0&0&0&0&0&0&0&0&0&0&0&0&0&0\\
0&-1&0&1&0&0&0&0&0&0&0&0&0&0&0&0&0&0&0&0\\
0&0&-1&0&0&0&1&0&0&0&0&0&0&0&0&0&0&0&0&0\\
0&0&0&-1&0&0&0&0&0&1&0&0&0&0&0&0&0&0&0&0\\
0&0&0&0&-1&0&0&0&0&0&0&0&1&0&0&0&0&0&0&0\\
0&0&0&0&0&-1&0&0&0&0&0&0&0&0&0&1&0&0&0&0\\
0&0&0&0&0&0&-1&0&0&0&0&0&0&0&0&0&0&0&1&0\\
-1&0&1&-1&-1&1&-1&1&0&0&-1&-1&0&0&-1&-1&1&-1&-1&1\\
1&-1&-1&1&0&-1&1&-1&-1&-1&0&0&-1&-1&0&0&1&1&-1&-1\\
1&1&-1&1&1&-1&1&-1&0&0&0&-1&0&0&1&1&-1&0&-1&1\\
0&0&-1&-1&1&1&1&-1&0&-1&0&1&-1&1&1&1&1&-1&1&1\\
-1&-1&1&0&1&-1&-1&-1&0&1&1&0&1&1&0&0&1&1&-1&1\\
-1&1&-1&-1&-1&1&1&0&-1&1&-1&-1&0&0&1&0&-1&0&0&1\\
-1&0&0&-1&0&-1&1&-1&-1&-1&1&-1&-1&-1&0&1&-1&1&-1&-1\\
1&1&-1&-1&-1&0&-1&-1&0&1&0&1&0&1&-1&1&0&0&0&-1\\
1&0&0&1&-1&-1&0&-1&-1&1&1&0&0&-1&0&-1&-1&0&-1&0\\
1&0&-1&-1&1&-1&-1&0&-1&0&0&0&1&1&1&1&0&1&1&1\\
0&1&1&-1&1&1&1&1&-1&1&1&-1&0&-1&1&1&1&1&-1&0\\
-1&-1&-1&-1&-1&-1&-1&1&-1&0&1&-1&-1&0&-1&0&1&0&1&-1\\
-1&0&-1&0&1&1&1&0&1&1&1&0&-1&0&-1&0&0&0&-1&-1
\end{pmatrix}$$

Again, if $(g_1 \quad \cdots \quad g_{20})$ is a row vector over GF(3) which is in the null space of $\mathbf{Q}_2 - \mathbf{I}$, we obtain certain equations in the g_i which finally lead to the relations

$$\begin{aligned} g_2 &= -g_3 = -g_5 = -g_6 = -g_7 = g_9 = -g_{11} = -g_{12} \\ &= g_{13} = g_{14} = g_{17} = g_{20} \end{aligned}$$

$$g_4 = g_8 = g_{10} = g_{15} = g_{16} = g_{18} = g_{19} = 0$$

Calculating the HCF of

$$x^{19} + x^{16} + x^{13} + x^{12} - x^{11} - x^{10} + x^8 - x^6 - x^5 - x^4 - x^2 + x + 1$$

and $h(x)$, gives one irreducible factor of $h(x)$ as

$$h_1(x) = x^{10} - x^9 + x^8 - x^7 + x^5 - x^3 + x^2 - x + 1$$

Then, by the process of actual division we obtain the other irreducible factor of $h(x)$ as

$$h_2(x) = x^{10} - x^8 + x^7 - x^6 + x^5 - x^4 + x^3 - x^2 + 1$$

Let

$$F = \mathrm{GF}(3)[x]/\langle h_2(x)\rangle$$

where $\langle h_2(x)\rangle$ denotes the ideal of GF(3)[x] generated by $h_2(x)$. Let

$$\alpha = x + \langle h_2(x)\rangle$$

Then α is a 61st root of unity in F. The elements α, α^4 being in different cyclotomic cosets, these have distinct minimal polynomials. Also α^4 does not satisfy $f_1(x)$, $f_2(x)$ or $h_1(x)$. Therefore, the minimal polynomial of α^4 has to be an irreducible factor of $g(x)$.

Let s_n denote the symmetric polynomial of $\{c^i | i \in C_4\}$ taken n elements at a time. We find that $s_1 = s_9$, $s_2 = s_8$, $s_3 = s_7$, $s_4 = s_6$. A process of direct calculations shows that

$$s_1 = s_4 = 0 \qquad s_2 = s_3 = 1$$

and then

$$\alpha^{40} + \alpha^{32} - \alpha^{28} - \alpha^{20} + \alpha^{12} + \alpha^8 + 1 = 0$$

It thus follows that the minimal polynomial of α^4 is

$$g_1(x) = x^{10} + x^8 - x^7 - x^5 - x^3 + x^2 + 1$$

Dividing $g(x)$ by $g_1(x)$ gives the other factor of $g(x)$ as

$$g_2(x) = x^{10} + x^7 + x^6 + x^5 + x^4 + x^3 + 1$$

We have thus obtained the factorization

$$x^{61} - 1 = (x-1) f_1(x) f_2(x) g_1(x) g_2(x) h_1(x) h_2(x)$$

of $x^{61} - 1$ as a product of irreducible factors over GF(3).

8

Quadratic residue codes

8.1 INTRODUCTION

Let p be an odd prime. Let Q denote the set of all quadratic residues mod p and N the set of all quadratic non-residues mod p. Let s be another prime which is a quadratic residue mod p. Then $s \in Q$ and it follows (from Proposition 5.6) that Q is closed with respect to multiplication by s. Therefore, Q is partitioned as a disjoint union of cyclotomic cosets modulo p under multiplication by s. Similarly, N is partitioned as a union of cyclotomic cosets modulo p under multiplication by s. Let α be a primitive pth root of unity in some extension of the field GF(s). By Euler's theorem, there exists a positive integer m such that $s^m \equiv 1 \pmod{p}$. Let ρ be a primitive element of an extension GF(s^m) of GF(s) of degree m. We may then take

$$\alpha = \rho^{(s^m - 1)/p}$$

It follows from Theorem 7.3 that

$$q(x) = \prod_{i \in Q} (x - \alpha^i) \qquad n(x) = \prod_{j \in N} (x - \alpha^j) \tag{8.1}$$

are polynomials with coefficients in GF(s).

Lemma 8.1

$$x^p - 1 = (x - 1)q(x)n(x)$$

Proof
As every α^i is a pth root of unity, every root of $q(x)$ and every root of $n(x)$ is a root of $x^p - 1$. Therefore, $q(x)$ and $n(x)$ divide $x^p - 1$. Also $Q \cap N = \phi$ and $q(x)n(x) | x^p - 1$. Clearly $x - 1 | x^p - 1$ and 1 is neither a root of $q(x)$ nor of $n(x)$. Therefore

$$(x - 1)q(x)n(x) | (x^p - 1)$$

But both polynomials are monic and of the same degree p. Therefore

$$x^p - 1 = (x-1)q(x)n(x)$$

Set

$$\mathscr{R} = \mathrm{GF(s)}[x]/\langle x^p - 1\rangle$$

where $\langle x^p - 1\rangle$ denotes the ideal of GF(s)[x] generated by $x^p - 1$.

Definition 8.1

Quadratic residue codes $\mathscr{F}$, $\mathscr{N}$, $\bar{\mathscr{F}}$ and $\bar{\mathscr{N}}$ are the cyclic codes of length p over GF(s) generated by the polynomials $q(x)$, $n(x)$, $(x-1)q(x)$ and $(x-1)n(x)$ respectively, i.e. these are the ideals in $\mathscr{R}$ generated by the respective polynomials.

It is clear that

$$\bar{\mathscr{F}} \subseteq \mathscr{F} \quad \text{and} \quad \bar{\mathscr{N}} \subseteq \mathscr{N}$$

As

$$\text{degree of } q(x) = \text{degree of } n(x) = \frac{p-1}{2}$$

both $\mathscr{F}$ and $\mathscr{N}$ are linear codes over GF(s) of dimension

$$p - \frac{p-1}{2} = \frac{p+1}{2}$$

each. Similarly, $\bar{\mathscr{F}}$ and $\bar{\mathscr{N}}$ are linear codes over GF(s) of dimension $(p-1)/2$ each.

Let $s = 2$ and $p = 7$. Then, 1, 2, 4 are quadratic residues mod 7 and 3, 5, 6 are quadratic non-residues mod 7. Consider the field

$$\mathbb{B}[x]/\langle x^3 + x + 1\rangle$$

of order 8. Then

$$\alpha = x + \langle x^3 + x + 1\rangle$$

is a primitive 7th root of unity having $x^3 + x + 1$ as its minimal polynomial over $\mathbb{B}$. Now

$$\begin{aligned} q(x) &= (x+\alpha)(x+\alpha^2)(x+\alpha^4) \\ &= x^3 + x^2(\alpha + \alpha^2 + \alpha^4) + x(\alpha^3 + \alpha^5 + \alpha^6) + 1 \\ &= x^3 + x[\alpha^3 + \alpha^2(\alpha+1) + (\alpha+1)^2] + 1 \\ &= x^3 + x(\alpha^2 + \alpha^2 + 1) + 1 \\ &= x^3 + x + 1 \end{aligned}$$

So, the quadratic residue code $\mathscr{F}$ of length 7 over $\mathbb{B}$ is generated by

$$x^3 + x + 1 + \langle x^7 + 1\rangle = q(x) + \langle x^7 + 1\rangle$$

Observe that

$$x^7 + 1 = (x + 1)(x^3 + x + 1)(x^3 + x^2 + 1)$$

and, therefore, the other quadratic residue code $\mathcal{N}$ is generated by

$$x^3 + x^2 + 1 = n(x)$$

Now

$$\begin{aligned}
q(x^3) + \langle x^7 + 1 \rangle &= (x^3)^3 + x^3 + 1 + \langle x^7 + 1 \rangle \\
&= x^3 + x^2 + 1 + \langle x^7 + 1 \rangle \\
&= n(x) + \langle x^7 + 1 \rangle \in \mathcal{N} \\
q(x^5) + \langle x^7 + 1 \rangle &= (x^5)^3 + x^5 + 1 + \langle x^7 + 1 \rangle \\
&= x^5 + x + 1 + \langle x^7 + 1 \rangle \\
&= (x^5 + x^4 + x^2) + (x^4 + x^3 + x) \\
&\quad + x^3 + x^2 + 1 + \langle x^7 + 1 \rangle \\
&= (x^3 + x^2 + 1)(x^2 + x + 1) + \langle x^7 + 1 \rangle \\
q(x^6) + \langle x^7 + 1) &= x^4 + x^6 + 1 + \langle x^7 + 1 \rangle \\
&= (x^3 + x^2 + 1)^2 + \langle x^7 + 1 \rangle
\end{aligned}$$

Thus

$$q(x^n) + \langle x^7 + 1 \rangle \in \mathcal{N}$$

whenever n is any quadratic non-residue mod 7.

Observe that the map $x \to x^5$ maps

$$x^6 + x^2 + 1 + \langle x^7 + 1 \rangle = (x^3 + x + 1)^2 + \langle x^7 + 1 \rangle \in \mathcal{F}$$

onto

$$\begin{aligned}
(x^5)^6 + (x^5)^2 + 1 + \langle x^7 + 1 \rangle &= x^2 + x^3 + 1 + \langle x^7 + 1 \rangle \\
&= x^3 + x^2 + 1 + \langle x^7 + 1 \rangle
\end{aligned}$$

which generates $\mathcal{N}$ and the map $x \to x^6$ maps

$$x^5 + x^4 + 1 + \langle x^7 + 1 \rangle = (x^3 + x + 1)(x^2 + x + 1) + \langle x^7 + 1 \rangle \in \mathcal{F}$$

onto the generator

$$x^3 + x^2 + 1 + \langle x^7 + 1 \rangle$$

of $\mathcal{N}$. Thus for every non-residue n modulo 7, there is an element

$$a(x) + \langle x^7 + 1 \rangle \in \mathcal{F}$$

which maps onto the generator

$$x^3 + x^2 + 1 + \langle x^7 + 1 \rangle$$

of $\mathcal{N}$. Also the map $x \to x^n$ determines a permutation

$$\boldsymbol{\sigma}: \{0, 1, 2, \ldots, 6\} \to \{0, 1, 2, \ldots, 6\}$$

as g.c.d.$(n, 7) = 1$. For example, for $n = 3$, the permutation determined is

$$\sigma = \begin{pmatrix} 0 & 1 & 2 & 3 & 4 & 5 & 6 \\ 0 & 3 & 6 & 2 & 5 & 1 & 4 \end{pmatrix} = (1 \quad 3 \quad 2 \quad 6 \quad 4 \quad 5)$$

In general

$$\sigma = \begin{pmatrix} 0 & 1 & 2 & 3 & 4 & 5 & 6 \\ 0 & \bar{1} & \bar{2} & \bar{3} & \bar{4} & \bar{5} & \bar{6} \end{pmatrix}$$

where for $1 \leq i \leq 6$, $\bar{i}$ denotes the least non-negative remainder on dividing ni by 7. The degrees of the generators of $\mathscr{F}$ and $\mathscr{N}$ being 3 each, the number of elements in the two codes are the same. Hence, it follows that the quadratic residue codes $\mathscr{F}$ and $\mathscr{N}$ are equivalent. Similarly, the expurgated quadratic residue codes $\bar{\mathscr{F}}$ and $\bar{\mathscr{N}}$ generated by

$$(x+1)(x^3+x+1) \quad \text{and} \quad (x+1)(x^3+x^2+1)$$

respectively are also equivalent.

Theorem 8.1
The quadratic residue codes $\mathscr{F}$ and $\mathscr{N}$ of length p over GF(s) generated by $q(x)$ and $n(x)$ are equivalent. Also the expurgated quadratic residue codes $\bar{\mathscr{F}}$ and $\bar{\mathscr{N}}$ of length p over GF(s) are equivalent.

Proof
Let n be a fixed quadratic non-residue mod p. Then there exists a positive integer r such that

$$nr \equiv 1 \pmod{p}$$

As 1 is always a quadratic residue, nr is a quadratic residue. But, then n being a non-residue it follows that r is a quadratic non-residue mod p. For any $i \in Q$, it follows that ir is a non-residue mod p. Now

$$q(x^n) = \prod_{i \in Q} (x^n - \alpha^i)$$

where α is a primitive pth root of unity in some extension of GF(s). Also

$$nr \equiv 1 \pmod{p} \Rightarrow \alpha^i = \alpha^{nri} = (\alpha^{ri})^n$$

so that α^{ir} is a root of $q(x^n)$. This is so for every $i \in Q$ and so

$$n(x) \mid q(x^n)$$

Hence the map induced by $x \to x^n$ maps code words from $\mathscr{F}$ onto code words in $\mathscr{N}$. Again

$$1 = nr + pt$$

for some integer t and, therefore,

$$x + \langle x^p - 1 \rangle = x^{nr+pt} + \langle x^p - 1 \rangle$$

If $t < 0$, then $-t > 0$ and

$$\begin{aligned} x^{-pt} + \langle x^p - 1 \rangle &= (x^p)^{-t} + \langle x^p - 1 \rangle \\ &= 1 + \langle x^p - 1 \rangle \end{aligned}$$

and so

$$\begin{aligned} x + \langle x^p - 1 \rangle &= (x^{nr+pt} + \langle x^p - 1 \rangle)(x^{-pt} + \langle x^p - 1 \rangle) \\ &= x^{nr} + \langle x^p - 1 \rangle \\ &= (x^n)^r + \langle x^p - 1 \rangle \end{aligned}$$

It follows that the map

$$F[x]/\langle x^p - 1 \rangle \to F[x]/\langle x^p - 1 \rangle$$

induced by $x \to x^n$ is onto and hence one–one as well. Therefore, the restriction of this map to

$$:\mathscr{F} \to \mathscr{N}$$

is also one–one and the two spaces being of the same dimension, the map is onto as well. Furthermore, the map $x \to x^n$ determines the permutation $\boldsymbol{\sigma}$ of the set $\{0, 1, 2, \ldots, p-1\}$ given by

$$\boldsymbol{\sigma} = \begin{pmatrix} 0 & 1 & 2 & \cdots & p-1 \\ 0 & \bar{1} & \bar{2} & \cdots & \overline{p-1} \end{pmatrix}$$

where for $1 \le i \le p-1$, $\bar{i}$ denotes the least non-negative remainder where ni is divided by p. Thus $\mathscr{F}$ and $\mathscr{N}$ are equivalent.

Equivalence of $\bar{\mathscr{F}}$ and $\bar{\mathscr{N}}$ follows similarly.

8.2 SOME EXAMPLES OF QUADRATIC RESIDUE CODES

Consider the $[7, 4, 3]$ binary Hamming code $\mathscr{C}$. A generator matrix of this code is (see p. 115)

$$\mathbf{G} = \begin{pmatrix} 1 & 1 & 1 & 0 & 0 & 0 & 0 \\ 1 & 0 & 0 & 1 & 1 & 0 & 0 \\ 0 & 1 & 0 & 1 & 0 & 1 & 0 \\ 1 & 1 & 0 & 1 & 0 & 0 & 1 \end{pmatrix}$$

Consider the field

$$\mathrm{GF}(2^3) = \mathbb{B}[X]/\langle x^3 + x + 1 \rangle$$

Then,

$$\alpha = x + \langle x^3 + x + 1 \rangle$$

is a primitive 7th root of unity. Now $1, 2, 4$ are quadratic residues modulo 7 and so $q(x)$ is a cubic polynomial having α as a root. Also, the minimal polynomial of α is of degree 3. Thus $q(x)$ must equal the minimal polynomial $x^3 + x + 1$ of α. The quadratic residue code being the polynomial code of length 7 generated by $1 + x + x^3$ it has a generator matrix

$$\mathbf{G}_1 = \begin{pmatrix} 1 & 1 & 0 & 1 & 0 & 0 & 0 \\ 0 & 1 & 1 & 0 & 1 & 0 & 0 \\ 0 & 0 & 1 & 1 & 0 & 1 & 0 \\ 0 & 0 & 0 & 1 & 1 & 0 & 1 \end{pmatrix}$$

The code generated by $\mathbf{G}_1$ is the same as the code generated by

$$\mathbf{G}_2 = \begin{pmatrix} 1 & 0 & 1 & 1 & 1 & 0 & 0 \\ 0 & 1 & 1 & 0 & 1 & 0 & 0 \\ 0 & 0 & 1 & 1 & 0 & 1 & 0 \\ 0 & 0 & 0 & 1 & 1 & 0 & 1 \end{pmatrix}$$

or the same as that generated by (interchanging the first and fourth rows)

$$\mathbf{G}_3 = \begin{pmatrix} 0 & 0 & 0 & 1 & 1 & 0 & 1 \\ 0 & 1 & 1 & 0 & 1 & 0 & 0 \\ 0 & 0 & 1 & 1 & 0 & 1 & 0 \\ 1 & 0 & 1 & 1 & 1 & 0 & 0 \end{pmatrix}$$

The permutation $\boldsymbol{\sigma} = (1 \quad 7 \quad 3 \quad 4 \quad 2 \quad 5)$ applied to the columns of $\mathbf{G}_3$ gives the generator matrix $\mathbf{G}$ of the Hamming code. Hence the $[7, 4, 3]$ binary Hamming code is equivalent to the binary quadratic residue code of length 7 generated by $x^3 + x + 1$.

Let $p = 23$. As $5^2 \equiv 2 (\mathrm{mod}\, 23)$, 2 is a quadratic residue modulo 23. As seen in Case (i) of Examples 7.4,

$$\begin{aligned} x^{23} + 1 &= (x + 1)(x^{11} + x^{10} + x^6 + x^5 + x^4 + x^2 + 1) \\ &\quad \times (x^{11} + x^9 + x^7 + x^6 + x^5 + x + 1) \end{aligned}$$

and each of the two factors of degree 11 is irreducible. Therefore, either may be taken as $q(x)$ (Theorem 8.1). Let

$$q(x) = x^{11} + x^{10} + x^6 + x^5 + x^4 + x^2 + 1$$

If d denotes the minimum distance of the quadratic residue code generated by $q(x)$, then $d \leq 7$. We shall prove two theorems about the minimum distance of quadratic residue codes from which it will follow that $d = 7$. Thus we have a $[23, 12, 7]$ code.

Given a code word $c = c_1 c_2 \cdots c_{23}$, the number of binary words of length 23 which are at distance at most 3 from c is

$$\begin{aligned} \binom{23}{0} + \binom{23}{1} + \binom{23}{2} + \binom{23}{3} &= 1 + 23(1 + 11 + 11 \times 7) \\ &= 1 + 23 \times 89 = 2048 = 2^{11} \end{aligned}$$

i.e.

$$O(\mathscr{S}(c,3)) = 2^{11}$$

and so

$$O\left(\bigcup_{c\in\mathscr{F}} \mathscr{S}(c,3)\right) = 2^{12} \times 2^{11} = 2^{23} = O(V(23,2))$$

Hence the code under consideration is perfect.

Definition 8.2

The binary quadratic residue code of length 23 is called the **Golay code** and is denoted by $\mathscr{G}_{23}$.

Consider the case when $s = 3$ and $p = 11$. As $5^2 \equiv 3 \pmod{11}$, 3 is a quadratic residue modulo 11. We have proved in Case (ii) of Examples 7.4 that

$$x^{11} - 1 = (x-1)(x^5 + x^4 - x^3 + x^2 - 1)(x^5 - x^3 + x^2 - x - 1)$$

and both the factors of degree 5 are irreducible. Therefore, either of the two may be taken as $q(x)$ (Theorem 8.1). Let

$$q(x) = x^5 + x^4 - x^3 + x^2 - 1$$

Then the minimum distance d of the code is at most 5. Again, as an application of the two theorems to be proved, we shall find that $d = 5$.

Let c be a fixed code word of the code. Then the number of ternary words which are at distance at most 2 from c is

$$1 + \binom{11}{1} \times 2 + \binom{11}{2} \times 2^2 = 1 + 22 + 55 \times 4 = 243 = 3^5$$

i.e.

$$O(\mathscr{S}(c,2)) = 3^5$$

Therefore

$$O\left(\bigcup_{C\in\mathscr{F}} \mathscr{S}(c,2)\right) = 3^6 \times 3^5 = 3^{11} = O(V(11,3))$$

Hence the code under consideration is perfect.

Definition 8.3

The $[11, 6, 5]$ ternary quadratic residue code is the Golay code $\mathscr{G}_{11}$.

Let $p = 13$, $s = 3$. Observe that 3 is a quadratic residue modulo 13. We have seen in Case (iii) of Examples 7.4 that

$$x^{13} - 1 = (x-1)(x^3 - x^2 - x + 1)(x^3 + x^2 + x - 1)(x^3 + x^2 - 1)(x^3 - x - 1)$$

where all the four factors of degree 3 are irreducible over $GF(3) = F$. Consider the extension

$$F[x]/\langle x^3 - x - 1\rangle$$

of F and let

$$\alpha = x + \langle x^3 - x - 1\rangle$$

Then $x^3 - x - 1$ is the minimal polynomial of α. Now

$$\alpha^3 - \alpha - 1 = 0$$

or

$$\alpha^3 = \alpha + 1 \neq 0$$

and, so

$$\begin{aligned}\alpha^{12} &= \alpha^4 + \alpha^3 + \alpha + 1\\ &= \alpha(\alpha + 1) + 2(\alpha + 1)\\ &= \alpha^2 + 2\end{aligned}$$

then

$$\alpha^{13} = \alpha^3 - \alpha = 1$$

Thus, α is a primitive 13th root of unity in

$$F[x]/\langle x^3 - x - 1\rangle$$

Now

$$Q = \{1, 4, 9, 3, 12, 10\} \quad \text{and} \quad N = \{2, 5, 6, 7, 8, 11\}$$

As $\alpha, \alpha^3, \alpha^9$ have the same minimal polynomial, these are the roots of the polynomial $x^3 - x - 1$.

The minimal polynomial of α^4 is

$$\begin{aligned}&(x - \alpha^4)(x - \alpha^{10})(x - \alpha^{12})\\ &\quad = x^3 - x^2(\alpha^4 + \alpha^{10} + \alpha^{12}) + x(\alpha^{14} + \alpha^{16} + \alpha^{22}) - 1\\ &\quad = x^3 - x^2(\alpha^2 + \alpha + \alpha(\alpha + 1)^3 + (\alpha^2 + \alpha)^3) + x(\alpha + \alpha^3 + \alpha^9) - 1\\ &\quad = x^3 - x^2(\alpha^2 + \alpha + \alpha(\alpha^3 + 1) + \alpha^6 + \alpha^3) + x(2\alpha + 1 + (\alpha + 1)^3) - 1\\ &\quad = x^3 - x^2(\alpha^2 + \alpha + \alpha^2 + 2\alpha + (\alpha + 1)^2 + \alpha + 1) + x(2\alpha + 1 + \alpha^3 + 1) - 1\\ &\quad = x^3 - x^2(2\alpha^2 + \alpha + 1 + \alpha^2 + 2\alpha + 1) + x(2\alpha + 1 + \alpha + 1 + 1) - 1\\ &\quad = x^3 + x^2 - 1\end{aligned}$$

Hence

$$\begin{aligned}q(x) &= (x^3 - x - 1)(x^3 + x^2 - 1)\\ &= x^6 + x^5 - x^4 - x^2 + x + 1\end{aligned}$$

Then $\mathscr{F}$ is the code of length 13 generated by $q(x)$, its dimension is 7 and the minimum distance $d \le 6$. It will follow from Theorem 8.8 that $d = 5$ or 6. The code is, therefore, capable of correcting any two errors. Observe that

$$\mathrm{O}(\mathscr{S}(c,2)) = 1 + 13 \times 2 + 13 \times 6 \times 4 = 339$$

and, so

$$\mathrm{O}\left(\bigcup_{C \in \mathscr{F}} \mathscr{S}(c,2)\right) = 339 \times 3^7 \neq 3^{13} = \mathrm{O}(V(13,3))$$

Hence $\mathscr{F}$ is not perfect.

8.3 EXTENDED QUADRATIC RESIDUE CODES AND DISTANCE PROPERTIES

We are interested in finding some information about the minimum distance of quadratic and extended quadratic residue codes. For that we first need to obtain duals of quadratic residue codes.

Theorem 8.2

$$\mathscr{F}^{\perp} = \bar{\mathscr{F}}, \mathscr{N}^{\perp} = \bar{\mathscr{N}} \quad \text{if } p = 4k - 1$$

and

$$\mathscr{F}^{\perp} = \bar{\mathscr{N}}, \mathscr{N}^{\perp} = \bar{\mathscr{F}} \quad \text{if } p = 4k + 1$$

Moreover, $\mathscr{F}$ is always generated by $\bar{\mathscr{F}}$ and

$$\sum_{i=0}^{p-1} x^i$$

while $\mathscr{N}$ is always generated by $\bar{\mathscr{N}}$ and

$$\sum_{i=0}^{p-1} x^i$$

Proof

The check polynomial of $\mathscr{F}$ is

$$h(x) = (x - 1)n(x)$$

and, therefore, the dual code $\mathscr{F}^{\perp}$ is generated by

$$\rho(x) = x^{(p+1)/2}(x^{-1} - 1)n(x^{-1}) = (x - 1) \prod_{n \in N} (1 - x\alpha^n)$$

(Theorem 6.2) which is a constant multiple of

$$(x - 1) \prod_{n \in N} (x - \alpha^{-n})$$

Now if p is of the form $4k + 1$, then -1 is a quadratic residue mod p so that

$-n \in N \; \forall n \in N$. Hence,

$$(x-1)\prod_{n\in\mathcal{N}}(x-\alpha^{-n})=(x-1)n(x)$$

Therefore $\mathcal{F}^{\perp}$ is the code generated by $(x-1)n(x)$ and so is $\bar{\mathcal{N}}$.

On the other hand, if p is of the form $4k-1$, then -1 is a quadratic non-residue modulo p and $-n \in Q \; \forall n \in N$. Hence, in this case

$$(x-1)\prod_{n\in N}(x-\alpha^{-n})=(x-1)q(x)$$

and, therefore, $\mathcal{F}^{\perp}$ is $\bar{\mathcal{F}}$.

The proof for the dual of $\mathcal{N}$ follows on the same lines. As

$$\sum_{i=0}^{p-1} x^i = q(x)n(x)$$

the ideal in $F[x]/\langle x^p - 1\rangle$ generated by

$$\sum_{i=0}^{p-1} x^i \quad \text{and} \quad (x-1)q(x)$$

is contained in the ideal generated by $q(x)$. Also $x-1$ and $n(x)$ do not have a common root and so are relatively coprime. Therefore

$$1=(x-1)a(x)+n(x)b(x)$$

for some $a(x), b(x) \in F[x]$. But then

$$\begin{aligned} q(x) &= (x-1)q(x)a(x)+q(x)n(x)b(x) \\ &= (x-1)q(x)a(x)+\left(\sum_{i=0}^{p-1} x^i\right)b(x) \end{aligned}$$

From this, it follows that the ideal of $F[x]/\langle x^p - 1\rangle$ generated by $q(x)$ is contained in the ideal generated by

$$\sum_{i=0}^{p-1} x^i \quad \text{and} \quad (x-1)q(x)$$

As $\bar{\mathcal{F}}$ is generated by $(x-1)q(x)$, $\mathcal{F}$ is generated by $\bar{\mathcal{F}}$ and

$$\sum_{i=0}^{p-1} x^i$$

Corollary

(i) If $\bar{\mathbf{G}}$ is a generator matrix for $\bar{\mathcal{F}}$, then

$$\left(\begin{array}{cccc} & \bar{\mathbf{G}} & & \\ \hline 1 & 1 & \cdots & 1 \end{array}\right)$$

is a generator matrix for $\mathcal{F}$.

(ii) If $\bar{\mathbf{G}}_1$ is a generator matrix for $\bar{\mathcal{N}}$, then

$$\left(\frac{\bar{\mathbf{G}}_1}{1 \quad 1 \quad \cdots \quad 1}\right)$$

is a generator matrix for $\mathcal{N}$.

Remark 8.1

Let

$$(x-1)q(x) = a_0 + a_1 x + \cdots + a_m x^m$$

where $m = (p+1)/2$. The code $\bar{\mathcal{F}}$ being generated by $(x-1)q(x)$, it follows that

$$\bar{\mathbf{G}} = \begin{pmatrix} a_0 & a_1 & \cdots & a_m & 0 & \cdots & \cdots & 0 \\ 0 & a_0 & \cdots & a_{m-1} & a_m & 0 & \cdots & 0 \\ \hdashline 0 & 0 & \cdots & 0 & a_0 & \cdots & \cdots & a_m \end{pmatrix}$$

which is an $(m-1) \times p$ matrix is a generator matrix of $\bar{\mathcal{F}}$ and clearly every row of $\bar{\mathbf{G}}$ is orthogonal to the all ones vector. Similarly, $\bar{\mathcal{N}}$ being a polynomial code generated by $(x-1)n(x)$, every row of the corresponding generator matrix $\bar{\mathbf{G}}_1$ is orthogonal to the all ones vector.

Definition 8.4

Recall that the Legendre symbol $\chi(i)$ is defined by

$$\chi(i) = \begin{cases} 0 & \text{if } i \text{ a multiple of } p \\ 1 & \text{if } i \text{ is a quadratic residue mod } p \\ -1 & \text{if } i \text{ is a non-residue mod } p \end{cases}$$

Since the product of two residues or two non-residues is a residue while the product of a residue and a non-residue is a non-residue, we have

$$\chi(i)\chi(j) = \chi(ij)$$

We then define the **Gaussian sum** by

$$\theta = \sum_{i=1}^{p-1} \chi(i)\alpha^i \tag{8.2}$$

As s is a prime, $s \neq p$, and α belongs to an extension of the field GF(s), it follows that

$$\theta^s = \sum_{i=1}^{p-1} \chi(i)^s \alpha^{is}$$

Now

$$\chi(i)^s = \chi(i)$$

for s odd, while for $s = 2$, we may take every $\chi(i) = 1$, so

$$\begin{aligned}\theta^s &= \sum_{i=1}^{p-1} \chi(i)\alpha^{is} \\ &= \sum_{i=1}^{p-1} \chi(i)\alpha^{i} = \theta\end{aligned}$$

Therefore, $\theta \in GF(s)$ (Theorem 7.1 part (ii)).

Examples 8.1

Case (i)
Let $p = 5$. Then

$$\theta = \alpha - \alpha^2 - \alpha^3 + \alpha^4$$

and

$$\begin{aligned}\theta^2 &= \alpha \times \alpha^4 + \alpha^2 \times \alpha^3 + \alpha^3 \times \alpha^2 + \alpha^4 \times \alpha + (\alpha \times \alpha + \alpha^2 \times \alpha^2 + \alpha^3 \times \alpha^3 \\ &\quad + \alpha^4 \times \alpha^4) + \alpha(-\alpha^2 - \alpha^2) - \alpha^2(\alpha + \alpha^4) - \alpha^3(\alpha + \alpha^4) + \alpha^4(-\alpha^2 - \alpha^3) \\ &= 4 + (\alpha + \alpha^2 + \alpha^3 + \alpha^4) - [\alpha^3 + \alpha^4 + \alpha^3 + \alpha + \alpha^4 + \alpha^2 + \alpha + \alpha^2] \\ &= 4 - 1 - 2(\alpha + \alpha^2 + \alpha^3 + \alpha^4) \\ &= 4 - 1 + 2 = 5 = p\end{aligned}$$

Case (ii)
Let $p = 7$. Then

$$\theta = \alpha + \alpha^2 - \alpha^3 + \alpha^4 - \alpha^5 - \alpha^6$$

and

$$\begin{aligned}\theta^2 &= -\sum_{i+j=7} \alpha^{i+j} + \sum_{i=1}^{6} \alpha^{2i} + \alpha(\alpha^2 - \alpha^3 + \alpha^4 - \alpha^5) + \alpha^2(\alpha - \alpha^3 + \alpha^4 - \alpha^6) \\ &\quad - \alpha^3(\alpha + \alpha^2 - \alpha^5 - \alpha^6) + \alpha^4(\alpha + \alpha^2 - \alpha^5 - \alpha^6) - \alpha^5(\alpha - \alpha^3 + \alpha^4 - \alpha^6) \\ &\quad - \alpha^6(\alpha^2 - \alpha^3 + \alpha^4 - \alpha^5) \\ &= -6 + \sum_{i=1}^{6} \alpha^i + [\alpha^3 - \alpha^4 + \alpha^5 - \alpha^6 + \alpha^3 - \alpha^5 + \alpha^6 - \alpha] \\ &\quad + [-\alpha^4 - \alpha^5 + \alpha + \alpha^2 + \alpha^5 + \alpha^6 - \alpha^2 - \alpha^3] \\ &\quad + [-\alpha^6 + \alpha - \alpha^2 + \alpha^4 - \alpha + \alpha^2 - \alpha^3 + \alpha^4] \\ &= -6 - 1 + [2\alpha^3 - \alpha^4 - \alpha] + [\alpha - \alpha^3 - \alpha^4 + \alpha^6] + [-\alpha^3 + 2\alpha^4 - \alpha^6] \\ &= -7 = -p\end{aligned}$$

Case (iii)
Let $p = 13$. Then

$$\theta = \alpha - \alpha^2 + \alpha^3 + \alpha^4 - \alpha^5 - \alpha^6 - \alpha^7 - \alpha^8 + \alpha^9 + \alpha^{10} - \alpha^{11} + \alpha^{12}$$

and

$$\theta^2 = \sum_{i+j=13} \alpha^{i+j} + \sum_{i=1}^{12} \alpha^{2i} + \sum_{k=1}^{12} \left(\sum_{i=1, i\neq k, 2i\neq k}^{12} \chi(i(k-i)) \right) \alpha^k$$

$$= 12 - 1 + \sum_{k=1}^{12} \left(\sum_{i=1, i\neq k, 2i\neq k}^{12} (i(k-i)) \right) \alpha^k$$

The coefficient of α in the above summation is

$$= 2[\chi(2(12)) + \chi(3 \times 11) + \chi(4 \times 10) + \chi(5 \times 9) + \chi(6 \times 8)]$$

Among the pairs (2, 12), (3, 11), (4, 10), (5, 9), (6, 8), the pairs (2, 12), (3, 11), (5, 9) are such that one of the numbers is a quadratic residue mod 13, while the other is a non-residue. Also, the pairs (4, 10) and (6, 8) are pairs in which both the numbers are either residues or non-residues mod 13. Therefore, the coefficient of α is -2.

$$\text{coeff of } \alpha^2 = 2[\chi(3 \times 12) + \chi(4 \times 11) + \chi(5 \times 10) + \chi(6 \times 9) + \chi(7 \times 8)]$$

Pairs (4, 11), (5, 10), (6, 9) are of numbers one of which is a residue and the other a non-residue while the pairs (3, 12), (7, 8) have both their numbers either residues or non-residues. Therefore

$$\text{coeff of } \alpha^2 = -2$$

$$\text{coeff of } \alpha^3 = 2[\chi(1 \times 2) + \chi(4 \times 12) + \chi(5 \times 11) + \chi(6 \times 10) + \chi(7 \times 9)] = -2$$
$$\text{coeff of } \alpha^4 = 2[\chi(1 \times 3) + \chi(5 \times 12) + \chi(6 \times 11) + \chi(7 \times 10) + \chi(8 \times 9)] = -2$$
$$\text{coeff of } \alpha^5 = 2[\chi(1 \times 4) + \chi(2 \times 3) + \chi(6 \times 12) + \chi(7 \times 11) + \chi(8 \times 10)] = -2$$
$$\text{coeff of } \alpha^6 = 2[\chi(1 \times 5) + \chi(2 \times 4) + \chi(7 \times 12) + \chi(8 \times 11) + \chi(9 \times 10)] = -2$$
$$\text{coeff of } \alpha^7 = 2[\chi(1 \times 6) + \chi(2 \times 5) + \chi(3 \times 4) + \chi(8 \times 12) + \chi(9 \times 11)] = -2$$
$$\text{coeff of } \alpha^8 = 2[\chi(1 \times 7) + \chi(2 \times 6) + \chi(3 \times 5) + \chi(9 \times 12) + \chi(10 \times 11)] = -2$$
$$\text{coeff of } \alpha^9 = 2[\chi(1 \times 8) + \chi(2 \times 7) + \chi(3 \times 6) + \chi(4 \times 5) + \chi(10 \times 12)] = -2$$
$$\text{coeff of } \alpha^{10} = 2[\chi(1 \times 9) + \chi(2 \times 8) + \chi(3 \times 7) + \chi(4 \times 6) + \chi(11 \times 12)] = -2$$
$$\text{coeff of } \alpha^{11} = 2[\chi(1 \times 10) + \chi(2 \times 9) + \chi(3 \times 8) + \chi(4 \times 7) + \chi(5 \times 6)] = -2$$
$$\text{coeff of } \alpha^{12} = 2[\chi(1 \times 11) + \chi(2 \times 10) + \chi(3 \times 9) + \chi(4 \times 8) + \chi(5 \times 7)] = -2$$

therefore

$$\sum_{k=1}^{12} \left(\sum_{i=1, i\neq k, 2i\neq k}^{12} \chi(i(k-i)) \right) \alpha^k = -2 \sum_{k=1}^{12} \alpha^k = -2 \times (-1) = 2$$

Hence

$$\theta^2 = 11 + 2 = 13 = p$$

We now prove a general result about θ^2, but for that we need a result of Perron about quadratic residues which we state without proof.

Theorem 8.3 (Perron)

(i) Suppose $p = 4k - 1$. Let $r_1, \ldots, r_{2k}$ be the $2k$ quadratic residues mod p together with 0, and let a be a number relatively prime to p. Then among the $2k$ numbers $r_i + a$, there are k residues (possibly including 0) and k non-residues.

(ii) Suppose $p = 4k - 1$. Let $n_1, n_2, \ldots, n_{2k-1}$ be the $2k - 1$ non-residues, and let a be prime to p. Then among the $2k - 1$ numbers $n_i + a$, there are k residues (possibly including 0) and $k - 1$ non-residues.

(iii) Suppose $p = 4k + 1$. Among the $2k + 1$ numbers $r_i + a$ are, if a is itself a residue, $k + 1$ residues (including 0) and k non-residues; and, if a is a non-residue, k residues (not including 0) and $k + 1$ non-residues.

(iv) Suppose $p = 4k + 1$. Among the $2k$ numbers $n_i + a$ are, if a is itself a residue, k residues (not including 0) and k non-residues; and, if a is a non-residue, $k + 1$ residues (including 0) and $k - 1$ non-residues.

Theorem 8.4
If $p = 4l + 1$, then $\theta^2 = p$.

Proof

$$\theta^2 = \sum_{i=1}^{p-1} \sum_{j=1}^{p-1} \chi(i)\chi(j)\alpha^{i+j}$$

As $p = 4k + 1$, -1 is a residue mod p and, therefore, i and $p - i$ are either both residues or both non-residues. Therefore, $p - 1$ terms in the sum with $i + j = p$ have coefficient $\chi(i)\chi(j) = 1$ each. Therefore

$$\begin{aligned}
\theta^2 &= p - 1 + \sum_{i=1}^{p-1} \chi(i)^2\alpha^{2i} + \sum_{\substack{i+j \neq p \\ i \neq j}}\sum \chi(i)\chi(j)\alpha^{i+j} \\
&= p - 1 + \sum_{i=1}^{p-1} \alpha^{2i} + \sum_{k=1}^{p-1} \left(\sum_{i=1.i \neq k, 2i \neq k}^{p-1} \chi(i(k-i)) \right)\alpha^k \\
&= p - 1 + \sum_{i=1}^{p-1} \alpha^i + \sum_{k=1}^{p-1} \psi(k)\alpha^k \\
&= p - 2 + \sum_{k=1}^{p-1} \psi(k)\alpha^k
\end{aligned}$$

where

$$\psi(k) = \sum_{i=1, i \neq k, 2i \neq k}^{p-1} \chi(i(k-i))$$

Observe that, in the summation for $\psi(k)$, there are $p - 3$ terms. Let

$$M_k = \{t/t = i(k-i) \text{ for some } i,\ 1 \leq i \leq p-1,\ i \neq k,\ 2i \neq k\}$$

Then M_k has $(p-3)/2$ elements. If i, $i \leq i \leq p-1$ is one choice for which $t = i(k-i)$, then $j = k - i$ is another choice with $1 \leq j \leq p-1$ for which the given t arises as

$$j(k-j) = (k-i)i = i(k-i)$$

Therefore

$$\psi(k) = 2 \sum_{t \in M_k} \chi(t)$$

Now, if $t = i(k-i)$, then $i^2 - ki + t = 0$. Therefore,

$$k^2 - 4t = \left(\frac{i^2+t}{i}\right)^2 - 4t = \left(\frac{i^2-t}{i}\right)^2$$

which being a square is in Q (as it is non-zero as well). Thus

$$k^2 - 4t \in Q \quad \text{or} \quad -4t = r - k^2$$

for some $r \in Q$. As $0 \notin M_k$, we have $r \neq k^2$. Therefore

$$-4M_k = \{r - k^2 / r \in Q,\ r \neq k^2\}$$

Let $\bar{Q} = Q \cup \{0\}$. As $k^2 \in Q$ and p being of the form $4l+1$, -1 is also a residue mod p, $-k^2$ is a residue mod p. It then follows (from Perron's Theorem 8.3 part (iii)) that among $\{r - k^2 / r \in \bar{Q}\}$ there are $k+1$ residues (including 0) and k non-residues. But

$$-4M_k = \{r - k^2 / r \in \bar{Q}\} \backslash \{0, -k^2\}$$

and, so, in the set $-4M_k$ there are $k-1$ residues and k non-residues mod p. Again -4 is a residue mod p and hence M_k contains $k-1$ residues and k non-residues mod p. Therefore,

$$\psi(k) = 2(-1) = -2$$

and

$$\theta^2 = p - 2 - 2\sum_{k=1}^{p-1} \alpha^k = p - 2 + 2 = p$$

Using part (i) of Theorem 8.3 and the fact that -1 is a non-residue mod p when $p \equiv -1 \pmod 4$, we can prove the following theorem.

Theorem 8.5
If $p = 4k - 1$, then $\theta^2 = -p$.

8.3.1 Extended QR codes

We are now in a position to extend QR codes by adding an overall parity check. For a code $\mathscr{C}$, let $\hat{\mathscr{C}}$ denote the extended code of $\mathscr{C}$. We like to extend $\mathscr{F}$ and $\mathscr{N}$ in such a way that dual of $\hat{\mathscr{F}}$ is either $\hat{\mathscr{F}}$ or $\hat{\mathscr{N}}$. Similarly for $\hat{\mathscr{N}}$.

Case (i): p = 4k − 1

If $(a_0, a_1, \ldots, a_{p-1})$ is a code word in $\mathscr{F}$ (or $\mathscr{N}$), the extended code $\hat{\mathscr{F}}$ (or $\hat{\mathscr{N}}$) is formed by taking

$$a_p = -y \sum_{i=0}^{p-1} a_i$$

where $1 + y^2 p = 0$. Then

$$(yp)^2 = -p = \theta^2$$

so that $y = \pm\theta/p$. As already seen, $\theta \in \mathrm{GF}(s)$. Also s and p being distinct primes, p is invertible in $\mathrm{GF}(s)$ and so $\pm\theta/p \in \mathrm{GF}(s)$. Hence a choice of y in $\mathrm{GF}(s)$ with the above condition is possible.

Case (ii): p = 4k + 1

If $(a_0, a_1, \ldots, a_{p-1})$ is a code word in $\mathscr{F}$, the extended code $\hat{\mathscr{F}}$ is formed by taking

$$a_p = y \sum_{i=0}^{p-1} a_i$$

where $1 - y^2 p = 0$ and if $a = (a_0, a_1, \ldots, a_{p-1})$ is in $\mathscr{N}$, the extended code $\hat{\mathscr{N}}$ is formed by taking

$$a_p = -y \sum_{i=0}^{p-1} a_i$$

where y, as before, satisfies $1 - y^2 p = 0$. Observe that

$$y^2 p^2 = p = \theta^2$$

(Theorem 8.4) and, so, $yp = \pm\theta$. The number p is invertible in $\mathrm{GF}(s)$ and, therefore, y satisfying the given condition can be obtained in $\mathrm{GF}(s)$.

Theorem 8.6

If $p = 4k + 1$, the extended QR codes $\hat{\mathscr{F}}$ and $\hat{\mathscr{N}}$ defined above satisfy

$$(\hat{\mathscr{F}})^{\perp} = \hat{\mathscr{N}}$$

Proof

Let $\bar{\mathbf{G}}$ be a generator matrix for $\hat{\mathscr{F}}$. Then

$$\mathbf{G} = \left(\begin{array}{cccc} & \bar{\mathbf{G}} & & \\ \hline 1 & 1 & \cdots & 1 \end{array}\right)$$

is a generator matrix for $\mathscr{F}$. A generator matrix for $\hat{\mathscr{F}}$ is then given by

$$\hat{\mathbf{G}} = \left(\begin{array}{cccc|c} & \bar{\mathbf{G}} & & & \mathbf{0} \\ \hline 1 & 1 & \cdots & 1 & yp \end{array}\right)$$

Let $\bar{\mathbf{H}}$ be a generator matrix for $\bar{\mathscr{N}}$ so that

$$\mathbf{H} = \left(\begin{array}{cccc} & \bar{\mathbf{H}} & & \\ \hline 1 & 1 & \cdots & 1 \end{array}\right)$$

is a generator matrix for $\mathscr{N}$. Generator matrix for $\hat{\mathscr{N}}$ is then given by

$$\hat{\mathbf{H}} = \left(\begin{array}{cccc|c} & \bar{\mathbf{H}} & & & \mathbf{0} \\ \hline 1 & 1 & \cdots & 1 & -yp \end{array}\right)$$

By Theorem 8.2

$$\mathscr{F}^{\perp} = \mathscr{N}$$

Therefore, every row of $\bar{\mathbf{H}}$ is orthogonal to every row of $\mathbf{G}$ and, then, every row of $(\bar{\mathbf{H}} \quad \mathbf{0})$ is orthogonal to every row of $(\mathbf{G} \quad \mathbf{0})$ and, hence, every row of $(\bar{\mathbf{H}} \quad \mathbf{0})$ is orthogonal to every row of $\hat{\mathbf{G}}$. Now the last row of $\hat{\mathbf{H}}$ is orthogonal to the last row of $\hat{\mathbf{G}}$ iff

$$p - y^2p^2 = 0 \quad \text{or} \quad (yp)^2 = \theta^2$$

Since y has been chosen to satisfy this condition, every row of $\hat{\mathbf{H}}$ is orthogonal to every row of $\hat{\mathbf{G}}$. Hence

$$\hat{\mathscr{N}} \leq (\hat{\mathscr{F}})^{\perp} \tag{8.3}$$

Now $\hat{\mathscr{N}}$ is a code of length $p+1$ and dimension $(p+1)/2$. Also

$$\dim(\hat{\mathscr{F}})^{\perp} = p + 1 - \dim \hat{\mathscr{F}} = p + 1 - \frac{p+1}{2} = \frac{p+1}{2}$$

It, therefore, follows from relation (8.3) that

$$(\hat{\mathscr{F}})^{\perp} = \hat{\mathscr{N}}$$

Using the case $p = 4k - 1$ of Theorem 8.2, we can similarly prove the following theorem.

Theorem 8.7

The extended QR codes $\hat{\mathscr{F}}$ and $\hat{\mathscr{N}}$ defined as above are self dual in the case $p = 4k - 1$.

Corollary

The extended (i) binary Golay code $\mathscr{G}_{24} = \hat{\mathscr{G}}_{23}$ and (iii) ternary Golay code $\mathscr{G}_{12} = \hat{\mathscr{G}}_{11}$ are self dual.

Corollary

If $p = 4k - 1$, the weight of every non-zero code word in the extended QR codes $\hat{\mathscr{F}}$ and $\hat{\mathscr{N}}$ is divisible by s while the weight of every non-zero code word in the QR codes $\mathscr{F}$ and $\mathscr{N}$ is congruent to 0 or $s - 1$ modulo s.

We now come to the main theorem of this section.

Theorem 8.8

If d is the minimum distance between code words of the augmented QR code $\mathscr{F}$ (or $\mathscr{N}$) neither of which is in the expurgated QR code $\bar{\mathscr{F}}$ (respectively $\bar{\mathscr{N}}$) except the 0 word, then $d^2 \geq p$. If $p = 4k - 1$, this minimum distance satisfies

$$d^2 - d + 1 \geq p$$

Proof

Observe that this minimum distance d equals the weight of a non-zero code word $a(x)$ in $\mathscr{F}$ which is not in $\bar{\mathscr{F}}$. Then

$$x - 1 \nmid a(x)$$

Let n be a quadratic non-residue mod p. Set

$$\bar{a}(x) = a(x^n)$$

As α^r, $r \in Q$, are among the roots of $a(x), \alpha^{r/n}$ are among the roots of $\bar{a}(x)$. Moreover, n is a non-residue implies $1/n$ is a non-residue. Therefore, α^m, $m \in N$ are among the roots of $\bar{a}(x)$. Thus $\bar{a}(x)$ is divisible by $n(x)$. As 1 is not a root of $a(x)$, $1^{1/n} = 1$ is not a root of $\bar{a}(x)$. Hence $\bar{a}(x) \in \mathscr{N}$ but is not in the expurgated code $\bar{\mathscr{N}}$. The number of non-zero terms in $\bar{a}(x)$ is precisely equal to the number of non-zero terms of $a(x)$, i.e.

$$\mathrm{wt}(\bar{a}(x)) = d$$

Therefore, the number of non-zero terms in $a(x)\bar{a}(x)$ is at most d^2. Also $a(x)\bar{a}(x)$ is divisible by

$$q(x)n(x) = \sum_{i=0}^{p-1} x^i$$

so that $a(x)\bar{a}(x)$ is non-zero constant multiple of

$$\sum_{i=0}^{p-1} x^i$$

and

$$\mathrm{wt}(a(x)\bar{a}(x)) = p$$

Hence $d^2 \geq p$.

If $p = 4k - 1$, then -1 is a quadratic non-residue mod p and we may take

$$\bar{a}(x) = a(x^{-1})$$

Then d terms in $a(x)\bar{a}(x)$ are each equal to 1 and so the number of terms in $a(x)\bar{a}(x)$ is at most $d^2 - d + 1$. Hence, in this case

$$d^2 - d + 1 \geq p$$

Example 8.2
Consider the case $p = 17$. As

$$6^2 \equiv 2 \pmod{17}$$

we can take $s = 2$. The cyclotomic cosets relative to 2 modulo 17 are:

$$C_0 = \{0\}$$

$$C_1 = \{1, 2, 4, 8, 16, 15, 13, 9\}$$

$$C_3 = \{3, 6, 12, 7, 14, 11, 5, 10\}$$

To factorize $x^{17} - 1$ as a product of irreducible polynomials, we find the HCF of

$$x^{16} + x^{15} + x^{13} + x^9 + x^8 + x^4 + x^2 + x + 1$$

and

$$\sum_{i=0}^{16} x^i$$

$$x^{16} + x^{15} + x^{13} + x^9 + x^8 + x^4 + x^2 + x + 1 \overline{\big)\, x^{16} + x^{15} + \cdots + x^4 + x^3 + x^2 + x + 1}$$

$$x^{16} + x^{15} + x^{13} + x^9 + x^8 + x^4 + x^2 + x + 1$$

$$\overline{x^{14} + x^{12} + x^{11} + x^{10} + x^7 + x^6 + x^5 + x^3}$$

$$x^{14} + x^{12} + x^{11}x^{10} + x^7 + x^6 + x^5 + x^3 \overline{\big)\, x^{16} + x^{15} + x^{13} + x^9 + x^8 + x^4 + x^2 + x + 1}$$

$$x^{16} + x^{14} + x^{13} + x^{12} + x^9 + x^8 + x^7 + x^5$$

$$\overline{x^{15} + x^{14} + x^{12} + x^7 + x^5 + x^4 + x^2 + x + 1}$$

$$x^{15} + x^{13} + x^{12} + x^{11} + x^8 + x^7 + x^6 + x^4$$

$$\overline{x^{14} + x^{13} + x^{11} + x^8 + x^6 + x^5 + x^2 + x + 1}$$

$$x^{14} + x^{12} + x^{11} + x^{10} + x^7 + x^6 + x^5 + x^3$$

$$\overline{x^{13} + x^{12} + x^{10} + x^8 + x^7 + x^3 + x^2 + x + 1}$$

$$x^{13} + x^{12} + x^{10} + x^8 + x^7 + x^3 + x^2 + x + 1 \overline{\big)\, x^{14} + x^{12} + x^{11} + x^{10} + x^7 + x^6 + x^5 + x^3}$$

$$x^{14} + x^{13} + x^{11} + x^9 + x^8 + x^4 + x^3 + x^2 + x$$

$$\overline{x^{13} + x^{12} + x^{10} + x^9 + x^8 + x^7 + x^6 + x^5 + x^4 + x^2 + x}$$

$$x^{13} + x^{12} + x^{10} + x^8 + x^7 + x^3 + x^2 + x + 1$$

$$\overline{\phantom{x^{13} + x^{12} + x^{10} + x^8 + x^7 + x^3 + x^2 + x + 1}}$$

$$
\begin{array}{l|l}
x^9+x^6+x^5+x^4+x^3+1 & x^{13}+x^{12}+x^{10}+x^8+x^7+x^3+x^2+x+1 \\
 & x^{13} \qquad\qquad x^{10}+x^8+x^7+x^4+x^9 \\ \hline
 & x^{12}+x^9+x^4+x^3+x^2+x+1 \\
 & x^{12}+x^9+x^8+x^3+x^7+x^6 \\ \hline
 & \quad x^8+x^7+x^6+x^4+x^2+x+1
\end{array}
$$

$$
\begin{array}{l|l}
x^8+x^7+x^6+x^4+x^2+x+1 & x^9+x^6+x^5+x^4+x^3+1 \\
 & x^9+x^8+x^5+x^7+x^3+x^2+x \\ \hline
 & x^8+x^7+x^6+x^4+x^2+x+1 \\
 & x^8+x^7+x^6+x^4+x^2+x+1 \\ \hline
 & \qquad 0
\end{array}
$$

Thus the required HCF is

$$x^8+x^7+x^6+x^4+x^2+x+1$$

Dividing

$$\sum_{i=0}^{16} x^i$$

by the HCF obtained we get the other factor as

$$x^8+x^5+x^4+x^3+1$$

Thus

$$x^{17}-1=(x-1)(x^8+x^5+x^4+x^3+1)(x^8+x^7+x^6+x^4+x^2+x+1)$$

Let

$$F=\mathbb{B}[x]/I \quad \text{and} \quad \alpha=x+I$$

where I is the ideal of $\mathbb{B}[x]$ generated by

$$x^8+x^5+x^4+x^3+1$$

The elements of C_1 being quadratic residues modulo 17 and those of C_3 being non-residues,

$$q(x)=x^8+x^5+x^4+x^3+1$$

and

$$n(x)=x^8+x^7+x^6+x^4+x^2+x+1$$

Let d be the minimum distance of the QR code $\mathscr{F}$. Then

$$d^2 \geq 17 \Rightarrow d \geq 5$$

Also $q(x)$ is a word of weight 5. Hence $d = 5$. The dimension of $\mathscr{F}$ is 9. A generator matrix of $\mathscr{F}$ is

$$\mathbf{G} = \begin{pmatrix} 1&0&0&1&1&1&0&0&1&0&0&0&0&0&0&0&0 \\ 0&1&0&0&1&1&1&0&0&1&0&0&0&0&0&0&0 \\ 0&0&1&0&0&1&1&1&0&0&1&0&0&0&0&0&0 \\ 0&0&0&1&0&0&1&1&1&0&0&1&0&0&0&0&0 \\ 0&0&0&0&1&0&0&1&1&1&0&0&1&0&0&0&0 \\ 0&0&0&0&0&1&0&0&1&1&1&0&0&1&0&0&0 \\ 0&0&0&0&0&0&1&0&0&1&1&1&0&0&1&0&0 \\ 0&0&0&0&0&0&0&1&0&0&1&1&1&0&0&1&0 \\ 0&0&0&0&0&0&0&0&1&0&0&1&1&1&0&0&1 \end{pmatrix}$$

and that of $\mathscr{N}$ is

$$\mathbf{H} = \begin{pmatrix} 1&1&1&0&1&0&1&1&1&0&0&0&0&0&0&0&0 \\ 0&1&1&1&0&1&0&1&1&1&0&0&0&0&0&0&0 \\ 0&0&1&1&1&0&1&0&1&1&1&0&0&0&0&0&0 \\ 0&0&0&1&1&1&0&1&0&1&1&1&0&0&0&0&0 \\ 0&0&0&0&1&1&1&0&1&0&1&1&1&0&0&0&0 \\ 0&0&0&0&0&1&1&1&0&1&0&1&1&1&0&0&0 \\ 0&0&0&0&0&0&1&1&1&0&1&0&1&1&1&0&0 \\ 0&0&0&0&0&0&0&1&1&1&0&1&0&1&1&1&0 \\ 0&0&0&0&0&0&0&0&1&1&1&0&1&0&1&1&1 \end{pmatrix}$$

By Theorem 8.2, we may also take

$$\mathbf{G}_1 = \begin{pmatrix} 1&1&0&1&0&0&1&0&1&1&0&0&0&0&0&0&0 \\ 0&1&1&0&1&0&0&1&0&1&1&0&0&0&0&0&0 \\ 0&0&1&1&0&1&0&0&1&0&1&1&0&0&0&0&0 \\ 0&0&0&1&1&0&1&0&0&1&0&1&1&0&0&0&0 \\ 0&0&0&0&1&1&0&1&0&0&1&0&1&1&0&0&0 \\ 0&0&0&0&0&1&1&0&1&0&0&1&0&1&1&0&0 \\ 0&0&0&0&0&0&1&1&0&1&0&0&1&0&1&1&0 \\ 0&0&0&0&0&0&0&1&1&0&1&0&0&1&0&1&1 \\ 1&1&1&1&1&1&1&1&1&1&1&1&1&1&1&1&1 \end{pmatrix}$$

as a generator matrix of $\mathscr{F}$ and

$$\mathbf{H}_1 = \begin{pmatrix} 1&0&0&1&1&1&1&0&0&1&0&0&0&0&0&0&0\\ 0&1&0&0&1&1&1&1&0&0&1&0&0&0&0&0&0\\ 0&0&1&0&0&1&1&1&1&0&0&1&0&0&0&0&0\\ 0&0&0&1&0&0&1&1&1&1&0&0&1&0&0&0&0\\ 0&0&0&0&1&0&0&1&1&1&1&0&0&1&0&0&0\\ 0&0&0&0&0&1&0&0&1&1&1&1&0&0&1&0&0\\ 0&0&0&0&0&0&1&0&0&1&1&1&1&0&0&1&0\\ 0&0&0&0&0&0&0&1&0&0&1&1&1&1&0&0&1\\ 1&1&1&1&1&1&1&1&1&1&1&1&1&1&1&1&1 \end{pmatrix}$$

as a generator matrix of $\mathscr{N}$. As

$$(x+1)q(x)(x+1)n(x) = (x+1)(x^{17}+1) = 0$$

in $\mathbb{B}[x]/I$, every one of the first eight rows of $\mathbf{G}_1$ is orthogonal to any one of the first eight rows of $\mathbf{H}_1$. If we add a column to $\mathbf{G}_1$ and a column to $\mathbf{H}_1$ such that each of the first eight entries of these columns is zero, and the last entries are λ and μ respectively, then the above orthogonality property stays good. Each one of the first eight rows of $\mathbf{G}_1$ and of $\mathbf{H}_1$ being of even weight, this is equivalent to saying that if

$$a = (a_0, \ldots, a_{16}) \in \mathscr{F}$$

then

$$a_{17} = \lambda \sum_{i=1}^{16} a_i$$

while if $a \in \mathscr{N}$, then

$$a_{17} = \mu \sum_{i=0}^{16} a_i$$

Taking $\lambda = \mu = 1$, we find that every row of $\hat{\mathbf{G}}_1$ is orthogonal to every row of $\hat{\mathbf{H}}_1$. Hence, the codes generated by $\hat{\mathbf{G}}_1$ and $\hat{\mathbf{H}}_1$ are orthogonal to each other. Therefore,

$$\hat{\mathscr{F}} \leq (\hat{\mathscr{N}})^{\perp} \quad \text{and} \quad \hat{\mathscr{N}} \leq (\hat{\mathscr{F}})^{\perp} \tag{8.4}$$

As on extending a code, the dimension remains unchanged, $\hat{\mathscr{F}}$ is an [18, 9, –] code and so is $\hat{\mathscr{N}}$. For a code $\mathscr{C}$ of dimension k

$$\dim \mathscr{C}^{\perp} = n - k$$

therefore

$$\dim (\hat{\mathscr{F}})^{\perp} = 9$$

and

$$\dim (\hat{\mathscr{N}})^{\perp} = 9$$

It then follows from the relation (8.4) that

$$(\hat{\mathscr{F}})^{\perp} = \hat{\mathscr{N}}$$

8.4 IDEMPOTENTS OF QUADRATIC RESIDUE CODES

We have studied idempotents of cyclic codes earlier. Recall that the idempotent in a cyclic code $\mathscr{C}$ is the unique element e in $\mathscr{C}$ which generates $\mathscr{C}$ and $e^2 = e$. Quadratic residue codes being cyclic codes, we study idempotents for binary and ternary quadratic residue codes here.

Lemma 8.2

Let p be a prime congruent to $\pm 1 \pmod 8$. Then there exists a primitive pth root α of unity in some extension field of $\mathbb{B} = \mathrm{GF}(2)$ such that $E(\alpha) = 0$, where

$$E_q(x) = \sum_{r \in Q} x^r$$

Proof

Since Q is closed under multiplication by 2, $E_q(x)$ is an idempotent in the ring $\mathbb{B}[x]/\langle x^p - 1\rangle$. Therefore, for every primitive pth root α of unity

$$E_q(\alpha)^2 = E_q(\alpha)$$

showing that $E_q(\alpha) \in \mathbb{B}$. Thus, either $E_q(\alpha) = 0$ or $E_q(\alpha) = 1$.

Suppose that $E_q(\alpha) = 1$ for every primitive pth root α of unity. As the primitive pth roots of unity are precisely the roots of the polynomial

$$f(x) = 1 + x + \cdots + x^{p-1}$$

$$f(x) | E_q(x) + 1$$

The polynomial $E_q(x)$ being of degree at most $p - 1$, we must have

$$f(x) = E_q(x) + 1 \tag{8.5}$$

The right-hand side of this relation has $(p + 1)/2$ non-zero terms while the left-hand side has $p - 1$ non-zero terms. The relation (8.5) is as such not possible. Hence there is at least one primitive pth root α of unity such that

$$E_q(\alpha) = 0$$

Theorem 8.9

If $p \equiv -1 \pmod 8$, then the primitive pth root α of unity in (8.1) can be suitably chosen so that the idempotents of the binary quadratic residue codes $\mathscr{F}$, $\mathscr{N}$, $\bar{\mathscr{F}}$ and $\bar{\mathscr{N}}$ are respectively

$$E_q(x) = \sum_{r \in Q} x^r \qquad E_n(x) = \sum_{n \in N} x^n$$

$$1 + E_n(x) \qquad \text{and} \qquad 1 + E_q(x)$$

Proof

Since Q and N are closed under multiplication by 2, $E_q(x)$, $E_n(x)$ are idempotents in $\mathbb{B}[x]/\langle x^p - 1\rangle$. But then $1 + E_n(x)$ and $1 + E_q(x)$ are also idempotents. As seen in Lemma 8.2, $E_q(\alpha)\in\mathbb{B}$. Similarly $E_q(\alpha)\in\mathbb{B}$. For any $i\in Q$

$$E_q(\alpha^i) = \sum_{r\in Q} \alpha^{ir} = \sum_{r\in Q} \alpha^r = E_q(\alpha)$$

while for any non-residue t

$$E_n(\alpha^t) = \sum_{n\in N} \alpha^{tn} = \sum_{r\in Q} \alpha^r = E_q(\alpha) \tag{8.6}$$

Choose α so that $E_q(\alpha) = 0$. Then α^i, $\forall i\in Q$, is a root of $E_q(x)$ so that

$$q(x)|E_q(x)$$

As $(p-1)/2$ is odd, 1 is not a root of $E_q(x)$ and

$$(x-1)\nmid E_q(x)$$

Now

$$1 + \alpha + \alpha^2 + \cdots + \alpha^{p-1} = 0$$

so that

$$E_q(\alpha) + E_n(\alpha) = 1$$

Therefore

$$E_n(\alpha) + 1 = 0 \qquad \text{and} \qquad E_q(\alpha^t) = 1$$

for every non-residue t. Thus

$$n(x)|(1 + E_q(x))$$

Also 1 is a root of $1 + E_q(x)$ and

$$(x-1)|(1 + E_q(x))$$

Let

$$1 + E_q(x) = n(x)(x-1)f_1(x)$$

then

$$1 = E_q(x) + n(x)(x-1)f_1(x) \tag{8.7}$$

Multiplying both sides of this relation by $q(x)$ gives

$$\begin{aligned} q(x) &= q(x)E_q(x) + q(x)n(x)(x-1)f_1(x) \\ &\equiv q(x)E_q(x) \quad \text{in } \mathbb{B}[x]/\langle x^p - 1\rangle \end{aligned}$$

Thus, the ideal generated by $q(x)$ in $\mathbb{B}[x]/\langle x^p - 1\rangle$ is contained in the ideal generated by $E_q(x)$. The reverse inclusion follows as $q(x)|E_q(x)$. Hence $E_q(x)$ is the idempotent of the cyclic code $\mathscr{F}$ (Theorem 6.1).

Let

$$E_q(x) = q(x) f_2(x)$$

Multiplying the relation (8.7) by $n(x)(x-1)$, gives

$$\begin{aligned} n(x)(x-1) &= q(x)n(x)(x-1)f_2(x) + n(x)(x-1)(1+E_q(x)) \\ &= n(x)(x-1)(1+E_q(x)) \quad \text{in } \mathbb{B}[x]/\langle x^p - 1\rangle \end{aligned}$$

From this relation and the fact that

$$n(x)(x-1) \mid (1+E_q(x))$$

it follows that $1+E_q(x)$ is the idempotent of the cyclic code $\bar{\mathcal{N}}$.

Again, it follows from (8.6) and that $E_q(\alpha)=0$ that

$$n(x) \mid E_n(x)$$

and the number of terms in $E_n(x)$ being odd

$$x-1 \nmid E_n(x)$$

The number of terms in $1+n(x)$ being even

$$(x-1) \nmid (1+n(x))$$

Also, for any quadratic residue i,

$$E_n(\alpha^i) = E_n(\alpha) = E_q(\alpha) + 1 = 1$$

Therefore, every α^i is a root of $1+E_n(x)$. Hence

$$q(x) \mid (1+E_n(x))$$

Let

$$E_n(x) = n(x) g_1(x)$$

and

$$1 + E_n(x) = q(x)(x-1) g_2(x)$$

Then

$$1 = E_n(x) + q(x)(x-1)g_2(x) = n(x)g_1(x) + (1+E_n(x)) \tag{8.8}$$

implies

$$n(x) \equiv E_n(x) n(x) (\text{mod } x^p - 1)$$

Therefore, the cyclic code generated by $n(x)$ is contained in the cyclic code generated by $E_n(x)$. That $n(x) \mid E_n(x)$ shows that the cyclic code generated by $E_n(x)$ is contained in the cyclic code $\mathcal{N}$. Hence $E_n(x)$ is the idempotent of the cyclic code $\mathcal{N}$.

On multiplying the relation (8.8) by $(x-1)q(x)$, we can prove that $1+E_n(x)$ is the idempotent of the cyclic code $\bar{\mathcal{F}}$.

Using a similar argument, we can prove the following theorem.

Theorem 8.10
If $p \equiv 1 \pmod 8$, then the primitive pth root α in (8.1) can be suitably chosen so that the idempotents of the binary quadratic residue codes $\mathscr{F}, \bar{\mathscr{F}}, \mathscr{N}, \bar{\mathscr{N}}$ are $1 + E_q(x)$, $E_n(x)$, $1 + E_n(x)$ and $E_q(x)$ respectively.

Proposition 8.10
If $s = 2$ and $p \equiv -1 \pmod 4$, the weight of every code word in $\hat{\mathscr{F}}$ is divisible by 4, and the weight of every code word in $\mathscr{F}$ is congruent to 0 or 3 modulo 4.

Proof
For quadratic residue codes s is a quadratic residue mod p. Thus 2 is a quadratic residue mod p and so

$$p = \pm 1 \pmod 8$$

But

$$p \equiv -1 \pmod 4$$

Therefore,

$$p = 8k - 1$$

for some natural number k. The number of residues or non-residues is then $4k - 1$. The idempotent of the expurgated QR code $\bar{\mathscr{F}}$ is

$$1 + \sum_{n \in N} x^n$$

which has $4k$ non-zero terms. In the corresponding generator matrix $\bar{\mathbf{G}}$ of this code, every row has $4k$ non-zero terms. A generator matrix for $\mathscr{F}$ is then

$$\mathbf{G} = \left(\frac{\bar{\mathbf{G}}}{1 \quad 1 \quad \cdots \quad 1} \right)$$

Since $\hat{\mathscr{F}}$ is obtained from $\mathscr{F}$ by adding an overall parity check, the weight of every row of the generator matrix $\hat{\mathbf{G}}$ corresponding to $\mathbf{G}$ of $\hat{\mathscr{F}}$ is divisible by 4. The code $\hat{\mathscr{F}}$ being self dual, any two rows of $\hat{\mathbf{G}}$ agree in an even number of terms. Therefore, the sum of any two rows of $\hat{\mathbf{G}}$ again has weight divisible by 4 and the result follows. ■

Let p be a prime of the form $12n + b$. Then $b = 1, 5, 7$ or 11. By the law of quadratic reciprocity

$$\left(\frac{3}{12n+b}\right)\left(\frac{12n+b}{3}\right) = (-1)^{(12n+b-1)/2} = (-1)^{(b-1)/2}$$

$$(-1)^{(b-1)/2} = \begin{cases} 1 & \text{if } b = 4k+1 \Rightarrow b = 1 \text{ or } b = 5 \\ -1 & \text{if } b = 4k-1 \Rightarrow b = 7 \text{ or } b = 11 \end{cases}$$

Thus when $b = 1$ or $b = 5$

$$\left(\frac{3}{12n+b}\right) = \left(\frac{b}{3}\right) = \begin{cases} 1 & \text{if } b = 1 \\ -1 & \text{if } b = 5 \end{cases}$$

When $b=7$ or $b=11$

$$\left(\frac{3}{12n+b}\right)=-\left(\frac{b}{3}\right)=\begin{cases}-1 & \text{if } b=7\\ 1 & \text{if } b=11\end{cases}$$

Hence, 3 is a quadratic residue for primes of the form $12n\pm 1$ and non-residue for primes of the form $12n\pm 5$.

We can thus talk of ternary quadratic residue codes of length p when p is of the form $12n\pm 1$. Also, -1 is a non-residue for $p=12n-1$ and a residue for $p=12n+1$.

Theorem 8.11
Suppose that $p\equiv -1 \pmod{12}$ and that

$$c(x)=\sum_{i=1}^{d} c_i x^{e_i}$$

is a code word of weight d in the ternary QR code of length p. Then

$$d\equiv 2 \pmod 4 \quad \text{or} \quad d\equiv 3 \pmod 4$$

Proof
Let $\hat{c}(x)$ be the word obtained from $c(x)$ by replacing x by x^{-1}. The product

$$c(x)\hat{c}(x)=\sum_{i=1}^{d} c_i^2+\sum_{\substack{i\neq j\\ i,j}} c_i c_j x^{e_i-e_j}$$

will have less than d^2-d+1 terms if some of the terms cancel, i.e. if

$$e_i-e_j=e_k-e_l$$

for some i, j, k, l. The number of such terms is of the form $4t$ and so

$$d^2-d+1-4t=p$$

which implies that

$$d\equiv 2 \pmod 4 \quad \text{or} \quad d\equiv 3 \pmod 4$$

If we recall that 2 is a quadratic residue for primes of the form $8n\pm 1$, we have the following theorem.

Theorem 8.12
If $p\equiv 1 \pmod 8$ and c is a code word of odd weight d in the binary QR code with generator $q(x)$, then $d\equiv 3 \pmod 4$.

Remark 8.2
Observe that the conclusion of Theorem 8.11 is valid for any quadratic residue code of prime length p provided $p\equiv -1 \pmod 4$.

Remark 8.3

A close look at the proof of Theorem 8.9 suggests the following working rule for obtaining the idempotents of the quadratic residue codes of $\mathscr{F}$, $\mathscr{N}$, $\bar{\mathscr{F}}$ and $\bar{\mathscr{N}}$ (not necessarily binary).

If $f(x)$ is a polynomial over GF(s) and

$$q(x) | f(x)$$

$x-1$ does not divide it and

$$(x-1)n(x) | (1-f(x))$$

then $f(x)$ is the idempotent of the QR code $\mathscr{F}$ and $1-f(x)$ is the idempotent of the QR code $\bar{\mathscr{N}}$.

When $s=2$, $p=4k\pm 1$ and, so, in either case $\theta^2=1$. When $s=3$, $p=12k\pm 1$ and again, in either case $\theta^2=1$. Since $\theta \in \mathrm{GF}(s)$, in both the cases $\theta=1$ or $\theta=-1$ (for $s=2$ it is always $\theta=1$).

Lemma 8.3

In

$$\mathscr{R} = F[x]/\langle x^p - 1\rangle$$

where $F=\mathrm{GF}(3)$ and $p=12k-1$

$$\left(\sum_{r\in Q} x^r\right)^2 = -\sum_{r\in Q} x^r$$

$$\left(\sum_{n\in N} x^n\right)^2 = -\sum_{n\in N} x^n$$

and

$$\left(\sum_{r\in Q} x^r\right)\left(\sum_{n\in N} x^n\right) = -(1+x+\cdots+x^{p-1})$$

Proof

For any residue t, it follows (from Perron's Theorem 8.3) that $\{r+t/r\in Q\}$ contains $3k-1$ residues and $3k$ non-residues. Therefore in $\{r+t/r, t\in Q\}$ every residue appears $3k-1$ times and every non-residue appears $3k$ times. Hence

$$\left(\sum_{r\in Q} x^r\right)^2 = (3k-1)\sum_{r\in Q} x^r + 3k\sum_{n\in N} x^n = -\sum_{r\in Q} x^r$$

Thus

$$-\sum_{r\in Q} x^r$$

is an idempotent. That

$$-\sum_{n\in N} x^n$$

is an idempotent follows as above using the observation that for any non-residue t, $\{t+n/n\in N\}$ contains $3k$ residues and $3k-1$ non-residues.

In the present case, i.e. $p=12k-1$, -1 is a non-residue and so $\forall n\in N$ there is an $r\in Q$ such that $r+n=0$ and for every $r\in Q$, there is an $n\in N$ such that $r+n=0$. Therefore for every non-residue n, $\{r+n/r\in Q\}$ has $3k$ residues (including 0) and $3k-1$ non-residues. Therefore, in $\{r+n/r\in Q,\ n\in N\}$ every residue or non-residue occurs $3k-1$ times and 0 also occurs $3k-1$ times. Hence

$$\left(\sum_{r\in Q} x^r\right)\left(\sum_{n\in N} x^n\right)=(3k-1)\sum_{i=0}^{p-1} x^i=-\sum_{i=0}^{p-1} x^i$$

Using parts (iii) and (iv) of Theorem 8.3, we can similarly prove the following Lemma.

Lemma 8.4

In

$$\mathcal{R}=F[x]/\langle x^p-1\rangle$$

where $F=\mathrm{GF}(3)$ and $p=12k+1$

$$\left(\sum_{r\in Q} x^r\right)^2=-\sum_{r\in Q} x^r$$

$$\left(\sum_{n\in N} x^n\right)^2=-\sum_{n\in N} x^n$$

and

$$\left(\sum_{r\in Q} x^r\right)\left(\sum_{n\in N} x^n\right)=0$$

Let

$$E_q(x)=-\sum_{r\in Q} x^r$$

$$E_n(x)=-\sum_{n\in N} x^n$$

$$F_q(x)=1-E_n(x)$$

and

$$F_n(x)=1-E_q(x)$$

Using the above lemma and proceeding as in the proof of Lemma 8.2, we can prove the next lemma.

Lemma 8.5

Let p be a prime congruent to $\pm 1 \pmod{12}$. Then there exists a primitive pth root α of unity in some extension field of $F=\mathrm{GF}(3)$ such that $E_n(\alpha)=0$.

Theorem 8.13
If $p \equiv 1 \pmod{12}$, then the primitive pth root α in (8.1) can be suitably chosen so that the idempotents of the ternary quadrative residue codes $\mathscr{F}$, $\bar{\mathscr{F}}$, $\mathscr{N}$ and $\bar{\mathscr{N}}$ are $1 - E_q(x)$, $E_n(x)$, $1 - E_n(x)$ and $E_q(x)$ respectively.

Proof
Choose α such that $E_n(\alpha) = 0$. Then

$$1 - E_q(\alpha) - E_n(\alpha) = 0 \Rightarrow 1 - E_q(\alpha) = 0$$

For $t \in Q$

$$E_q(\alpha^t) = -\sum_{r \in Q} \alpha^{rt} = -\sum_{r \in Q} \alpha^r = E_q(\alpha) = 1$$

Therefore

$$q(x) | (1 - E_q(x))$$

Also

$$E_q(1) = \frac{p-1}{2} \equiv 0 \pmod 3)$$

Therefore

$$(x-1) | E_q(x)$$

For any $n \in N$

$$E_q(\alpha^n) = -\sum_{r \in Q} \alpha^{rn} = -\sum_{t \in N} \alpha^t = E_n(\alpha) = 0$$

It then follows that

$$n(x) | E_q(x)$$

Thus

$$(x-1)n(x) | E_q(x)$$

and it follows from Remark 8.3 that $1 - E_q(x)$ is the idempotent of $\mathscr{F}$, while $E_q(x)$ is the idempotent of $\bar{\mathscr{N}}$. We can similarly prove that $1 - E_n(x)$ is the idempotent of $\mathscr{N}$ and $E_n(x)$ is the idempotent of $\bar{\mathscr{F}}$. ■

Proceeding similarly, we have the following theorem.

Theorem 8.14
If $p \equiv -1 \pmod{12}$, then the primitive pth root α in (8.1) can be suitably chosen so that the idempotents of the ternary quadratic residue codes $\mathscr{F}$, $\bar{\mathscr{F}}$, $\mathscr{N}$ and $\bar{\mathscr{N}}$ are $E_q(x)$, $1 - E_n(x)$, $E_n(x)$ and $1 - E_q(x)$ respectively.

Remark 8.4

If $p = 12k - 1$, we have observed that the minimum distance d of a QR code satisfies

$$d \equiv 2 (\text{mod } 4) \quad \text{or} \quad d \equiv 3 (\text{mod } 4)$$

and

$$d \equiv 0 (\text{mod } 3) \quad \text{or} \quad d \equiv 2 (\text{mod } 3)$$

Then

$$d \equiv 4m + 2 \quad \text{or} \quad d \equiv 4m + 3$$

and so

$$d \equiv m + 2 \quad \text{or} \quad d \equiv m (\text{mod } 3)$$

For $d \equiv 0 (\text{mod } 3)$ then shows that

$$m \equiv 1 (\text{mod } 3) \quad \text{or} \quad m \equiv 0 (\text{mod } 3)$$

But then

$$d \equiv 3 (\text{mod } 12) \quad \text{or} \quad d \equiv 6 (\text{mod } 12)$$

For $d \equiv 2 (\text{mod } 3)$ shows that

$$m \equiv 0 (\text{mod } 3) \quad \text{or} \quad m \equiv 2 (\text{mod } 3)$$

so that

$$d \equiv 2 (\text{mod } 12) \quad \text{or} \quad d \equiv 11 (\text{mod } 12)$$

Thus the minimum distance of a ternary QR code of length $p = 12k - 1$ is always congruent to 2, 3, 6 or 11 modulo 12 and hence the minimum distance of extended ternary QR code is always congruent to 0 or 3 or 6 modulo 12.

8.5 SOME EXAMPLES

Case (i)

In Case (v) of Examples 7.4, we obtained the irreducible factors of $x^{37} - 1$ over GF(3):

$$x^{37} - 1 = (x - 1) f(x) g(x)$$

where $f(x)$, $g(x)$ are irreducible factors of degree 18 each. We may take either of these as a generator polynomial of the QR code. The weight of the polynomial (word) $g(x)$ being 10, it follows that the minimum distance d of the code is at most 10. As $d^2 \geq 37$, we have

$$7 \leq d \leq 10$$

Case (ii) – ternary QR code of length 61

Let $\alpha = x + \langle h_2(x) \rangle$ be the primitive 61st root of unity in the field F as constructed in Case (vi) of Examples 7.4. A generator polynomial of the QR code is

$$q(x) = \prod_{i \in Q} (x - \alpha^i)$$

where $Q = C_1 \cup C_4 \cup C_5$ is the set of all quadratic residues modulo 61. Here C_0, C_1, C_2, C_4, C_5, C_8 and C_{10} are the cyclotomic cosets relative to 3 modulo 61 as obtained in Case (vi) of Examples 7.4. Thus $q(x)$ is the product of the minimal polynomials of α, α^4 and α^5. Factorization of $x^{61} - 1$ as a product of irreducible polynomials over GF(3) has been obtained in Case (vi) of Examples 7.4. We find that α^5 satisfies the irreducible factor $f_2(x)$ and so $f_2(x)$ is its minimal polynomial. We have already proved in Case (vi) of Examples 7.4 that $g_1(x)$ is the minimal polynomial of α^4. Therefore,

$$\begin{aligned} q(x) &= f_2(x) g_1(x) h_2(x) \\ &= x^{30} + x^{29} - x^{28} + x^{27} - x^{25} - x^{21} - x^{20} - x^{19} - x^{15} - x^{11} \\ &\quad - x^{10} - x^9 - x^5 + x^3 - x^2 + x + 1 \end{aligned}$$

which is a word of weight 17. Let d denote the minimum distance of the QR code. Then $d^2 \geq 61$ and so we have $8 \leq d \leq 17$. Observe that

$$\begin{aligned} (x^6 - x^5 - x^4 - x^3 + x^2 + 1)q(x) &= x^{36} - x^{27} + x^{23} + x^{18} + x^{17} + x^{13} \\ &\quad + x^9 - x^8 - x^5 + x^3 + x + 1 \end{aligned}$$

which is a word of weight 12. Hence $d \leq 12$.

Taking

$$n(x) = \prod_{i \in N} (x - \alpha^i)$$

where $N = C_2 \cup C_8 \cup C_{10}$ is the set of all quadratic non-residues modulo 61, we find that

$$\begin{aligned} n(x) &= f_1(x) g_2(x) h_1(x) \\ &= x^{30} - x^{28} + x^{27} - x^{26} - x^{25} + x^{21} - x^{19} + x^{18} + x^{17} - x^{16} \\ &\quad + x^{15} - x^{14} + x^{13} + x^{12} - x^{11} + x^9 - x^5 - x^4 + x^3 - x^2 + 1 \end{aligned}$$

Also, on direct computation we find that

$$\begin{aligned} (x^9 + x^7 - x^6 - x^5 + x^4 + x^3 - 1)n(x) &= x^{39} - x^{27} + x^{25} - x^{21} - x^{20} - x^{18} \\ &\quad + x^{14} + x^6 - x^5 + x^2 - 1 \end{aligned}$$

which is a word of weight 11. As the codes generated by $q(x)$ and $n(x)$ are equivalent and equivalent codes have the same minimum distance, the minimum distance of the ternary quadratic residue code of length 61 is at most 11.

Case (iii)–QR code of length 11 over GF(5)

We have obtained the factorization of $x^{11}-1$ over GF(5) as a product of irreducible polynomials in Case (iv) of Examples 7.4

$$x^{11}-1=(x-1)f(x)g(x)$$

where

$$f(x)=x^5-x^4-x^3+x^2-2x-1$$

and

$$g(x)=x^5+2x^4-x^3+x^2+x-1$$

As one of the equivalent QR codes $\mathscr{F}$ and $\mathscr{N}$ is generated by $f(x)$ and the other by $g(x)$, we find that the minimum distance ∂ of the QR code $\mathscr{F}$ is at most 6. Also

$$(x+1)g(x)=x^6+3x^5+x^4+2x^2-1$$

and, so, $\partial\leq 5$. As $\partial^2\geq 11$, we have $4\leq\partial\leq 5$.

Theorem 8.15

The minimum distance of the code is 5.

Proof

An arbitrary code word is

$$\begin{aligned}a,\quad 2a+b,\quad -a+2b+c,\quad a-b+2c+d,\quad a+b-c+2d+e,\\ -a-b+c-d+2e+1,\quad -b+c+d-e+2,\\ -c+d+e-1,\quad -d+e+1,\quad -e+1,\quad -1\end{aligned}$$

where $a, b, c, d, e\in\mathrm{GF}(5)$.

To prove that $\partial=5$, we consider the various possible cases:

Case A: $\boldsymbol{a=0, 2a+b=0}$ ***so that*** $\boldsymbol{b=0}$ ***as well***

The word becomes

$$\begin{aligned}0,\quad 0,\quad c,\quad 2c+d,\quad -c+2d+e,\quad c-d+2e+1,\quad c+d-e+2,\\ -c+d+e-1,\quad -d+e+1,\quad -e+1,\quad -1\end{aligned}$$

If $c=d=0$, it is fairly easy to see that the word is of weight at least 5.

Case A(i): $c=0$ *but* $2c+d\neq 0$

Then $d\neq 0$ and the word takes the form

$$\begin{aligned}0,\quad 0,\quad 0,\quad d,\quad 2d+e,\quad -d+2e+1,\quad d-e+2,\quad d+e-1,\\ -d+e+1,\quad -e+1,\quad -1\end{aligned}$$

If $2d + e = 0$, then among the entries

$$-d + 2e + 1 = 1 \qquad d - e + 2 = 3d + 2 \qquad d + e - 1 = -d - 1$$
$$-d + e + 1 = -3d + 1 \qquad 2d + 1$$

at least three are non-zero so that the word is of weight at least 5. If $2d + e \neq 0$, but $-d + 2e + 1 = 0$, then among the entries

$$d - e + 2 = e + 3 \qquad 3e - 2 \qquad -e \qquad -e + 1$$

at least two are non-zero.

If $2d + e \neq 0$, $-d + 2e + 1 \neq 0$ but $d - e + 2 = 0$, then $-d + e + 1 = 3 \neq 0$ and so again the word is of weight at least 5.

Case A(ii): $c \neq 0$ but $2c + d = 0$
The word becomes

$$0, \quad 0, \quad c, \quad 0, \quad b, \quad 3c + 2e + 1, \quad -c - e + 2, \quad -3c + e - 1, \quad 2c + e,$$
$$-e + 1, \quad -1$$

For $e = 0$, the entries $3c + 1$, $4c + 2$, $-3c - 1$, $2c, 1, -1$ have at least four non-zero terms.

For $e \neq 0$, but $3c + 2e + 1 = 0$, we have $c = e - 2$ and among

$$-c - e + 2 = -2e + 4 \qquad -3c + e - 1 = -2e \qquad 3e - 4$$
$$-e + 1 \qquad -1$$

at least three terms are non-zero.

Case A(iii): $c \neq 0$, $2c + d \neq 0$ but $-c + 2d + e = 0$
Then $c = 2d + e$ and the last six terms of the word are (among the first five there are two non-zero terms):

$$d + 3e + 1 \qquad 3d - 2e + 2 \qquad -d - 1 \qquad -d + e + 1 \qquad -e + 1 \qquad -1$$

For $e = 0$, among the terms

$$-d - 1, \quad -d + e + 1 = -d + 1, \quad -e + 1 = 1, \quad -1$$

at least three are non-zero.

For $e \neq 0$, but $d + 3e + 1 = 0$, we have $d = 2e - 1$ and the last five terms are

$$4e - 1, \quad -2e, \quad -e + 2, \quad -e + 1, \quad -1$$

out of which at least four are non-zero.

Case A(iv): $c \neq 0$, $2c + d \neq 0$, $-c + 2d + e \neq 0$ but $c - d + 2e + 1 = 0$
Then $c = d - 2e - 1$ and among the last five terms

$$2d - 3e + 1, \quad 3e, \quad -d + e + 1, \quad -e + 1, \quad -1$$

at least two are non-zero.

Thus, in Case (A) we always have a word of weight at least 5.

***Case B:* $a = 0$, $2a + b \neq 0$ *so that* $b \neq 0$**
Then the word becomes

$$0, \quad b, \quad 2b + c, \quad -b + 2c + d, \quad b - c + 2d + e, \quad -b + c - d + 2e + 1,$$
$$-b + c + d - e + 2, \quad -c + d + e - 1, \quad -d + e + 1, \quad -e + 1, \quad -1$$

Case B(i): $2b + c = 0$ *so that* $c = -2b$
Then the word is

$$0, \quad b, \quad 0, \quad d, \quad 3b + 2d + e, \quad 3b - d + 2e + 1, \quad 3b + d - e + 2,$$
$$2b + d + e - 1, \quad -d + e + 1, \quad -e + 1, \quad -1$$

If $d = 0$, the word is

$$0, \quad b, \quad 0, \quad 0, \quad 3b + e, \quad 3b + 2e + 1, \quad 3b - e + 2, \quad 2b + e - 1,$$
$$e + 1, \quad -e + 1, \quad -1$$

If $3b + e = 0$ so that $e = 2b$, the word becomes

$$0, \quad b, \quad 0, \quad 0, \quad 0, \quad 2b + 1, \quad b + 2, \quad 4b - 1, \quad 2b + 1, \quad -2b + 1, \quad -1$$

which is of weight at least 5.
A similar argument also shows that the word is of weight at least 5 in the cases:

(a) $d \neq 0$, $3b + 2d + e = 0$ and
(b) $d \neq 0$, $3b + 2d + e \neq 0$ but $3b - d + 2e + 1 = 0$.

Case B(ii): $2b + c \neq 0$ *but* $-b + 2c + d = 0$
Then $b = 2c + d$. The word then is

$$0, \quad b, \quad 2b + c, \quad 0, \quad c + 3d + e, \quad -c - 2d + 2e + 1, \quad -3c - e + 2,$$
$$-c + d + e - 1, \quad -d + e + 1, \quad -e + 1, \quad -1$$

Here b, $2b + c$ and -1 are three non-zero entries and by considering the cases

(a) $c + 3d + e = 0$
(b) $c + 3d + e \neq 0$ but $-c - 2d + 2e + 1 = 0$,

we can prove that there are at least two non-zero entries among the rest of the entries.

***Case C:* $a \neq 0$, $2a + b = 0$**
Then the word is

$$a, \quad 0, \quad c, \quad 3a + 2c + d, \quad -a - c + 2d + e, \quad a + c - d + 2e + 1,$$
$$2a + c + d - e + 2, \quad -c + d + e - 1, \quad -d + e + 1, \quad -e + 1, \quad -1$$

Again considering the subcases

(i) $c = 0$
(ii) $c \neq 0$ but $3a + 2c + d = 0$ and
(iii) $c \neq 0$, $3a + 2c + d \neq 0$ but $-a - c + 2d + e = 0$

we can prove that the word is of weight at least 5.

In the remaining two cases:

Case D: $a \neq 0$, $2c + b \neq 0$ but $-a + 2b + c = 0$
Case E: $a \neq 0$, $2a + b \neq 0$, $-a + 2b + c \neq 0$ but $a - b + 2c + d = 0$

also we can prove similarly that there is no word of weight 4 in this code.

Hence the minimum distance of the code is 5.

Exercise 8.1

1. Prove that 2 is a quadratic residue modulo *a* prime p iff $p \equiv \pm 1 \pmod 8$.
2. Determine all primes p for which 5 is a quadratic residue mod p.
3. Let p be a prime congruent to $\pm 1 \pmod 8$. Then there exists a primitive pth root α of unity in some extension field of $\mathbb{B}$ such that $E_q(\alpha) = 1$, where

$$E_q(x) = \sum_{r \in Q} x^r$$

4. Prove Theorems 8.10, 8.12 and 8.14.
5. Determine, if possible, weight distributions of some of the codes constructed.

9

Maximum distance separable codes

9.1 NECESSARY AND SUFFICIENT CONDITIONS FOR MDS CODES

In this chapter, we study an interesting class of linear codes – interesting because these codes have the maximum possible error detection/correction possibility. Another point of interest is a question of existence of such codes which translates into a question purely on vector spaces.

We have seen earlier that if $\mathscr{C}$ is a linear $[n, k, d]$ code over a field F, then $d \leq n - k + 1$.

Definition 9.1

A linear $[n, k, d]$ code over F with $d = n - k + 1$ is called a **maximum distance separable (MDS) code**.

In this chapter, unless explicitly stated to the contrary, we do not insist that the first k columns of a generator matrix of a linear $[n, k, d]$ form the identity matrix or that the last $n - k$ columns of a parity check matrix form the identity matrix.

We begin our study with the following simple observation.

Proposition 9.1

Let $\mathscr{C}$ be a linear $[n, k, d]$ code over a field F of q elements, q a prime power with a parity check matrix $\mathbf{H}$. Then $\mathscr{C}$ has a code word of eight $\leq l$ iff l columns of $\mathbf{H}$ are linearly dependent.

Proof

Let $b = b_1 b_2 \cdots b_n$ be a code word in $\mathscr{C}$ with $\mathrm{wt}(b) = l$. Let $b_{i_1}, \ldots, b_{i_l}$ be the non-zero entries of b. Then

$$\mathbf{Hb}^t = \mathbf{0} \Rightarrow b_{i_1}\mathbf{H}_{i_1} + \cdots + b_{i_l}\mathbf{H}_{i_l} = 0$$

where $\mathbf{H}_1, \mathbf{H}_2, \ldots, \mathbf{H}_n$ denote the columns of $\mathbf{H}$. Thus, l columns

$$\mathbf{H}_{i_1}, \ldots, \mathbf{H}_{i_l}$$

of $\mathbf{H}$ are linearly dependent.

Conversely, suppose that l columns of $\mathbf{H}$, say

$$\mathbf{H}_{i_1}, \ldots, \mathbf{H}_{i_l}$$

are linearly dependent. Then there exist scalars

$$b_{i_1}, \ldots, b_{i_l}$$

not all zero such that

$$b_{i_1}\mathbf{H}_{i_1} + \cdots + b_{i_l}\mathbf{H}_{i_l} = 0$$

Take $c = c_1 \cdots c_n$ with

$$c_{i_j} = b_{i_j},\ 1 \leq j \leq l \quad \text{and} \quad c_i = 0 \text{ for every other } i$$

Then c is a word of weight at most l and $\mathbf{H}c^t = 0$. Thus c is a code word of weight at most l.

Theorem 9.1

Let $\mathscr{C}$ be a linear $[n, k, d]$ code over F with a parity check matrix $\mathbf{H}$. Then $\mathscr{C}$ is an MDS code iff every $n - k$ columns of $\mathbf{H}$ are linearly independent.

Proof

Suppose that $\mathscr{C}$ is an MDS code. Then $d = n - k + 1$ and so there is no non-zero code word of weight at most $n - k$. It follows from Proposition 9.1 that every $n - k$ columns of $\mathbf{H}$ are linearly independent.

Conversely, suppose that every $n - k$ columns of $\mathbf{H}$ are linearly independent. Then there is no non-zero code word of weight at most $n - k$. Therefore

$$d \geq n - k + 1$$

But

$$d \leq n - k + 1$$

always and, so

$$d = n - k + 1$$

Hence $\mathscr{C}$ is an MDS code.

Theorem 9.2

If a linear $[n, k, d]$ code $\mathscr{C}$ is MDS, then so is its dual $\mathscr{C}^\perp$.

Proof

As already seen $\mathscr{C}^\perp$ is a linear $[n, n-k, -]$ code. Let d_1 be the minimum distance of $\mathscr{C}^\perp$. Then

$$d_1 \leq n - (n-k) + 1 = k + 1$$

Let $\mathbf{H}$ be a parity check matrix of $\mathscr{C}$. The code $\mathscr{C}$ being MDS, every $n-k$ columns of $\mathbf{H}$ are linearly independent. Therefore, if any k columns of $\mathbf{H}$ are omitted, the remaining columns in that order (being linearly independent) form a square submatrix of $\mathbf{H}$ of rank $n-k$. Let a be a word of length $n-k$ and suppose that the code word $\mathbf{aH}$ of $\mathscr{C}^\perp$ has at least $n-k$ zeros. Let $\bar{\mathbf{H}}$ be the submatrix of $\mathbf{H}$ obtained by omitting k columns including those which correspond to the possible k non-zero entries of $\mathbf{aH}$. Then $\mathbf{a}\bar{\mathbf{H}}=\mathbf{0}$. As $\bar{\mathbf{H}}$ is a square matrix of order $n-k$ with rank $n-k$, $\bar{\mathbf{H}}$ is non-singular. It then follows from $\mathbf{a}\bar{\mathbf{H}}=\mathbf{0}$ that $\mathbf{a}=\mathbf{0}$ and, hence, $\mathbf{aH}=\mathbf{0}$. This proves that

$$d_1 \geq k+1$$

and, so

$$d_1 = k+1$$

Hence $\mathscr{C}^\perp$ is an MDS code.

Corollary

Let $\mathscr{C}$ be an $[n,k,d]$ linear code over $F=\mathrm{GF}(q)$. Then the following statements are equivalent:

(i) $\mathscr{C}$ is MDS.
(ii) Every k columns of a generator matrix $\mathbf{G}$ of $\mathscr{C}$ are linearly independent.
(iii) Every $n-k$ columns of a parity check matrix $\mathbf{H}$ of $\mathscr{C}$ are linearly independent.

Proof

Equivalence of (i) and (iii) has been proved in Theorem 9.1.

Let $\mathbf{G}$ be a generator matrix of $\mathscr{C}$. By Theorem 5.2 $\mathbf{G}$ is a parity check matrix of $\mathscr{C}^\perp$ which is an $[n, n-k, -]$ linear code. Therefore, by Theorem 9.1, $\mathscr{C}^\perp$ is an MDS code iff every k columns of $\mathbf{G}$ are linearly independent. As

$$(\mathscr{C}^\perp)^\perp = \mathscr{C}$$

it follows from the above theorem that $\mathscr{C}$ is MDS iff $\mathscr{C}^\perp$ is MDS. Hence $\mathscr{C}$ is MDS iff every k columns of $\mathbf{G}$ are linearly independent.

Examples 9.1

Case (i)

Let F be a field of q elements, q a prime power and e be the word with every entry equal to 1. Let $\mathscr{C}$ be the linear space over F generated by e. Then every non-zero element of $\mathscr{C}$ has weight n. Thus $\mathscr{C}$ is a linear code of dimension 1 and minimum distance $n=n-1+1$. Hence, $\mathscr{C}$ is an $[n,1,n]$ MDS code.

Case (ii)

Let F be any finite field and for i, $1 \leq i \leq n$, let e^i be the word of length n with 1 in the ith position and 0 everywhere else. Let $\mathscr{C}$ be the vector space over F generated by e^i, $1 \leq i \leq n$. Obviously the minimum distance of this code is 1. Also its dimension is n. Hence $\mathscr{C}$ is an $[n, n, 1]$ MDS code.

Case (iii)

Let F be any finite field and for any i, $1 \leq i \leq n-1$, let e^i be the word of length n with 1 in the ith and $(i+1)$th position and 0 everywhere else. These $n-1$ words are linearly independent over F and so generate a linear code $\mathscr{C}$ of dimension $n-1$. Clearly the minimum distance of this code is

$$2 = n - (n-1) + 1$$

Hence $\mathscr{C}$ is an $[n, n-1, 2]$ linear MDS code over F.

Case (iv)

For i, $1 \leq i \leq n-2$, let e^i be the binary word of length n with ith, $(i+1)$th and $(i+2)$th entry equal to 1 and every other entry equal to 0. These vectors are linearly independent and so generate a linear code of dimension $n-2$. Observe that $e^1 + e^2$ is a word of weight 2 and the minimum distance of this code is

$$2 < n - (n-2) + 1 = 3$$

This linear $[n, n-2, 2]$ code over any finite field F is not an MDS code.

Case (v)

The $[7, 4, 3]$ binary Hamming code is not MDS and then its dual also cannot be an MDS code.

Case (vi)

Let $F = \mathrm{GF}(3)$ – the field of three elements – and $\mathscr{C}$ be the code of length 3 generated by the matrix

$$\mathbf{G} = \begin{pmatrix} 1 & 0 & 1 \\ 0 & 1 & -1 \end{pmatrix}$$

It is a code of length 3 and dimension 2. Every two columns of the matrix $\mathbf{G}$ are linearly independent and, therefore, the code is MDS. Explicitly, all the code words of this code are:

$$0\ 0\ 0,\quad 1\ 0\ 1,\quad 0\ 1\ -1,\quad -1\ 0\ -1,\quad 0\ -1\ 1,\quad 1\ 1\ 0,\quad -1\ 1\ 1,$$
$$1\ -1\ -1,\quad -1\ -1\ 0$$

and the minimum distance is $2 = 3 - 2 + 1$.

The dual of this code is generated by

$$\mathbf{H} = (1 \quad -1 \quad 1)$$

and is of dimension 1. This minimum distance of this code is 3 as all the code words of this code are:

$$0\ 0\ 0, \quad 1\ -1\ 1, \quad -1\ 1\ -1$$

Exercise 9.1

1. Let α be a primitive cube root of unity and

$$F = \{0, 1, \alpha, \alpha^2\}$$

the field of 4 elements. Prove that the code $\mathscr{C}$ generated by the matrix

$$\mathbf{G} = \begin{pmatrix} 1 & 0 & 1 & 1 \\ 0 & 1 & \alpha & \alpha^2 \end{pmatrix}$$

is an MDS code. Also find its dual and verify that $\mathscr{C}^\perp$ is also MDS.
Find the weight enumerator of $\mathscr{C}$ as well as $\mathscr{C}^\perp$.

2. Let

$$F = \{0, 1, -1\}$$

be the field of 3 elements and $\mathscr{C}$ be the code generated by

$$\mathbf{G} = \begin{pmatrix} 1 & 0 & 1 & 1 \\ 0 & 1 & -1 & 1 \end{pmatrix}$$

Is the code $\mathscr{C}$ MDS? Find $\mathscr{C}^\perp$ also.

3. Find the duals of the codes of Cases (i) – (iii) of Examples 9.1.

Definition 9.2

We have shown that linear $[n, 1, n]$, $[n, n-1, 2]$ and $[n, n, 1]$ codes exist over any finite field F and these are MDS codes. These are called **trivial MDS codes**.

Proposition 9.2

The only binary MDS codes are the trivial codes.

Proof

Let $\mathscr{C}$ be a binary $[n, k, d]$ MDS code. If $k = 1$, then $\mathscr{C}$ is a trivial MDS code and so we may suppose that $k > 1$. Let $\mathbf{G}$ be a generator matrix of $\mathscr{C}$ with the first k columns of $\mathbf{G}$ forming the identity matrix. If $n > k + 1$, then $\mathscr{C}$ has a column, say jth, of weight less than k and greater than 1. Suppose that the ith entry of this column is 0. Then the first k columns of $\mathbf{G}$ except the ith together with the jth column are linearly dependent. This proves that $\mathscr{C}$ cannot be an MDS code. Hence

$$k \le n \le k + 1$$

and $\mathscr{C}$ is a trivial MDS code. ■

We shall come to a similar property of MDS codes when we discuss the existence of MDS codes.

Using Theorem 9.1, we now prove another useful criterion for MDS codes.

Theorem 9.3
Let $\mathscr{C}$ be an $[n, k, -]$ code with parity check matrix

$$\mathbf{H} = (\mathbf{A} \quad \mathbf{I}_{n-k})$$

Then $\mathscr{C}$ is an MDS code iff every square submatrix of $\mathbf{A}$ is non-singular.

Proof
Let $\mathbf{B}_r$ be a square submatrix of $\mathbf{A}$ constituted by parts of the i_1th, i_2th, ..., i_rth rows of $\mathbf{A}$ with

$$i_1 < i_2 < \cdots < i_r (\leq n - k)$$

Let $\mathbf{M}_r$ be the square submatrix of $\mathbf{H}$ of order $n-k$ constituted by the columns of $\mathbf{A}$ parts of which occur in $\mathbf{B}_r$ and the remaining $n-k-r$ columns from the identity matrix $\mathbf{I}_{n-k}$ which are different from i_1th, i_2th, ..., i_rth columns of $\mathbf{I}_{n-k}$. Then

$$\det \mathbf{M}_r = \pm \det \mathbf{B}_r$$

so that $\mathbf{B}_r$ is non-singular iff $\mathbf{M}_r$ is. Therefore, every $n-k$ columns of $\mathbf{H}$ are linearly independent iff every square submatrix of $\mathbf{A}$ is non-singular. The result then follows from Theorem 9.1. ■

The above theorem can equally well be stated in terms of generator matrix.

Theorem 9.4
Let $\mathscr{C}$ be an $[n, k, -]$ code with generator matrix

$$\mathbf{G} = (\mathbf{I}_k \quad \mathbf{A})$$

Then $\mathscr{C}$ is an MDS code iff every square submatrix of $\mathbf{A}$ is non-singular.

Examples 9.2

Case (i)
Consider the matrix

$$\mathbf{A} = \begin{pmatrix} 1 & 6 & 2 & 5 & 1 \\ 1 & 4 & 3 & 3 & 6 \\ 1 & 5 & 5 & 1 & 5 \end{pmatrix}$$

over GF(7). It is clear that every square submatrix of order 2 is non-singular.

There are ${}^5C_3 = 10$ square submatrices of order 3:

$$\det\begin{pmatrix}1 & 6 & 2\\ 1 & 4 & 3\\ 1 & 5 & 5\end{pmatrix} = 2 \qquad \det\begin{pmatrix}1 & 6 & 5\\ 1 & 4 & 3\\ 1 & 5 & 1\end{pmatrix} = 6 \qquad \det\begin{pmatrix}1 & 6 & 1\\ 1 & 4 & 6\\ 1 & 5 & 5\end{pmatrix} = 4$$

$$\det\begin{pmatrix}1 & 2 & 5\\ 1 & 3 & 3\\ 1 & 5 & 1\end{pmatrix} = 2 \qquad \det\begin{pmatrix}1 & 2 & 1\\ 1 & 3 & 6\\ 1 & 5 & 5\end{pmatrix} = 3 \qquad \det\begin{pmatrix}1 & 5 & 1\\ 1 & 3 & 6\\ 1 & 1 & 5\end{pmatrix} = 5$$

$$\det\begin{pmatrix}6 & 2 & 5\\ 4 & 3 & 3\\ 5 & 5 & 1\end{pmatrix} = 3 \qquad \det\begin{pmatrix}6 & 2 & 1\\ 4 & 3 & 6\\ 5 & 5 & 5\end{pmatrix} = 5 \qquad \det\begin{pmatrix}6 & 5 & 1\\ 4 & 3 & 6\\ 5 & 1 & 5\end{pmatrix} = 2$$

$$\det\begin{pmatrix}2 & 5 & 1\\ 3 & 3 & 6\\ 5 & 1 & 5\end{pmatrix} = 4$$

Thus every square submatrix of **A** is non-singular and **A** can be used to obtain two MDS codes:

(a) The $[8, 3, -]$ code over GF(7) with generator matrix $\mathbf{G} = (\mathbf{I}_3 \quad \mathbf{A})$ is an MDS code.

(b) The $[8, 5, -]$ code over GF(7) with parity check matrix $\mathbf{H} = (\mathbf{A} \quad \mathbf{I}_3)$ is an MDS code.

Case (ii)

Consider the matrix

$$\mathbf{A} = \begin{pmatrix}1 & 2 & 3 & 4\\ 1 & 3 & 1 & 2\end{pmatrix}$$

over GF(5). It is clear that every square submatrix of **A** is non-singular and, so, this matrix gives two MDS codes:

(a) The code with generator matrix $\mathbf{G} = (\mathbf{I}_2 \quad \mathbf{A})$ is a $[6, 2, -]$ MDS code over GF(5).

(b) The code with parity check matrix $\mathbf{H} = (\mathbf{A} \quad \mathbf{I}_2)$ is a $[6, 4, -]$ MDS code over GF(5).

Exercise 9.2

1. Does there exist a ternary MDS code of length $n \geq 5$ and dimension 2?
2. Construct a ternary MDS code of length 4 and dimension 2.
3. Does there exist a ternary MDS code of dimension 3 and length (i) 5? (ii) 6?
4. Does there exist an MDS code over GF(5) of length $n \geq 7$ and dimension 2?
5. Construct all possible MDS codes over GF(5) of dimension 2 and length (i) 4; (ii) 5; (iii) 6.

6. Give reasonable necessary and sufficient conditions for a polynomial code over GF(q) with generator polynomial $g(X)$ to be MDS.
7. Give reasonable necessary and sufficient conditions for a cyclic code of length n over GF(q), g.c.d.$(n, q) = 1$ to be MDS.

9.2 THE WEIGHT DISTRIBUTION OF MDS CODES

Proposition 9.3
Let $\mathscr{C}$ be an $[n, k, d]$ MDS code. Then any k symbols of the code words may be taken as message symbols.

Proof
Let $i_1, i_2, \ldots, i_k$ be the chosen k positions. Let $\mathbf{G}$ be a generator matrix of $\mathscr{C}$. Then $\mathbf{G}$ is a $k \times n$ matrix, every k columns of which are linearly independent. Take

$$\bar{\mathbf{G}} = (\mathbf{G}_{i_1} \quad \mathbf{G}_{i_2} \quad \cdots \quad \mathbf{G}_{i_k})$$

where $\mathbf{G}_1, \mathbf{G}_2, \ldots, \mathbf{G}_n$ are the columns of $\mathbf{G}$. Then $\bar{\mathbf{G}}$ is an invertible matrix and so

$$\mathbf{G}' = \bar{\mathbf{G}}^{-1}\mathbf{G}$$

is also a generator matrix of $\mathscr{C}$. Given a message word $a = a_1, a_1 \text{---} a_k$, let

$$\mathbf{a}' = \mathbf{a}\bar{\mathbf{G}}^{-1}$$

Then

$$\mathbf{a} = \mathbf{a}'\bar{\mathbf{G}} = (\mathbf{a}'\mathbf{G}_{i_1} \quad \cdots \quad \mathbf{a}'\mathbf{G}_{i_k})$$

Now the code word in the code $\mathscr{C}$ corresponding to the message word a (and its associated vector $\mathbf{a}$) is

$$\mathbf{a}\mathbf{G}' = \mathbf{a}\bar{\mathbf{G}}^{-1}\mathbf{G} = \mathbf{a}'\mathbf{G} = (\cdots \quad a'\mathbf{G}_{i_1} \quad \cdots \quad a'\mathbf{G}_{i_k} \quad \cdots)$$

in which the entries in the i_1th, i_2th, $\ldots$, i_kth positions are $a_1, a_2, \ldots, a_k$ respectively.

Theorem 9.5
Let $\mathscr{C}$ be an $[n, k, d]$ code over GF(q). Then $\mathscr{C}$ is an MDS code iff $\mathscr{C}$ has a minimum distance code word with non-zero entries in any d coordinates.

Proof
Let $\mathscr{C}$ be an MDS code. By the above proposition, any k coordinates can be taken as positions of the message symbols. Let $d = n - k + 1$ coordinate positions be given. Take one of these, say ith, and the complementary $k - 1$ coordinates as message symbols. Consider the message word which has 1 in

the position corresponding to the ith position in code words and 0 elsewhere in the remaining $k-1$ positions. The code word corresponding to this message word has a non-zero entry in one of the chosen d positions (the ith position) and 0 in every one of the $k-1$ complementary positions. The weight of this code word being $n-k+1$, each one of the remaining $n-k$ chosen positions must be occupied by a non-zero entry.

Conversely, suppose that $\mathscr{C}$ has a code word of weight d in any d coordinate positions. If $d=n-k+1$, we have nothing to prove. Suppose that $d \leq n-k$.

Let $\mathbf{G}$ be a generator matrix of $\mathscr{C}$. Then $\mathbf{G}$ is a $k \times n$ matrix of rank k. Therefore $\mathbf{G}$ has a set of k linearly independent columns. For the sake of simplicity (of notation), let us assume that the first k columns of $\mathbf{G}$ are linearly independent. Let $\bar{\mathbf{G}}$ be the submatrix of $\mathbf{G}$ formed by the first k columns of $\mathbf{G}$. Set

$$\mathbf{G}' = \bar{\mathbf{G}}^{-1}\mathbf{G}$$

Then $\mathbf{G}'$ is also a generator matrix of $\mathscr{C}$ and the first k columns of $\mathbf{G}'$ form the identity matrix of order k. Let $\mathbf{b}=\mathbf{a}\mathbf{G}'$ be the vector associated with a code word of weight d with the non-zero entries in the last d coordinate positions. If

$$b = b_1 b_2 \text{---} b_n \quad \text{and} \quad a = a_1 \text{---} a_k$$

then $b_i = a_i$ for $1 \leq i \leq k$ and $b_i = 0$ for $1 \leq i \leq n-d$. Since $d \leq n-k$, $k \leq n-d$ and, therefore, $a_i = b_i = 0$ for $1 \leq i \leq k$, i.e. $a=0$. But then $\mathbf{b}=\mathbf{a}\mathbf{G}'=\mathbf{0}-\mathbf{a}$ contradiction. Hence $d=n-k+1$ and $\mathscr{C}$ is an MDS code.

Corollary

The number of code words of weight $n-k+1$ in an $[n, k, d]$ MDS code over GF(q) is

$$(q-1)\binom{n}{n-k+1}$$

Proof

Let $n-k+1$ coordinate positions be given and let

$$b = b_1 b_2 \text{---} b_n$$

be a code word of weight $n-k+1$ with non-zero entries at the given coordinate positions. Since every non-zero multiple of b gives a new code word of weight $n-k+1$ with non-zero entries at the given positions, we obtain $(q-1)$ code words with this property. Let

$$c = c_1 c_2 \text{---} c_n$$

be a code word of weight $n-k+1$ with non-zero entries at the given positions. If

$$c \notin \{\alpha b \mid \alpha \in \mathrm{GF}(q), \alpha \neq 0\}$$

then there exists an i, $1 \le i \le n$, such that

$$0 \ne c_i \notin \{\alpha b_i | \alpha \in \mathrm{GF}(q), \alpha \ne 0\}$$

But then

$$0(\{c_i\} u \{\alpha b_i | \alpha \in \mathrm{GF}(q), \alpha \ne 0\}) = q$$

and every element of the set is non-zero – a contradiction. Hence c is a scalar multiple of b. Thus, given any $n-k+1$ positions, there are exactly $q-1$ code words with non-zero entries at these positions. Since $n-k+1$ positions can be chosen out of the n positions in

$$\binom{n}{n-k+1} \text{ ways}$$

the total number of code words of weight $n-k+1$ is

$$(q-1)\binom{n}{n-k+1}$$

Examples 9.3

Case (i)
In the code of Case (vi) in Examples 9.1, there are 6 code words of weight 2, 2 code words of weight 3 and 1 code word of weight 0. Therefore

$$W_{\mathscr{C}}(x, y) = x^3 + 6xy^2 + 2y^3$$

also

$$W_{\mathscr{C}^\perp} = x^3 + 2y^3$$

Case (ii)
Let $\mathscr{C}$ be the $[4, 2, -]$ code of question 2 in Exercise 9.1. Also, the code words of $\mathscr{C}$ are:

$$0\ 0\ 0\ 0,\quad 1\ 0\ 1\ 1,\quad -1\ 0\ -1\ -1,\quad 0\ 1\ -1\ 1,\quad 0\ -1\ 1\ -1,$$
$$1\ 1\ 0\ -1,\quad -1\ 1\ 1\ 0,\quad 1\ -1\ -1\ 0,\quad -1\ -1\ 0\ 1$$

Therefore

$$W_{\mathscr{C}} = x^4 + 8xy^3$$

Observe that

$$8 = (3-1)\binom{4}{4-2+1}$$

The dual $\mathscr{C}^\perp$ is generated by the matrix

$$\begin{pmatrix} -1 & 1 & 1 & 0 \\ -1 & -1 & 0 & 1 \end{pmatrix}$$

and all the code words of $\mathscr{C}^{\perp}$ are

$$0\ 0\ 0\ 0,\quad -1\ 1\ 1\ 0,\quad 1\ -1\ -1\ 0,\quad -1\ -1\ 0\ 1,\quad 1\ 1\ 0\ -1,$$
$$1\ 0\ 1\ 1,\quad 0\ \ 1\ -1\ 1,\quad 0\ -1\ 1\ -1,\quad -1\ 0\ -1\ -1$$

and

$$W_{\mathscr{C}^{\perp}} = x^4 + 8xy^3$$

Observe that $\mathscr{C}^{\perp} = \mathscr{C}$ and so this code is self dual.

Exercise 9.3

Find the number of minimum distance code words in the codes of question 1 in Exercise 9.1, and Case (iv) of Examples 9.1. Find also the weight enumerators of these codes.

9.3 AN EXISTENCE PROBLEM

In this section we throw some light on the following problems.

Problem 9.1
Given k and q, find the largest value of n for which an $[n, k, n-k+1]$ MDS code exists over GF(q). We denote this largest value of n by $m(k, q)$. When $q = 2$ and $k \neq 1$, it follows from Proposition 9.2 that

$$m(k, 2) = k + 1 \quad \text{or} \quad m(k, 2) = k$$

We shall obtain here a general theorem of which this is a particular case. But we first translate this problem into one of linear algebra.

In view of the corollary to Theorem 9.2, it follows that this problem is equivalent to the following.

Problem 9.2
Given k and q, find the largest n for which there is a $k \times n$ matrix over GF(q), every k columns of which are linearly independent.

Given a $k \times n$ matrix **G** over GF(q) every k columns of which are linearly independent, let V be a vector space generated by some k columns of **G**. The vector space V then has a set of n vectors (i.e. the set of all columns of **G**) such that every k vectors of these are linearly independent. On the other hand, if V is a k-dimensional vector space having a set of n vectors, every k elements of which are linearly independent, regarding these n vectors as columns of length k, we obtain a $k \times n$ matrix **G** over GF(q) such that every k columns of **G** are linearly independent. In view of this, Problem 9.2 translates into the following.

Problem 9.3
Given a k-dimensional vector space V over GF(q), what is the order of a largest subset of V with the property that every k of these vectors form a basis of V?

It has been conjectured that

$$m(k,q)=\begin{cases} q+1 & \text{for } 2\le k\le q-1 \\ k+1 & \text{for } q\le k \end{cases}$$

except that

$$m(3,q)=m(q-1,q)=q+2 \quad \text{if } q=2^m$$

Theorem 9.6

$$m(2,q)=q+1$$

for any prime power q.

Proof
Let V be a 2-dimensional vector space over GF(q) and let α be a primitive element of GF(q). Let S be a largest subset of V with the property that every two elements of S are linearly independent over GF(q). The set S contains at least two elements, say $\mathbf{e}_1$ and $\mathbf{e}_2$. Every other element of S is then of the form

$$\alpha^i\mathbf{e}_1+\alpha^j\mathbf{e}_2$$

for suitable non-negative integers i, j. Given i, $0\le i\le q-2$, let

$$S_i=\{\beta(\mathbf{e}_1+\alpha^i\mathbf{e}_2)\mid 0\ne\beta\in \mathrm{GF}(q)\}$$

Any two elements of S_i are linearly dependent, and at the most one element from each S_i, $0\le i\le q-2$, belongs to S. Also no element from an S_i is a scalar multiple of any element from S_j for $j\ne i$. Hence, at least one element from each S_i, $0\le i\le q-2$, belongs to S, for otherwise the set S can be enlarged without upsetting the linear independence of any two elements. Thus S contains $q+1$ elements and

$$m(2,q)=q+1$$

Theorem 9.7

$$m(k,q)=k+1 \quad \text{for } q\le k$$

Proof
Let V be a k-dimensional vector space over GF(q). Let S be a largest subset of V with the property that any k vectors from the set S are linearly independent. Choose any k vectors $\mathbf{e}_1,\mathbf{e}_2,\ldots,\mathbf{e}_k$ from the set S. Every other element of S is then of the form

$$\sum_{1\le i\le k} a_i\mathbf{e}_i$$

where a_i belong to the multiplicative group GF(q)* of GF(q). Suppose two elements of this form

$$\sum_{1 \le i \le k} a_i \mathbf{e}_i \quad \text{and} \quad \sum_{1 \le i \le k} b_i \mathbf{e}_i$$

where a_i, $b_j \in \mathrm{GF}(q)^*$ are in S.

Consider the k equations

$$a_i x_i = b_i \quad 1 \le i \le k$$

Each of these k equations has a unique solution in $\mathrm{GF}(q)^*$. Since the order of $\mathrm{GF}(q)^*$ is $q-1<k$, at least two of these equations have the same solution. Without loss of generality, we may assume that

$$a_1 x = b_1 \quad \text{and} \quad a_2 x = b_2$$

have the same solution, say x. Then the vector

$$x\left(\sum_{1 \le i \le k} a_i x_i\right) - \sum_{1 \le i \le k} b_i x_i$$

is a linear combination of the $k-2$ vectors $\mathbf{e}_3, \ldots, \mathbf{e}_k$. The vectors

$$\mathbf{e}_3, \ldots, \mathbf{e}_k \quad \sum_{1 \le i \le k} a_i \mathbf{e}_i \quad \text{and} \quad \sum_{1 \le i \le k} b_i \mathbf{e}_i$$

are thus linearly dependent. This is a contradiction and hence S can contain at the most one element of the form

$$\sum_{1 \le i \le k} a_i \mathbf{e}_i \quad a_i \in \mathrm{GF}(q)^*$$

Thus

$$m(k, q) \le k+1$$

Moreover, S must contain at least one element of the form

$$\sum_{1 \le i \le k} a_i \mathbf{e}_i \quad a_i \in \mathrm{GF}(q)^*$$

for otherwise S can be enlarged to a set S^* in which any k vectors are linearly independent. Therefore

$$m(k, q) = k+1$$

9.4 REED–SOLOMON CODES

A class of examples of MDS codes is provided by Reed–Solomon codes. Let $F = \mathrm{GF}(q)$ be a field of order q where q is a prime power. A Reed–Solomon code is a BCH code of length $n = q-1$ over F.

Recall that, for the construction of a BCH code of length n over F, we need to find a primitive element α in an extension field K of F with the degree $[K:F]$ equal to r, where r is the least positive integer satisfying

$$q^r \geq n+1$$

Therefore, for a Reed–Solomon code $r=1$ and α is a primitive element of F itself. Then $X-\alpha^i$ is the minimal polynomial of α^i. Generator polynomial of the Reed–Solomon code $\mathscr{C}$ with minimum distance d is given by

$$g(X)=(X-\alpha)(X-\alpha^2)\cdots(X-\alpha^{d-1})$$

which is of degree $d-1$. Hence the dimension k of $\mathscr{C}$ is

$$k=n-(d-1)=n-d+1 \quad \text{and} \quad d=n-k+1.$$

Therefore, we have the following theorem.

Theorem 9.8
A Reed–Solomon code is an MDS code.

Examples 9.4

Case (i)
Consider $F=\mathrm{GF}(7)$ in which 3 is a primitive element. The Reed–Solomon code with minimum distance $d=5$ has generator polynomial

$$\begin{aligned} g(X) &= (X-3)(X-2)(X-6)(X-4) \\ &= (X^2+2X+6)(X^2-3X+3) \\ &= X^4+6X^3+3X^2+2X+4 \end{aligned}$$

All the code words of this code are given by

$$\begin{aligned} &(aX+b)(X^4+6X^3+3X^2+2X+4) \\ &= aX^5+(6a+b)X^4+(3a+6b)X^3+(2a+3b)X^2+(4a+2b)X+4b \end{aligned}$$

where $a, b \in \mathrm{GF}(7)$.

Case (ii)
Consider again $F=\mathrm{GF}(7)$ in which 3 is a primitive element. The Reed–Solomon code with minimum distance $d=6$ has generator polynomial

$$\begin{aligned} g(X) &= (X-3)(X-2)(X-6)(X-4)(X-5) \\ &= X^5+X^4+X^3+X^2+X+1 \end{aligned}$$

and the code words are $aaaaaa$, where $a \in \mathrm{GF}(7)$.

Case (iii)
The polynomial $X^2 + X + 1$ is irreducible over $\mathbb{B}$ and so

$$F = \mathbb{B}[X]/\langle X^2 + X + 1 \rangle$$

is a field of 4 elements. If

$$\alpha = X + \langle X^2 + X + 1 \rangle$$

then α is a primitive element of F and we can take

$$F = \{0, 1, \alpha, \alpha + 1\}$$

A generator polynomial of the Reed–Solomon code $\mathscr{C}$ of length 3 over F and minimum distance $d = 2$ is

$$g(X) = X + \alpha$$

All the elements of $\mathscr{C}$ are

$$(aX + b)(X + \alpha) = aX^2 + (b + a\alpha)X + \alpha b$$

where a, $b \in F$ or

$$000 \quad \alpha 10 \quad \alpha^2\alpha 0 \quad 110 \quad 0\alpha 1 \quad \alpha\alpha^2 1 \quad \alpha^2 01 \quad 111 \quad 0\alpha^2\alpha \quad \alpha 1\alpha \quad \alpha^2 1\alpha$$
$$10\alpha \quad 01\alpha^2 \quad \alpha 0\alpha^2 \quad \alpha^2 1\alpha^2 \quad 1\alpha\alpha^2$$

(Remember! $\alpha^2 = \alpha + 1$.)

Case (iv)
Let $F = \mathrm{GF}(3) = \{0, 1, 2\}$. Then 2 is a primitive element of F and a generator polynomial of the Reed–Solomon code with minimum distance 2 is

$$g(X) = X - 2 = X + 1$$

The code words of this code are given by $(aX + b)(X + 1)$, where $a, b \in F$ or

$$000 \quad 110 \quad 220 \quad 011 \quad 121 \quad 201 \quad 022 \quad 102 \quad 212$$

10

Automorphism group of a code

10.1 AUTOMORPHISM GROUP OF A BINARY CODE

Let $\mathscr{C}$ be a binary code of length n. If σ is a permutation of the set $S=\{1,2,\ldots,n\}$, then $\mathscr{C}'=\{\sigma(c)|c\in\mathscr{C}\}$ is a code equivalent to $\mathscr{C}$. If, however, $\mathscr{C}'=\mathscr{C}$ then σ is called an **automorphism** of the code $\mathscr{C}$. Let Aut($\mathscr{C}$) denote the set of all automorphisms of $\mathscr{C}$. Observe that if σ, τ are in Aut($\mathscr{C}$), then so is $\sigma\tau$. The set S_n of all permutations of S being a finite group, it follows that Aut($\mathscr{C}$) is a subgroup of S_n.

Definition 10.1
The subgroup Aut($\mathscr{C}$) of S_n is called the **automorphism group** of the code $\mathscr{C}$.

Remark 10.1
To every permutation σ of S corresponds a permutation matrix $\mathbf{P}$ of order n such that $\sigma(\mathbf{c})=\mathbf{cP}$ for $\mathbf{c}$, the vector associated with $c\in\mathscr{C}$ and conversely. Writing $(\mathbf{c})\sigma$ for $\sigma(\mathbf{c})$, we find that the map $\sigma\to\mathbf{P}$ gives an isomorphism between the symmetric group S_n of degree n and the group of all permutation matrices of order n. We may thus have (up to isomorphism)

$$\text{Aut}(\mathscr{C})=\{\mathbf{P}|\mathbf{P}\text{ is a permutation matrix of order } n \text{ with } \mathbf{cP}\in\mathscr{C}\ \forall c\in\mathscr{C}\}$$

It is, in general, not easy to determine the automorphism group of a code. We consider some examples.

Examples 10.1

Case (i)
Consider first the repetition code

$$\mathscr{C} = \{00\cdots0 \quad 11\cdots1\}$$

of length n. Every transposition $(1 \quad i) \in \text{Aut}(\mathscr{C})$ and, so,

$$\text{Aut}(\mathscr{C}) = S_n$$

Case (ii)
Let $\mathscr{C}$ be the $(n, n+1)$ parity check code. Then $\mathscr{C}$ is obtained from the set of all words of length n by adding an overall parity check. $\mathscr{C}$ is then a linear code of dimension n with a basis consisting of the n elements $c_1c_2\text{---}c_{n+1}$ of weight 2 with $c_{n+1} = 1$. The transpositions

$$(1 \quad 2), (1 \quad 3), \ldots, (1 \quad n)$$

leave the basis elements of $\mathscr{C}$ unchanged and are, therefore, in $\text{Aut}(\mathscr{C})$. Let

$$c = c_1c_2\text{---}c_nc_{n+1} \in \mathscr{C}$$

Then

$$c_{n+1} = \sum_{i=1}^{n} c_i$$

Applying the transposition $\sigma = (1, n+1)$ to $\mathbf{c}$, gives

$$\sigma(\mathbf{c}) = (c_{n+1} \quad c_2 \quad \cdots \quad c_nc_1)$$

and

$$c_{n+1} = \sum_{i=1}^{n} c_i$$

shows that

$$c_i = \sum_{i=2}^{n+1} c_i$$

Thus $\sigma(\mathbf{c}) \in \mathscr{C}$. Therefore

$$(1, n+1) \in \text{Aut}(\mathscr{C})$$

As the transpositions

$$(12), \ldots, (1 \quad n+1)$$

generate the symmetric group S_{n+1} of degree $n+1$, we have

$$\text{Aut}(\mathscr{C}) = S_{n+1}$$

Case (iii)
Let $\mathscr{C}$ be the code of length 4 generated by 1011, 1001. Then

$$\mathscr{C} = \{0000, 1011, 1001, 0010\}$$

Clearly

$$(1 \quad 4) \in \mathrm{Aut}(\mathscr{C})$$

but none of $(1 \quad 2), (1 \quad 3), (2 \quad 3), (2 \quad 4), (3 \quad 4)$ is in $\mathrm{Aut}(\mathscr{C})$. But then none of

$$(1 \quad 2 \quad 4) = (1 \quad 4)(1 \quad 2)$$

$$(1 \quad 3 \quad 4) = (1 \quad 4)(1 \quad 3)$$

$(1 \quad 4)(2 \quad 3)$ is in $\mathrm{Aut}(\mathscr{C})$. It is also clear that $(1 \quad 2 \quad 3)$, $(2 \quad 3 \quad 4)$, $(1 \quad 2)(3 \quad 4)$, $(1 \quad 3)(2 \quad 4)$ do not belong to $\mathrm{Aut}(\mathscr{C})$. A simple observation of the elements of $\mathscr{C}$ shows that none of the cycles of length 4 is in $\mathrm{Aut}(\mathscr{C})$. Hence

$$\mathrm{Aut}(\mathscr{C}) = \{1, (1 \quad 4)\}$$

Case (iv)
Let

$$\mathscr{C} = \{0000, 1011, 1001, 0010, 1100, 0111, 0101, 1110\}$$

Clearly

$$(1 \quad 2), (1 \quad 4), (2 \quad 4) \in \mathrm{Aut}(\mathscr{C})$$

but

$$(1 \quad 3), (2 \quad 3), (3 \quad 4) \notin \mathrm{Aut}(\mathscr{C})$$

Then

$$(1 \quad 2 \quad 3) = (1 \quad 3)(1 \quad 2)$$

$$(1 \quad 3 \quad 4) = (1 \quad 4)(1 \quad 3)$$

$$(2 \quad 3 \quad 4) = (2 \quad 4)(2 \quad 3)$$

$(1 \quad 2)(3 \quad 4), (1 \quad 3)(2 \quad 4), (1 \quad 4)(2 \quad 3)$ cannot be in $\mathrm{Aut}(\mathscr{C})$.

Since any cycle of length 4 is a product of the three transpositions $(1 \quad 2)$, $(1 \quad 3), (1 \quad 4)$ in some order two of which are in $\mathrm{Aut}(\mathscr{C})$ but one is not, it follows that none of these four cycles is in $\mathrm{Aut}(\mathscr{C})$. Hence

$$\mathrm{Aut}(\mathscr{C}) = \{1, (1 \quad 2), (1 \quad 4), (2 \quad 4), (1 \quad 2 \quad 4), (1 \quad 4 \quad 2)\}$$

Case (v)
Let $\mathscr{C}$ be a linear code and $\mathscr{C}_1$ be a code obtained from $\mathscr{C}$ by adding an overall parity check. Any element of $\mathscr{C}_1$ is of the form

$$c' = (c, c_{n+1})$$

where $c \in \mathscr{C}$ and

$$c_{n+1} = \sum_{i=1}^{n} c_i$$

If

$$\sigma \in \mathrm{Aut}(\mathscr{C}) \quad \text{as } c_{n+1} = \sum c_i = \sum c_{\sigma(i)}$$

then

$$(\sigma(c), c_{n+1}) \in \mathscr{C}_1$$

Thus $\sigma \in \mathrm{Aut}(\mathscr{C}_1)$ and, so, $\mathrm{Aut}(\mathscr{C}) \leq \mathrm{Aut}(\mathscr{C}_1)$.

Remark 10.2

Let $\mathscr{C}$ be a linear code of length n, $\{c^1, c^2, \ldots, c^k\}$ a set of linearly independent elements and σ a permutation of the set $\{1, 2, \ldots, n\}$. Then the elements

$$\sigma(c^1), \sigma(c^2), \ldots, \sigma(c^k)$$

are again linearly independent. Thus if $\mathscr{C}$ is a linear code of dimension k then

$$\sigma(\mathscr{C}) = \{\sigma(c) | c \in \mathscr{C}\}$$

is again a linear code of dimension k. In particular, if $\sigma \in \mathrm{Aut}(\mathscr{C})$, then $\sigma(\mathscr{C}) = \mathscr{C}$.

Proposition 10.1

If $\mathscr{C}$ is a linear code, then

$$\mathrm{Aut}(\mathscr{C}) = \mathrm{Aut}(\mathscr{C}^{\perp})$$

Proof

Let $\sigma \in \mathrm{Aut}(\mathscr{C})$. For $c' \in \mathscr{C}^{\perp}$, $\mathbf{c}(\mathbf{c}')^{\mathrm{t}} = 0 \forall c$ in $\mathscr{C}$ (here $\mathbf{a}^{\mathrm{t}}$ denotes the transpose of the row vector $\mathbf{a}$ and $\mathbf{c}$ is the vector formed from the elements of the code word c) which then implies that

$$\sigma(\mathbf{c})\sigma(\mathbf{c}')^{\mathrm{t}} = 0 \forall c \in \mathscr{C}$$

As $\sigma(\mathscr{C}) = \mathscr{C}$, it follows that $\mathbf{c}\sigma(\mathbf{c}')^{\mathrm{t}} = 0 \forall c \in \mathscr{C}$. Thus $\sigma(c') \in \mathscr{C}^{\perp}$ and, so, $\sigma \in \mathrm{Aut}(\mathscr{C}^{\perp})$. Hence

$$\mathrm{Aut}(\mathscr{C}) \leq \mathrm{Aut}(\mathscr{C}^{\perp})$$

This then implies that ($\mathscr{C}^{\perp}$ being a linear code)

$$\mathrm{Aut}(\mathscr{C}^{\perp}) \leq \mathrm{Aut}((\mathscr{C}^{\perp})^{\perp}) = \mathrm{Aut}(\mathscr{C}) \qquad \blacksquare$$

The above result is not true for non-linear codes.

Examples 10.2

Case (i)

Let

$$\mathscr{C} = \{000, 100, 010, 001, 110, 111\}$$

It is clear that

$$\mathrm{Aut}(\mathscr{C}) = \{1, (1 \quad 2)\}$$

Also $\mathscr{C}^\perp = \{000\}$ and $\mathrm{Aut}(\mathscr{C}^\perp) = S_3$ – the symmetric group of degree 3.

Case (ii)
Let

$$\mathscr{C} = \{000, 110, 111, 101, 010\}$$

Clearly $\mathrm{Aut}(\mathscr{C}) = 1$. Now, $\mathscr{C}^\perp = \{000\}$ and so $\mathrm{Aut}(\mathscr{C}^\perp) = S_3$ – the symmetric group of degree 3.

Incidentally, we have also given an example of a code the automorphism group of which is trivial.

Case (iii)
The code

$$\mathscr{C} = \{000, 100, 010, 001, 110\}$$

is a non-linear code and $(1 \quad 2) \in \mathrm{Aut}(\mathscr{C})$ while $(1 \quad 3), (2 \quad 3)$ are not in $\mathrm{Aut}(\mathscr{C})$. Therefore

$$\mathrm{Aut}(\mathscr{C}) = \{1, (1 \quad 2)\}$$

is a group of order 2.

Proposition 10.2
Let $\mathscr{C}$ be a linear code and $\mathscr{C}_1$ be obtained from $\mathscr{C}$ by adding the all one vector **1**. Then

$$\mathrm{Aut}(\mathscr{C}) \leq \mathrm{Aut}(\mathscr{C}_1)$$

while equality holds if $\mathscr{C}$ is of odd length and the code words have only even weights.

Proof
We need to consider only the case when $\mathbf{1} \notin \mathscr{C}$. Then

$$\mathscr{C}_1 = \mathscr{C} \cup \{\mathbf{c} + \mathbf{1} \mid \mathbf{c} \in \mathscr{C}\}$$

As

$$\sigma(\mathbf{c} + \mathbf{1}) = \sigma(\mathbf{c}) + \mathbf{1} \in \mathscr{C}_1 \, \forall \sigma \in \mathrm{Aut}(\mathscr{C})$$

we have $\mathrm{Aut}(\mathscr{C}) \leq \mathrm{Aut}(\mathscr{C}_1)$.

Now suppose that $\mathscr{C}$ is of odd length and its words have only even weights. Then every word of the form $\mathbf{c} + \mathbf{1}$, $\mathbf{c} \in \mathscr{C}$, has odd weight. For any $\sigma \in \mathrm{Aut}(\mathscr{C}_1)$ and $\mathbf{c} \in \mathscr{C}$, $\sigma(\mathbf{c})$ is in $\mathscr{C}_1$ having even weight and, so, $\sigma(\mathbf{c}) \in \mathscr{C}$. Thus $\mathrm{Aut}(\mathscr{C}_1) \leq \mathrm{Aut}(\mathscr{C})$.

The automorphism group of a cyclic code

The automorphism group of a cyclic code contains all cycles of length n (i.e. cyclic permutations of the set $\{1, 2, \ldots, n\}$) and their powers.

Now n being odd, there exist integers r and s such that

$$1 = 2r + ns$$

Then, with $I = \langle X^n - 1 \rangle$ the ideal of $\mathbb{B}[X]$ generated by $X^n - 1$

$$\begin{aligned} X + I &= X^{2r} \cdot X^{ns} + I \\ &= X^{2r} + I \end{aligned}$$

as $X^{ns} - 1 \in I$. If $r < 0$, let t be the least positive integer such that

$$2r + 2nt > 0$$

Then

$$X^{2r} + I = X^{2(r+nt)} + I$$

Thus

$$\sigma_2 : \{1 + I, X + I, \ldots, X^{n-1} + I\} \to \{1 + I, X + I, \ldots, X^{n-1} + I\}$$

given by

$$\sigma_2(X + I) = X^2 + I$$

is an onto map and hence a permutation. For any polynomial $a(X) \in \mathbb{B}[X]$ of degree at most $n - 1$

$$\sigma_2(a(X) + I) = a(X^2) + I = (a(X) + I)^2$$

so that whenever $a(X) + I$ is in a cyclic code $\mathscr{C}$ of length n (i.e. an ideal of $\mathbb{B}[X]/I$), then so is $\sigma_2(a(X) + I)$. Therefore, σ_2 is in Aut($\mathscr{C}$). Clearly, the order of the permutation σ_2 is the number of elements in the cyclotomic coset C_1 modulo n relative to 2 determined by 1.

Examples 10.3

Case (i)

Let $\mathscr{C} = \langle X + 1 + I \rangle$, where $I = \langle X^3 - 1 \rangle$, be the cyclic code of length 3 generated by $1 + X$. Here the cyclotomic coset $C_1 = \{1, 2\}$ and σ_2 is a transposition. Therefore,

$$\text{Aut}(\mathscr{C}) \geq \{1, \sigma_2, (1\ \ 2\ \ 3), (1\ \ 3\ \ 2)\}$$

and so it is S_3–the symmetric group of degree 3.

Alternatively, observe that

$$\mathscr{C} = \{000, 110, 011, 101\}$$

and it is clear that the transpositions (1 2), (1 3) which generate S_3 are in Aut($\mathscr{C}$) and, therefore, Aut($\mathscr{C}$) = S_3.

Case (ii)
The cyclic code of length 5 generated by $1 + X$ is

$$\begin{aligned}\mathscr{C} &= \{(a_0, a_0 + a_1, a_1 + a_2, a_2 + a_3, a_3) \mid a_i \in \mathbb{B}\} \\ &= \{00000, 11000, 01100, 00110, 00011, 10100, 11110, \\ &\quad 11011, 01111, 01010, 00101, 10010, 10111, 11101, \\ &\quad 01001, 10001\}\end{aligned}$$

The permutation σ_2 maps

$$1 \to 1,\ X \to X^2,\ X^2 \to X^4,\ X^3 \to X,\ X^4 \to X^3$$

and so

$$\sigma_2 = (2\quad 3\quad 5\quad 4)$$

which is a cycle of length 4.

However, a simple observation shows that

$$(1\quad 2), (1\quad 3), (1\quad 4), (1\quad 5) \in \mathrm{Aut}(\mathscr{C})$$

and, therefore, Aut($\mathscr{C}$) = S_5.

Exercise 10.1
Determine the automorphism group of the (4, 7) binary Hamming code.

10.2 AUTOMORPHISM GROUP OF A NON-BINARY CODE

Definition 10.2
A **monomial matrix** over a field F is a square matrix with exactly one non-zero entry in every row and in every column.

For example

$$\begin{pmatrix} 0 & 2 & 0 \\ 3 & 0 & 0 \\ 0 & 0 & 1 \end{pmatrix}$$

is a monomial matrix of order 3 while

$$\begin{pmatrix} 0 & 2 & 0 \\ 3 & 0 & 0 \\ 1 & 0 & 0 \end{pmatrix}$$

is not a monomial matrix.

Observe that a monomial matrix over $\mathbb{B}$ is just a permutation matrix. On the other hand, if **M** is a monomial matrix of order n over a field F with d_i, $1 \le i \le n$, being the non-zero entry in the ith row of **M**, then $\mathbf{M} = \mathbf{DP}$, where

$$\mathbf{D} = \mathrm{diag}(d_1, d_2, \ldots, d_n)$$

and **P** is the permutation matrix obtained from **M** by replacing every non-zero entry of **M** by 1. Alternatively, we can also write $\mathbf{M} = \mathbf{PD}'$ with

$$\mathbf{D}' = \mathrm{diag}(d'_1, d'_2, \ldots, d'_n)$$

where, for $1 \le i \le n$, d'_i denotes the non-zero entry of **M** in the ith column.

Every monomial matrix of order n over a field F is invertible and the set of all monomial matrices of order n forms a group under multiplication. This group is called the **monomial group** of degree n.

Definition 10.3

The **automorphism group** $\mathrm{Aut}(\mathscr{C})$ of a linear code $\mathscr{C}$ over $\mathrm{GF}(q)$, q a prime, is the set of all monomial matrices **M** over $\mathrm{GF}(q)$ such that $c\mathbf{M} \in \mathscr{C}\ \forall c \in \mathscr{C}$.

The product of two elements in $\mathrm{Aut}(\mathscr{C})$ is again in $\mathrm{Aut}(\mathscr{C})$ and the monomial group over $\mathrm{GF}(q)$ being finite, $\mathrm{Aut}(\mathscr{C})$ is indeed a group.

Theorem 10.1

If $\mathscr{C}$ is a linear $[n, 1, -]$ code over $F = \mathrm{GF}(q)$, q a prime, then order of $\mathrm{Aut}(\mathscr{C})$ is $(q-1)^{n-m+1}\,(m!)$ where m is the number of non-zero components in a basis vector of $\mathscr{C}$. ($m!$ denotes the product of $1, 2, \ldots, m$.)

Proof

Let

$$\mathbf{x} = (x_1 \quad x_2 \quad \cdots \quad x_n)$$

be a basis vector of $\mathscr{C}$. If

$$\mathbf{y} = (y_1 \quad y_2 \quad \cdots \quad y_n)$$

is another element of $\mathscr{C}$ which also generates $\mathscr{C}$, then **y** is a multiple of **x**. Therefore

$$y_i = 0 \quad \text{iff} \quad x_i = 0$$

i.e. the positions of non-zero components in any vector forming a basis of $\mathscr{C}$ remain unchanged. Let $\mathbf{M} = \mathbf{PD}$ where **P** is a permutation matrix of order n and **D** is the diagonal matrix

$$\mathrm{diag}(d_1, d_2, \ldots, d_n)$$

with $d_i \neq 0$, $1 \le i \le n$. Let σ be the permutation of the set $\{1, 2, \ldots, n\}$ corresponding to the permutation matrix **P**. Then

$$\mathbf{xPD} = (d_1 x_{\sigma(1)} \quad \cdots \quad d_n x_{\sigma(n)})$$

Therefore, $\mathbf{xPD} = ax$ for some $a \neq 0$ in F iff

$$d_i x_{\sigma(i)} = a x_i \forall i, \quad 1 \leq i \leq n$$

This, in particular, shows that

$$x_{\sigma(i)} \neq 0 \quad \text{iff} \quad x_i \neq 0$$

i.e. σ is effectively a permutation of the non-zero component positions in $\mathbf{x}$. Thus $\mathbf{xPD} \in \mathscr{C}$ iff:

(i) σ is effectively a permutation of the non-zero component positions in x and
(ii) $d_i x_{\sigma(i)} = a x_i$ for some $a \neq 0$ in F.

The number of permutations σ which are effectively permutations of the non-zero component positions in x is $m!$ and the number of choices for a is $q-1$. Also, every diagonal entry d_i corresponding to $x_i = 0$ has $q-1$ choices. Hence, the total number of choices for $\mathbf{D}$ is $(q-1)^{n-m+1}$ and, therefore, the number of choices for $\mathbf{PD}$ in Aut($\mathscr{C}$) is $(q-1)^{n-m+1}$ $(m!)$, i.e.

$$\text{order of Aut}(\mathscr{C}) = (q-1)^{n-m+1}(m!)$$

Remark 10.3

Every monomial matrix over $\mathbb{B}$ being a permutation matrix and every permutation matrix may be regarded as a permutation of the set $\{1, 2, \ldots, n\}$, Aut($\mathscr{C}$) for a binary linear code as defined earlier is identical with Aut($\mathscr{C}$) as defined above. We could, as such, have avoided giving separate definitions for Aut($\mathscr{C}$) for binary and non-binary codes but the procedure adopted is more instructive, especially for binary codes.

Examples 10.4

Case (i)

For the [3, 1, 2] ternary linear code

$$\mathscr{C} = \{000, 110, 220\}$$

we have

$$\text{Aut}(\mathscr{C}) = \left\{ \mathbf{I}, \begin{pmatrix} 0 & 1 & 0 \\ 1 & 0 & 0 \\ 0 & 0 & d \end{pmatrix}, 2\mathbf{I}, \begin{pmatrix} 0 & 2 & 0 \\ 2 & 0 & 0 \\ 0 & 0 & d \end{pmatrix}; \text{ where } d = 1, 2 \right\}$$

Case (ii)

The automorphism group of the [3, 1, 2] ternary code

$$\mathscr{C} = \{000, 101, 202\}$$

is

$$\left\{ \mathbf{I}, \begin{pmatrix} 0 & 0 & 1 \\ 0 & d & 0 \\ 1 & 0 & 0 \end{pmatrix}, 2\mathbf{I}, \begin{pmatrix} 0 & 0 & 2 \\ 0 & d & 0 \\ 2 & 0 & 0 \end{pmatrix}; \text{ where } d = 1, 2 \right\}$$

Case (iii)

Next, consider the [3, 1, 3] ternary code

$$\mathscr{C} = \{000, 111, 222\}$$

The basis word 111 is left invariant by every element of S_3 and so the order of Aut($\mathscr{C}$) is $(2 \times 3!) = 12$. Also

$$\mathrm{Aut}(\mathscr{C}) = \left\{ \mathbf{I}, \begin{pmatrix} 0 & 1 & 0 \\ 1 & 0 & 0 \\ 0 & 0 & 1 \end{pmatrix}, \begin{pmatrix} 0 & 0 & 1 \\ 0 & 1 & 0 \\ 1 & 0 & 0 \end{pmatrix}, \begin{pmatrix} 1 & 0 & 0 \\ 0 & 0 & 1 \\ 0 & 1 & 0 \end{pmatrix}, \begin{pmatrix} 0 & 1 & 0 \\ 0 & 0 & 1 \\ 1 & 0 & 0 \end{pmatrix}, \begin{pmatrix} 0 & 0 & 1 \\ 1 & 0 & 0 \\ 0 & 1 & 0 \end{pmatrix}, \right.$$

$$\left. 2\mathbf{I}, \begin{pmatrix} 0 & 2 & 0 \\ 2 & 0 & 0 \\ 0 & 0 & 2 \end{pmatrix}, \begin{pmatrix} 0 & 0 & 2 \\ 0 & 2 & 0 \\ 2 & 0 & 0 \end{pmatrix}, \begin{pmatrix} 2 & 0 & 0 \\ 0 & 0 & 2 \\ 0 & 2 & 0 \end{pmatrix}, \begin{pmatrix} 0 & 2 & 0 \\ 0 & 0 & 2 \\ 2 & 0 & 0 \end{pmatrix}, \begin{pmatrix} 0 & 0 & 2 \\ 2 & 0 & 0 \\ 0 & 2 & 0 \end{pmatrix} \right\}$$

Case (iv)

Consider the [3, 1, 3] code over GF(5) generated by $(1 \quad 2 \quad 3)$. Let $\mathbf{P}$ be a permutation matrix of order 3 with σ as its corresponding permutation and

$$\mathbf{D} = \mathrm{diag}(d_1, d_2, d_3),\ d_1 d_2 d_3 \neq 0$$

Then

$$(1 \quad 2 \quad 3)\mathbf{PD} = a(1 \quad 2 \quad 3)$$

for some $a \neq 0$ in GF(5) iff

$$(d_1\sigma(1), d_2\sigma(2), d_3\sigma(3)) = a(1 \quad 2 \quad 3)$$

i.e. iff

$$d_1 = a\sigma(1)^{-1}$$
$$d_2 = a2\sigma(2)^{-1}$$
$$d_3 = a3\sigma(3)^{-1}$$

Therefore

$$(1 \quad 2 \quad 3)\mathbf{PD} = a(\sigma(1)^{-1} \quad 2\sigma(2)^{-1} \quad 3\sigma(3)^{-1})$$

Giving all possible values to σ we find

$$\mathrm{Aut}(\mathscr{C}) = \left\{ a\begin{pmatrix} 1 & 0 & 0 \\ 0 & 1 & 0 \\ 0 & 0 & 1 \end{pmatrix}, a\begin{pmatrix} 0 & 2 & 0 \\ 3 & 0 & 0 \\ 0 & 0 & 1 \end{pmatrix}, a\begin{pmatrix} 0 & 0 & 3 \\ 0 & 1 & 0 \\ 2 & 0 & 0 \end{pmatrix}, \right.$$

$$\left. a\begin{pmatrix} 1 & 0 & 0 \\ 0 & 0 & 4 \\ 0 & 4 & 0 \end{pmatrix}, a\begin{pmatrix} 0 & 0 & 3 \\ 3 & 0 & 0 \\ 0 & 4 & 0 \end{pmatrix}, a\begin{pmatrix} 0 & 2 & 0 \\ 0 & 0 & 4 \\ 2 & 0 & 0 \end{pmatrix} \right\}$$

where a runs over all the non-zero elements of GF(5).

Case (v)

Let $\mathscr{C}$ be the linear code of length 3 over GF(5) generated by 102, 201. The two elements being linearly independent, $\mathscr{C}$ is of dimension 2 and

$$\mathscr{C} = \{000, 102, 204, 301, 403, 201, 402, 103, 304, 303, 004, 200, 401, 400, 101, 302, 003, 002, 203, 404, 100, 104, 300, 001, 202\}$$

Let $\mathbf{M}$ be a monomial matrix of order 3 with $(1, 2)$ entry or $(3, 2)$ entry non-zero. Then the second column of $\mathbf{M}$ is

$$\begin{pmatrix} 0 \\ 0 \\ a \end{pmatrix} \quad \text{or} \quad \begin{pmatrix} a \\ 0 \\ 0 \end{pmatrix}$$

where $a \neq 0$. The products

$$(1 \quad 0 \quad 2) \begin{pmatrix} a \\ 0 \\ 0 \end{pmatrix} = a$$

$$(2 \quad 0 \quad 1) \begin{pmatrix} a \\ 0 \\ 0 \end{pmatrix} = 2a$$

$$(1 \quad 0 \quad 2) \begin{pmatrix} 0 \\ 0 \\ a \end{pmatrix} = 2a$$

$$(2 \quad 0 \quad 1) \begin{pmatrix} 0 \\ 0 \\ a \end{pmatrix} = a$$

show that in $(1 \quad 0 \quad 2)\mathbf{M}$ and $(2 \quad 0 \quad 1)\mathbf{M}$ the middle entries are non-zero. Therefore $\mathbf{M} \notin \mathrm{Aut}(\mathscr{C})$. Hence if $\mathbf{M} \in \mathrm{Aut}(\mathscr{C})$, then

$$\mathbf{M} = \begin{pmatrix} 0 & 0 & a \\ 0 & b & 0 \\ c & 0 & 0 \end{pmatrix}$$

or

$$\mathbf{M} = \begin{pmatrix} a & 0 & 0 \\ 0 & b & 0 \\ 0 & 0 & c \end{pmatrix}$$

where $abc \neq 0$. Observe that

$$(1 \quad 0 \quad 2) \begin{pmatrix} 0 & 0 & a \\ 0 & b & 0 \\ c & 0 & 0 \end{pmatrix} = x(102) + y(201)$$

where $x = c - a$, $y = 3a + 3c$. Thus

$$(1 \quad 0 \quad 2) \quad \begin{pmatrix} 0 & 0 & a \\ 0 & b & 0 \\ c & 0 & 0 \end{pmatrix} \in \mathscr{C}$$

Similarly, we can show that

$$(2 \quad 0 \quad 1) \quad \begin{pmatrix} 0 & 0 & a \\ 0 & b & 0 \\ c & 0 & 0 \end{pmatrix} \in \mathscr{C}$$

Hence

$$\begin{pmatrix} 0 & 0 & a \\ 0 & b & 0 \\ c & 0 & 0 \end{pmatrix} \in \mathrm{Aut}(\mathscr{C})$$

Similarly, we can prove that

$$\begin{pmatrix} a & 0 & 0 \\ 0 & b & 0 \\ 0 & 0 & c \end{pmatrix} \in \mathrm{Aut}(\mathscr{C})$$

We thus have

$$\mathrm{Aut}(\mathscr{C}) = \left\{ \begin{pmatrix} a & 0 & 0 \\ 0 & b & 0 \\ 0 & 0 & c \end{pmatrix}, \begin{pmatrix} 0 & 0 & a \\ 0 & b & 0 \\ c & 0 & 0 \end{pmatrix} \middle| \, a, b, c \in \mathrm{GF}(5),\ abc \neq 0 \right\}$$

Exercise 10.2

1. Find Aut($\mathscr{C}$), when $\mathscr{C}$ is the linear code
   (i) of length 3 over GF(5) generated by 120, 210;
   (ii) of length 3 over GF(5) generated by 013, 031;
   (iii) of length 3 over GF(3) generated by 120, 110;
   (iv) of length 3 over GF(3) generated by 102, 101;
   (v) of length 3 over GF(3) generated by 102, 201;
   (vi) of length 3 over GF(5) generated by 112; and
   (vii) of length 3 over GF(3) generated by 110 and 101.
2. Prove that every monomial matrix of order n over GF(q) is invertible.
3. Prove that the set of all monomial matrices of order n over GF(q) forms a group under multiplication.

10.3 AUTOMORPHISM GROUP – ITS RELATION WITH MINIMUM DISTANCE

We prove here only one result of Sloane and Thompson (1983) showing the relevance of the automorphism group of a code in connection with its minimum distance. For this, we need a few observations about permutation groups and these are available with their proofs in W. R. Scott (1964).

Let G be a permutation group defined on a non-empty set M. Mark the deviation from earlier notation: so far we have used $\mathbf{G}$ to denote a generator matrix. An **orbit** of G is a subset S of M such that there exists an element $\mathbf{a} \in M$ for which $S = \mathbf{a}G = \{\sigma(\mathbf{a}) | \sigma \in G\}$. The group G is called **transitive** if it has only one orbit, i.e. if $\forall \mathbf{a}, \mathbf{b} \in M$, there exists $\sigma \in G$ such that $\sigma(\mathbf{a}) = \mathbf{b}$. For $\mathbf{a} \in M$, let

$$G_{\mathbf{a}} = \{\sigma \in G | \sigma(\mathbf{a}) = \mathbf{a}\}$$

i.e. the subgroup of G fixing the element $\mathbf{a}$ of M.

Proposition 10.3
If S is an orbit of G, and $\mathbf{a} \in S$, then

(i) $O(G) = O(G_{\mathbf{a}})O(S)$;
(ii) if G is transitive, then $O(G) = O(G_{\mathbf{a}}) \deg G$.

As an immediate consequence of this we have the following lemma.

Lemma 10.1
If $O(G)$ is odd while $O(M)$ is even, then G is **not transitive**.

Definition 10.4
Let G be transitive. A proper subset B of M is called a **block** of G if:

(i) $O(B) > 1$;
(ii) for any $\sigma \in G$, either $B = B\sigma$ or $B \cap B\sigma = \emptyset$.

Definition 10.5
A transitive group without blocks is called **primitive** and a transitive group with blocks is called **imprimitive**.

By a block system of an imprimitive permutation group G, we mean a set S of blocks of G such that:

(i) M is the disjoint union of all the blocks of G in S;
(ii) if $B \in S$ and $\sigma \in G$, then $B\sigma \in S$.

Proposition 10.4
Order of every block of G divides the order of M.

Theorem 10.2
If the permutation group G is transitive and has a non-trivial normal subgroup H which is intransitive, then the set of orbits for H is a block system for G.

Recall that if G is a finite group (not necessarily a permutation group) of order $p^r m$, where p is a prime not dividing m, then any subgroup of G of order p^r is called a **Sylow** p-subgroup of G. Sylow p-subgroups in G always exist. We need the next proposition.

Proposition 10.5
Let p be a prime divisor of the order $O(G)$ of a finite group G. If G contains a cyclic Sylow p-subgroup P of G, then G contains a normal subgroup N with $G/N \cong P$.

Next, we recall the definition of a projective special linear group over GF(p), p a prime.

Let p be an odd prime and $M = \{0, 1, \ldots, p-1, \infty\}$ where ∞ is the symbol introduced to represent any element of the form $a/0$, $a \neq 0$. It is fairly easy to see that if $a, b, c, d \in \mathrm{GF}(p)$, $y, z \in M$ such that

$$ay + b \neq 0 \qquad cy + d \neq 0$$

and

$$\frac{ay+b}{cy+d} = \frac{az+b}{cz+d}$$

then $y = z$. Thus

$$y \to \frac{ay+b}{cy+d}$$

for $ad - bc = 1$, $a, b, c, d \in \mathrm{GF}(p)$ is a one–one map: $M \to M$ and, hence, it is a permutation of M. If σ, σ' are permutations of M given by

$$\sigma(y) = \frac{ay+b}{cy+d}$$

$$\sigma'(y) = \frac{a'y+b'}{c'y+d'}$$

$$ad - bc = a'd' - b'c' = 1$$

then

$$\sigma'\sigma(y) = \frac{(aa' + b'c)y + (a'b + b'd)}{(ac' + cd')y + (bc' + dd')}$$

and

$$(aa' + b'c)(bc' + dd') - (a'b + b'd)(ac' + cd') = (ad - bc)(a'd' - b'c') = 1$$

Therefore a product of two permutations of M of the form described is again a permutation of the same form. Hence, the set of all such permutations of M is a group called the projective special linear group and is denoted by $\mathrm{PSL}_2(p)$. We recall the following result of Assmus and Mattson (1969) without proof.

Theorem 10.3
The automorphism groups of the two extended quadratic residue codes each contain a subgroup of which the permutation part is precisely $\mathrm{PSL}_2(p)$.

Using this theorem they then deduce the following corollary.

Corollary
The minimum distance in the augmented code $\hat{\mathscr{F}}$ is one less than that in $\mathscr{F}$.

For some other applications of the automorphism group we may refer to Assmus and Mattson (1972). We next recall the following theorem.

Theorem 10.4
If all the characteristic roots of a linear transformation T of a vector space V of dimension n are equal, each equal to a (say), then there exists a basis of V w.r.t. which the matrix of T is the square matrix (Jordan normal form) of order n

$$\begin{pmatrix} a & 0 & 0 & \cdots & 0 & 0 & 0 \\ 1 & a & 0 & \cdots & 0 & 0 & 0 \\ 0 & 1 & a & \cdots & 0 & 0 & 0 \\ \vdots & & \ddots & \cdots & \ddots & & \vdots \\ 0 & 0 & 0 & \cdots & 1 & a & 0 \\ 0 & 0 & 0 & \cdots & 0 & 1 & a \end{pmatrix}$$

We now recall the following theorem.

Theorem 10.5
Suppose $\mathscr{C}$ is a binary self-dual code of length n and is fixed (setwise) by a group of permutations H with $\mathrm{O}(H)$ odd. Let

$$(\mathbb{B}^n)_0 = \{v \in \mathbb{B}^n \mid vh = v, \forall h \in H\} \quad \text{and} \quad \mathscr{C}_0 = \mathscr{C} \cap (\mathbb{B}^n)_0.$$

Then

$$\dim(\mathbb{B}^n)_0 = 2 \dim \mathscr{C}_0$$

Proposition 10.6
Let V be a finite dimensional vector space of dimension n over a field F and T a linear transformation of V all the characteristic roots of which are equal to 1. For every k, $1 \leq k \leq n$, V has exactly one T-invariant subspace of dimension k.

Proof

Since all the characteristic roots of T are equal to 1, there exists a basis $e_1, e_2, \ldots, e_n$ with respect to which T is represented by its Jordan normal form

$$\mathbf{A} = \begin{pmatrix} 1 & 0 & 0 & \cdots & 0 & 0 & 0 \\ 1 & 1 & 0 & \cdots & 0 & 0 & 0 \\ 0 & 1 & 1 & \cdots & 0 & 0 & 0 \\ \vdots & & \ddots & \cdots & \ddots & & \vdots \\ 0 & 0 & 0 & \cdots & 1 & 1 & 0 \\ 0 & 0 & 0 & \cdots & 0 & 1 & 1 \end{pmatrix}$$

which is a square matrix of order n. then we have

$$\left.\begin{aligned} T(e_1) &= e_1 \\ T(e_i) &= e_{i-1} + e_i \end{aligned}\right\} \quad \text{for } 2 \leq i \leq n$$

Let W be a k-dimensional T-invariant subspace of V and let T_1 be the endomorphism of W induced by T. Then the characteristic polynomial of T_1 divides the characteristic polynomial of T so that all the characteristic roots of T_1 are also equal to 1. Then there exists a basis $w_1, w_2, \ldots, w_k$ of W such that

$$\left.\begin{aligned} T_1(w_1) &= w_1 \\ T_1(w_i) &= w_{i-1} + w_i \end{aligned}\right\} \quad \text{for } 2 \leq i \leq k$$

For $1 \leq i \leq k$, let

$$w_i = \sum_{j=1}^{n} \alpha_{ij} e_j$$

Then

$$\begin{aligned} w_1 = T_1(w_1) &= \sum_{j=1}^{n} \alpha_{ij} T(e_j) \\ &= \alpha_{11} e_1 + \sum_{j=2}^{n} \alpha_{ij}(e_{j-1} + e_j) \\ &= \sum_{j=1}^{n-1} (\alpha_{1j} + \alpha_{1j+1}) e_j + \alpha_{1n} e_n \end{aligned}$$

Therefore

$$\alpha_{1j} = \alpha_{1j} + \alpha_{1j+1} \qquad 1 \leq j \leq n-1$$

which show that

$$\alpha_{12} = \alpha_{13} = \cdots = \alpha_{1n} = 0$$

and

$$w_1 = \alpha_{11} e_1$$

We claim that

$$w_i = \alpha_{11}(e_1 + e_2 + \cdots + e_i) \quad 1 \leq i \leq k \tag{10.1}$$

Suppose that we have proved the relation up to $i < k$. Then

$$\begin{aligned} w_i + w_{i+1} &= T(w_{i+1}) \\ &= \alpha_{i+1,1} e_1 + \sum_{j=2}^{n} \alpha_{i+1,j}(e_{j-1} + e_j) \end{aligned}$$

which on comparison of coefficients of e_j gives

$$\alpha_{ij} + \alpha_{i+1,j} = \alpha_{i+1,j} + \alpha_{i+1,j+1} \quad 1 \leq j \leq n-1$$

or

$$\alpha_{ij} = \alpha_{i+1,j+1} \quad 1 \leq j \leq n-1 \tag{10.2}$$

The relations (10.1) for i and (10.2) together show that

$$\alpha_{i+1,j} = 0 \quad \text{for } j > i+1$$

and

$$\alpha_{i+1,j} = \alpha_{11} \quad \text{for } 1 \leq j \leq i+1$$

This proves that

$$w_{i+1} = \alpha_{11}(e_1 + e_2 + \cdots + e_{i+1})$$

Thus (10.1) holds $\forall i$, $1 \leq i \leq k$. Since $\alpha_{11} \neq 0$ (otherwise T is singular), W is spanned by

$$e_1, e_1 + e_2, \ldots, e_1 + e_2 + \cdots + e_k$$

which is the same as the space spanned by $e_1, \ldots, e_k$.

Theorem 10.6

Suppose that $\mathscr{C}$ is a binary self dual code of length $n = 2^a b$, $a \geq 1$, $b \geq 1$ and b odd, that is fixed (setwise) by a permutation group G satisfying the conditions

(a) G is transitive on the n coordinate positions;
(b) G has a Sylow 2-subgroup which is cyclic of order 2^a.

Then $\mathscr{C}$ contains code words of weight congruent to 2 modulo 4.

Proof

Let P be the cyclic Sylow 2-subgroup of G with generator π. Since $O(P)=2^a$, $O(G)=2^a e$, where e is odd and divisible by b. Then G contains a normal subgroup H with $G/H \cong P$, $O(H)=e$ (Proposition 10.5).

Let

$$\mathscr{C}_0=\{u\in\mathscr{C}\,|\,uh=u\,\forall\, h\in H\}$$

Since $O(H)$ is odd and n is even, H is not transitive (by Lemma 10.1). Then it follows (from Theorem 10.2) that G is imprimitive with the orbits of H forming a complete block system of G. All the blocks of G have the same length, l (say), and suppose that there are m blocks. Then

$$lm=n=2^a b$$

Each block being an orbit of H, H is transitive on each block and, therefore, $l|O(H)$ (Proposition 10.3(i)). Then l is odd. Therefore $2^a|m$. From the definition of complete block system, it follows that π is transitive on the blocks so that $m\leq 2^a$. Thus $m=2^a$ and the orbits of H consist of 2^a blocks of length b each.

Therefore, the fixed subspace $(\mathbb{B}^n)_0$ has dimension 2^a with one generator for each block. Relabel the n coordinates in such a way that the elements in every orbit of H are consecutively numbered. Then the generator matrix of $(\mathbb{B}^n)_0$ becomes

$$\begin{pmatrix} \overset{b}{\overline{111}} & 000 & \cdots & 000 \\ 000 & 111 & \cdots & 000 \\ \vdots & \cdots & \ddots & \vdots \\ 000 & 000 & \cdots & 111 \end{pmatrix}$$

It follows from Theorem 10.5 that $\dim\mathscr{C}_0$ is 2^{a-1}. Also the action of the generator π of the cyclic group P on the blocks is represented by the square matrix

$$\mathbf{A}=\begin{pmatrix} 0 & 1 & 0 & \cdots & 0 \\ 0 & 0 & 1 & \cdots & 0 \\ 0 & 0 & 0 & \ddots & 1 \\ 1 & 0 & 0 & \cdots & 0 \end{pmatrix}$$

of order 2^a. Since the determinant of a matrix remains unchanged except for a possible change of sign when some rows are interchanged, the characteristic roots of $\mathbf{A}$ are the same as the characteristic roots of the identity matrix $\mathbf{I}$ of order n. Hence the characteristic roots of $\mathbf{A}$ are all 1. Therefore, there is a basis

$$v_1, v_2, \ldots, v_2 a$$

for $\mathbb{B}^{2^a}$ w.r.t. which π is represented by its Jordan normal form (Theorem 10.4), which is the square matrix

$$\mathbf{B} = \begin{pmatrix} 1 & 0 & 0 & \cdots & 0 & 0 & 0 \\ 1 & 1 & 0 & \cdots & 0 & 0 & 0 \\ 0 & 1 & 1 & \cdots & 0 & 0 & 0 \\ \vdots & \ddots & & \cdots & & \ddots & \vdots \\ 0 & 0 & 0 & \cdots & 1 & 1 & 0 \\ 0 & 0 & 0 & \cdots & 0 & 1 & 1 \end{pmatrix}$$

of order 2^a. Then, $\forall k$, $1 \leq k \leq 2^a$, $\mathbb{B}^n$ has exactly one subspace of dimension k. The rows of $2^{a-1} \times n$ matrix

$$\begin{pmatrix} \overset{b}{\overline{111}} & 000 & \cdots & 000 & \overset{b}{\overline{111}} & 000 & \cdots & 000 \\ 000 & 111 & \cdots & 000 & 000 & 111 & \cdots & 000 \\ \vdots & \ddots & & & \vdots & & \ddots & \vdots \\ 000 & 000 & \cdots & 111 & 000 & 000 & \cdots & 111 \end{pmatrix}$$

in which each row has two blocks of b ones as shown are linearly independent. Therefore, the rows of this matrix generate a subspace of dimension 2^{a-1} which must be the unique subcode $\mathscr{C}_0$. Since b is odd, weight of each row of $\mathbf{A}$ is equivalent to 2(mod 4), i.e. $\mathscr{C}_0$ contains words of weight equivalent to 2(mod 4).

Corollary

No binary cyclic self dual code has all its weights divisible by 4.

Proof

Let $\mathscr{C}$ be a binary cyclic self dual code of length $n = 2^a b$, where b is odd. Since the length of a self dual code is even, $a \geq 1$. Let σ be a cyclic permutation fixing $\mathscr{C}$, and let $G = \langle \sigma \rangle$ be the cyclic group generated by σ. Then $P = \langle \sigma^b \rangle$ is a cyclic Sylow 2-subgroup of G with order 2^a. The result then follows from the above theorem.

11

Hadamard matrices and Hadamard codes

11.1 HADAMARD MATRICES

Definition 11.1

A **Hadamard matrix** $\mathbf{M}$ of order n is a square matrix of order n with every entry equal to 1 or -1 such that $\mathbf{MM}^t = n\mathbf{I}$. (Here $\mathbf{M}^t$ denotes the transpose of the matrix $\mathbf{M}$.)

Remarks 11.1

Note (i)

Let $\mathbf{M}$ be a Hadamard matrix of order n. Then

$$\mathbf{MM}^t = n\mathbf{I} \Rightarrow (\det \mathbf{M})^2 = n^n$$

so that $\det \mathbf{M} \neq 0$ and hence $\mathbf{M}$ is non-singular.

Also

$$\mathbf{MM}^t = n\mathbf{I} \Rightarrow \mathbf{M}^{-1}\left(\frac{1}{n}\mathbf{MM}^t\right) = \mathbf{M}^{-1}$$

i.e.

$$\mathbf{M}^{-1} = \frac{1}{n}\mathbf{M}^t$$

Therefore, $\mathbf{M}^t\mathbf{M} = n\mathbf{I}$. Hence $\mathbf{M}^t$ is also a Hadamard matrix of order n.

Note (ii)

Let $\mathbf{M}$ be a Hadamard matrix of order n. Let $\mathbf{R}_1, \mathbf{R}_2, \ldots, \mathbf{R}_n$ denote the rows of $\mathbf{M}$. Let $\mathbf{M}_1$ be the matrix obtained from $\mathbf{M}$ by multiplying every entry of the ith row of $\mathbf{M}$ by -1. Since $\mathbf{MM}^t = n\mathbf{I}$

$$\mathbf{R}_j\mathbf{R}_k^t = \begin{cases} 0 & \text{if } j \neq k,\ 1 \leq j,\ k \leq n \\ n & \text{if } j = k,\ 1 \leq j \leq n \end{cases} \tag{11.1}$$

Let $\mathbf{S}_1, \ldots, \mathbf{S}_n$ denote the rows of $\mathbf{M}_1$ so that $\mathbf{S}_j = \mathbf{R}_j$ if $j \neq i$ and $\mathbf{S}_i = -\mathbf{R}_i$. Then $\mathbf{M}_1\mathbf{M}_1^t = (\lambda_{jk})$, where

$$\begin{aligned} \lambda_{jk} &= \mathbf{S}_j\mathbf{S}_k^t \\ &= \begin{cases} \mathbf{R}_j\mathbf{R}_k^t & \text{if } j \neq i,\ k \neq i \\ -\mathbf{R}_i\mathbf{R}_k^t & \text{if } j = i,\ k \neq i \\ -\mathbf{R}_j\mathbf{R}_i^t & \text{if } j \neq i,\ k = i \\ \mathbf{R}_i\mathbf{R}_i^t & \text{if } j = k = i \end{cases} \\ &= \begin{cases} 0 & \text{if } j \neq k \\ n & \text{if } j = k \end{cases} \end{aligned}$$

Hence $\mathbf{M}_1\mathbf{M}_1^t = n\mathbf{I}$ and $\mathbf{M}_1$ is a Hadamard matrix of order n. Thus if every entry of some row of $\mathbf{M}$ is multiplied by -1, then the resulting matrix is a Hadamard matrix of order n. Similarly, if every entry of a column of $\mathbf{M}$ is multiplied by -1, the resulting matrix is again a Hadamard matrix of order n.

Note (iii)

Given a Hadamard matrix of order n, by a repeated application of Note (ii) above, we can obtain a Hadamard matrix of order n in which every entry in the first row and in the first column is $+1$.

Note (iv)

If any two rows or any two columns are interchanged in a Hadamard matrix, then the resulting matrix is again Hadamard.

Definition 11.2

A Hadamard matrix of order n in which every entry in the first row and in the first column is $+1$ is called a **normalized Hadamard matrix** of order n.

In view of Note (iii) above, observe that if a Hadamard matrix of order n exists, then so does a normalized Hadamard matrix of n.

Examples 11.1

Case (i)

If

$$\begin{pmatrix} 1 & 1 \\ 1 & a \end{pmatrix}^2 = 2\mathbf{I}$$

then $1+a=0$ so that $a=-1$ and

$$\begin{pmatrix} 1 & 1 \\ 1 & -1 \end{pmatrix}$$

is a normalized Hadamard matrix of order 2.

We observe that

$$\begin{pmatrix} -1 & -1 \\ 1 & -1 \end{pmatrix} \begin{pmatrix} -1 & 1 \\ -1 & -1 \end{pmatrix} \begin{pmatrix} 1 & 1 \\ -1 & 1 \end{pmatrix} \begin{pmatrix} 1 & -1 \\ 1 & 1 \end{pmatrix}$$

are some of the other Hadamard matrices of order 2.

Case (ii)

Let

$$\mathbf{M} = \begin{pmatrix} 1 & 1 & 1 \\ 1 & a & b \\ 1 & c & d \end{pmatrix}$$

be a normalized Hadamard matrix of order 3. Then $\mathbf{MM}^t = 3\mathbf{I}$ and $1+a+b=0$.

But this relation is not possible with $a=\pm 1$ and $b=\pm 1$. Thus, there is no normalized Hadamard matrix of order 3 and, hence, there does not exist any Hadamard matrix of order 3.

We next give a procedure for obtaining a Hadamard matrix of order $2n$ from a given Hadamard matrix of order n.

Proposition 11.1

If $\mathbf{M}$ is a Hadamard matrix of order n, then

$$\begin{pmatrix} \mathbf{M} & \mathbf{M} \\ \mathbf{M} & -\mathbf{M} \end{pmatrix}$$

is a Hadamard matrix of order $2n$.

Proof

$$\begin{aligned} \begin{pmatrix} \mathbf{M} & \mathbf{M} \\ \mathbf{M} & -\mathbf{M} \end{pmatrix} \begin{pmatrix} \mathbf{M} & \mathbf{M} \\ \mathbf{M} & -\mathbf{M} \end{pmatrix}^t &= \begin{pmatrix} \mathbf{M} & \mathbf{M} \\ \mathbf{M} & -\mathbf{M} \end{pmatrix} \begin{pmatrix} \mathbf{M}^t & \mathbf{M}^t \\ \mathbf{M}^t & -\mathbf{M}^t \end{pmatrix} \\ &= \begin{pmatrix} \mathbf{MM}^t + \mathbf{MM}^t & \mathbf{MM}^t - \mathbf{MM}^t \\ \mathbf{MM}^t - \mathbf{MM}^t & \mathbf{MM}^t + \mathbf{MM}^t \end{pmatrix} \\ &= \begin{pmatrix} 2n\mathbf{I}_n & \mathbf{0} \\ \mathbf{0} & 2n\mathbf{I}_n \end{pmatrix} \\ &= 2n \begin{pmatrix} \mathbf{I}_n & \mathbf{0} \\ \mathbf{0} & \mathbf{I}_n \end{pmatrix} \\ &= 2n\mathbf{I}_{2n} \end{aligned}$$

Hence

$$\begin{pmatrix} \mathbf{M} & \mathbf{M} \\ \mathbf{M} & -\mathbf{M} \end{pmatrix}$$

is a Hadamard matrix of order $2n$. Note: $\mathbf{I}_k$ denotes the identity matrix of order k. ■

We can also similarly prove that if $\mathbf{M}$ is a Hadamard matrix of order n, then

$$\begin{pmatrix} \mathbf{M} & -\mathbf{M} \\ \mathbf{M} & \mathbf{M} \end{pmatrix} \quad \begin{pmatrix} \mathbf{M} & \mathbf{M} \\ -\mathbf{M} & \mathbf{M} \end{pmatrix} \quad \text{and} \quad \begin{pmatrix} -\mathbf{M} & \mathbf{M} \\ \mathbf{M} & \mathbf{M} \end{pmatrix}$$

are Hadamard matrices of order $2n$.

Using the above procedure, we find that

$$\begin{pmatrix} 1 & 1 & 1 & 1 \\ 1 & -1 & 1 & -1 \\ 1 & 1 & -1 & -1 \\ 1 & -1 & -1 & 1 \end{pmatrix} \quad \text{and} \quad \begin{pmatrix} -1 & 1 & 1 & -1 \\ 1 & 1 & -1 & -1 \\ -1 & 1 & -1 & 1 \\ 1 & 1 & 1 & 1 \end{pmatrix}$$

are Hadamard matrices of order 4.

Exercise 11.1

1. Without using the procedure of Proposition 11.1 and the remark below it, obtain a normalized Hadamard matrix of order 4.
2. Obtain a Hadamard matrix of order 8.

Theorem 11.1
If a Hadamard matrix of order n exists, then $n = 1$, 2 or a multiple of 4.

Proof
The matrix (1) is trivially a (normalized) Hadamard matrix of order 1 and we have already obtained Hadamard matrices of order 2. Also, we have proved that there does not exist a Hadamard matrix of order 3. So, we suppose that $n \geq 4$ and that there exists a Hadamard matrix and, hence, a normalized Hadamard matrix $\mathbf{M}$ of order n.

Since every row of $\mathbf{M}$ from second row onward is orthogonal to the first, the number of $+1$s in any such row equals the number of -1s in it. This proves that n is even, say $n = 2m$. By permuting the columns of $\mathbf{M}$, if necessary, we can assume that the first m entries in the second row of $\mathbf{M}$ are $+1$. Among the first m entries of the third row, suppose that j of these are $+1$ and the remaining $m - j$ are -1. Then among the last m entries of the third row $m - j$ entries are $+1$ and j of them are -1. By the orthogonality of the second and third rows, we then find

$$j-(m-j)-(m-j)+j=0$$

giving $m=2j$. Hence $n=4j$.

Definition 11.3

Two Hadamard matrices are said to be **equivalent** if one of them can be obtained from the other by permuting rows or columns or by multiplying rows or columns by -1.

Remarks 11.2

Note (i)

Observe that the relation of two Hadamard matrices being equivalent is an equivalence relation.

Note (ii)

There are only two Hadamard matrices (1) and (-1) of order 1 and these are clearly equivalent.

Note (iii)

There is only one normalized Hadamard matrix of order 2 and as every Hadamard matrix of order n is equivalent to a normalized Hadamard matrix of order n, it follows that any two Hadamard matrices of order 2 are equivalent.

Note (iv)

Let

$$\mathbf{M}=\begin{pmatrix} 1 & 1 & 1 & 1 \\ 1 & a & b & c \\ 1 & d & e & f \\ 1 & g & h & i \end{pmatrix}$$

be a normalized Hadamard matrix of order 4. Then $\mathbf{MM}^t=4\mathbf{I}$ shows that

$$1+a+b+c=0$$

$$1+d+e+f=0$$

$$1+g+h+i=0$$

$$1+ad+be+cf=0$$

$$1+ag+bh+ci=0$$

$$1+dg+ch+fi=0$$

If $a=b=-1$, $c=1$, then $f=-1$, $i=-1$ and $d+e=0$, $g+h=0$, $dg+eh+2=0$. If $d=-1$, $e=1$, then $h=-1$, $g=1$ while if $d=1$, $e=-1$, then $g=-1$, $h=1$. Therefore, the choice of a, b, c as above gives two normalized matrices:

$$\mathbf{M}_1=\begin{pmatrix}1&1&1&1\\1&-1&-1&1\\1&-1&1&-1\\1&1&-1&-1\end{pmatrix}\qquad \mathbf{M}_2=\begin{pmatrix}1&1&1&1\\1&-1&-1&1\\1&1&-1&-1\\1&-1&1&-1\end{pmatrix}$$

Clearly $\mathbf{M}_2$ can be obtained from $\mathbf{M}_1$ by interchanging the third and fourth rows and so $\mathbf{M}_1$ and $\mathbf{M}_2$ are equivalent.

If $a=c=-1$, $b=1$, then $e=-1=h$ and $d+f=0=g+i$, $dg+fi+2=0$. If $d=-1$, $f=1$, then $g=1$, $i=-1$ while if $d=1$, $f=-1$, then $g=-1$, $i=1$. Thus the present choice of a, b, c gives two normalized matrices:

$$\mathbf{M}_3=\begin{pmatrix}1&1&1&1\\1&-1&1&-1\\1&-1&-1&1\\1&1&-1&-1\end{pmatrix}\qquad \mathbf{M}_4=\begin{pmatrix}1&1&1&1\\1&-1&1&-1\\1&1&-1&-1\\1&-1&-1&1\end{pmatrix}$$

But $\mathbf{M}_3$ is obtained from $\mathbf{M}_1$ by interchanging the second and third rows while $\mathbf{M}_4$ is obtained from $\mathbf{M}_1$ by applying the permutation $(\mathbf{R}_2\mathbf{R}_4\mathbf{R}_3)$ to the rows of $\mathbf{M}_1$. Thus, both $\mathbf{M}_3$ and $\mathbf{M}_4$ are equivalent to $\mathbf{M}_1$.

If $a=1$, $b=c=-1$, then $d=g=-1$, $e+f=0=h+i$ and $eh+fi+2=0$. If $e=1$, $f=-1$, then $h=-1$, $i=1$ while if $e=-1$, $f=1$, then $h=1$, $i=-1$. Thus, the two possible normalized matrices in this case are:

$$\mathbf{M}_5=\begin{pmatrix}1&1&1&1\\1&1&-1&-1\\1&-1&1&-1\\1&-1&-1&1\end{pmatrix}\quad\text{and}\quad \mathbf{M}_6=\begin{pmatrix}1&1&1&1\\1&1&-1&-1\\1&-1&-1&1\\1&-1&1&-1\end{pmatrix}$$

The matrix $\mathbf{M}_6$ follows from $\mathbf{M}_5$ by interchanging the third and fourth rows. Thus, $\mathbf{M}_5$ and $\mathbf{M}_6$ are equivalent. Also clearly $\mathbf{M}_1$ and $\mathbf{M}_5$ are equivalent.

This exhausts all possible choices for a, b, c and, therefore, up to equivalence there is only one normalized Hadamard matrix of order 4. Hence, there is only one equivalence class of Hadamard matrices of order 4.

The above could alternatively and in a simpler way be obtained as follows: For a normalized Hadamard matrix of order 4, the second row has two -1s and two $+1$s. Thus there are three choices for the second row. Then, with each choice of the second row, there are two choices for the third row and once the second and third rows have been chosen, there is only one choice for the fourth row. Hence, there are only six normalized Hadamard matrices of order 4 and these are the matrices $\mathbf{M}_1$ to $\mathbf{M}_6$ as above. Equivalence of these matrices needs the same argument as above.

11.2 HADAMARD CODES

Definition 11.4
A matrix obtained from a Hadamard matrix $\mathbf{M}_n$ of order n by changing 1s into 0s and -1s into 1s is called a **binary Hadamard matrix** of order n.

Let $\mathbf{M}_n$ be a normalized Hadamard matrix of order n and $\mathbf{A}_n$ be the binary Hadamard matrix of order n obtained from $\mathbf{M}_n$. Since any two rows of $\mathbf{M}_n$ are orthogonal, therefore, any two rows of $\mathbf{M}_n$ agree in $n/2$ places and differ in the remaining $n/2$ places. It follows that:

(i) the distance between any two rows of $\mathbf{A}_n$ is $n/2$;
(ii) the weight of every non-zero row of $\mathbf{A}_n$ is $n/2$.

Also, clearly, every row of $\mathscr{A}_n$ has first entry 0.

Let $\mathscr{A}_n$ denote the set of all the rows of $\mathscr{A}_n$ with first entry deleted. The set $\mathscr{A}_n$ has n elements of length $n-1$ and the distance between any two elements of $\mathscr{A}_n$ is $n/2$.

Let $\mathscr{C}_n$ denote the set of all rows of $\mathscr{A}_n$ together with their complements. Then elements of $\mathscr{C}_n$ are words of length n, are $2n$ in number and the minimum of the distance between any two of them is $n/2$.

Let $\mathscr{B}_n$ denote the set of all elements of $\mathscr{A}_n$ together with their complements. The elements of $\mathscr{B}_n$ are words of length $n-1$, are $2n$ in number and distance between any two of them is at least

$$\frac{n}{2}-1$$

(for $n > 2$).

$\mathscr{A}_n$, $\mathscr{B}_n$ and $\mathscr{C}_n$ are called **Hadamard codes**. These are binary codes but none of $\mathscr{A}_n$, $\mathscr{B}_n$ and $\mathscr{C}_n$ is in general a group. Thus, these are non-linear codes in general. Observe that the codes $\mathscr{B}_n$ and $\mathscr{C}_n$ satisfy the Plotkin bound (Theorem 6.7) as every code should but it is attained in the case of $\mathscr{A}_n$.

We end this section and also the chapter by constructing Hadamard codes for $n = 2, 4, 8$.

Examples 11.2

Case (i)
Consider

$$\begin{pmatrix} 1 & 1 \\ 1 & -1 \end{pmatrix}$$

It is a normalized Hadamard matrix of order 2 and gives the following three codes:

$$\mathscr{A}_2 = \{0, 1\} \qquad \mathscr{B}_2 = \{0, 1\} \qquad \mathscr{C}_2 = \{00, 01, 10, 11\}$$

Case (ii)
The normalized Hadamard matrix

$$\mathbf{M} = \begin{pmatrix} 1 & 1 & 1 & 1 \\ 1 & -1 & 1 & -1 \\ 1 & 1 & -1 & -1 \\ 1 & -1 & -1 & 1 \end{pmatrix}$$

gives the Hadamard codes

$$\mathscr{A}_4 = \{000, 101, 011, 110\}$$

$$\mathscr{B}_4 = \{000, 101, 011, 110, 111, 010, 100, 001\}$$

$$\mathscr{C}_4 = \{0000, 0101, 0011, 0110, 1111, 1010, 1100, 1001\}$$

Case (iii)
The matrix

$$\begin{pmatrix} 1 & 1 & 1 & 1 & 1 & 1 & 1 & 1 \\ 1 & -1 & 1 & -1 & 1 & -1 & 1 & -1 \\ 1 & 1 & -1 & -1 & 1 & 1 & -1 & -1 \\ 1 & -1 & -1 & 1 & 1 & -1 & -1 & 1 \\ 1 & 1 & 1 & 1 & -1 & -1 & -1 & -1 \\ 1 & -1 & 1 & -1 & -1 & 1 & -1 & 1 \\ 1 & 1 & -1 & -1 & -1 & -1 & 1 & 1 \\ 1 & -1 & -1 & 1 & -1 & 1 & 1 & -1 \end{pmatrix}$$

is a normalized Hadamard matrix of order 8 and gives the following Hadamard codes:

$$\mathscr{A}_8 = \begin{matrix} \{0000000, 0001111, \\ 1010101, 1011010, \\ 0110011, 0111100, \\ 1100110, 1101001\} \end{matrix}$$

$$\mathscr{B}_8 = \begin{matrix} \{0000000, 1110000, \\ 1010101, 0100101, \\ 0110011, 1000011, \\ 1100110, 0010110, \\ 1111111, 0001111, \\ 0101010, 1011010, \\ 1001100, 0111100, \\ 0011001, 1101001\} \end{matrix}$$

$$\mathscr{C}_8 = \begin{array}{l} \{00000000, 11111111, \\ 01010101, 10101010, \\ 00110011, 11001100, \\ 01100110, 10011001, \\ 00001111, 11110000, \\ 01011010, 10100101, \\ 00111100, 11000011, \\ 01101001, 10010110\} \end{array}$$

Exercise 11.2

1. Write another normalized Hadamard matrix of order (i) 4; (ii) 8, and obtain the corresponding Hadamard codes. Compare these codes with the codes obtained in Examples 11.2, Cases (ii) and (iii) above.
2. Determine the number of normalized Hadamard matrices of order 8. Also determine the number of equivalence classes of Hadamard matrices of order 8.
3. Do any two Hadamard matrices of order 8 give the same Hadamard codes? Justify your answer!

Bibliography

Assmus, E. F. and Mattson, H. F. (1963) Error correcting codes: an axiomatic approach. *Information and Control*, **6**, 315–30.

Assmus, E. F. Jr and Mattson, H. F. (1966) Perfect codes and Mathieu groups, *Arch. Math.*, **17**, 121–35.

Assmus, E. F. Jr and Mattson H. F. Jr (1967) On tactical configurations and error correcting codes. *J. Combinatorial Theory*, **2**, 243–57.

Assmus, E. F. Jr and Mattson, H. F. Jr (1969) New 5-designs. *J. Combinatorial Theory*, **6**, 122–51.

Assmus, E. F. Jr and Mattson, H. F. Jr (1972) On weights in quadratic residue codes. *Discrete Mathematics*, **3**, 1–20.

Assmus, E. F. Jr and Pless, V. (1983) On the covering radius of extremal self-dual codes. *IEEE Trans. Information Theory*, **29**, 359–63.

Beenker, G. F. M. (1984) A note on quadratic residue codes over GF(9) and their ternary images. *IEEE Trans. Information Theory*, **30**, 403–404.

Berlekamp, E. R. (1968) *Algebraic Coding Theory*, McGraw-Hill.

Berman, S. D. (1967) Semisimple cyclic and Abelian codes, II. *Cybernetics*, **3**, 17–23.

Birkhoff, G. and Bartee, T. C. (1970) *Modern Applied Algebra*, McGraw-Hill.

Blake, I. F. and Mullin, R. C. (1975) *The Mathematical Theory of Coding*, Academic Press.

Bose, R. C. and Ray-Choudhuri, D. K. (1960) On a class of error correcting binary group codes. *Information and Control*, **3**, 68–79.

Calderbank, R. (1983) A square root bound on the minimum weight in quasi-cyclic codes. *IEEE Trans. Information Theory*, **29**, 332–7.

Cohen, G. D., Karpovsky, M. G., Mattson, H. F. Jr and Schatz, J. R. (1985) Covering radius-survey and recent results. *IEEE Trans. Information Theory*, **31**, 328–43.

Coppersmith, D. and Seroussi, G. (1984) On the minimum distance of some quadratic residue codes. *IEEE Trans. Information Theory*, **30**, 407–11.

Dornhoff, L. L. and Hohn, F. E. (1978) *Applied Modern Algebra*, Macmillan.

Graham, R. L. and Sloane, N. J. A. (1985) On the covering radius of codes. *IEEE Trans. Information Theory*, **31**, 385–401.

Helleseth, T. (1978) All binary 3-error correcting BCH codes of length $2^m - 1$ have covering radius 5. *IEEE Trans. Information Theory*, **24**, 257–8.

Herstein, I. N. (1968) *Non-commutative Rings*, Carus Math. Monographs No. 15, Math. Assoc. Amer.

Herstein, I. N. (1976) *Topics in Algebra*, Vikas Publishing House, New Delhi.

Hocquenghem, A. (1959) Codes correcteurs d'erreurs. *Chiffres* (Paris), **2**, 147–56.

Kasami, T., Lin, S. and Peterson, W. W. (1968) Some results on cyclic codes which are invariant under the affine group and their applications. *Information and Control*, **11**, 475–96.

Lambek, J. (1966) *Lectures on Rings and Modules*, Ginn (Blaisdell), Boston, Massachusetts.

Lidl, R. and Niederreiter, H. (1986) *Introduction to Finite Fields and their Applications*, Cambridge University Press.

MacWilliams, F. J. and Sloane, N. J. A. (1978) *Theory of Error Correcting Codes*, North-Holland.

Manju Pruthi (1992) Primitive idempotents in semisimple group algebras of finite cyclic groups and cyclic codes over finite fields. PhD thesis, M.D. University, Rohtak, India.

McEliece, R. J. (1977) *The Theory of Information and Coding, Encyclopaedia of Mathematics and its Applications*, Vol. 3, Addison-Wesley.

Niven, I. and Zuckerman, H. S. (1972) *An Introduction to Number Theory*, Wiley Eastern.

Remijn, J. C. C. M. and De Vroedt, C. (1984) The minimum distance of the [38, 19] ternary extended QR-code is 11. *IEEE Trans. Information Theory*, **30**, 405–7.

Roos, C. (1983) A new lower bound for minimum bound of a cyclic code. *IEEE Trans. Information Theory*, **29**, 330–2.

Scott, W. R. (1964) *Group Theory*, Prentice-Hall.

Shaughnessy, E. P. (1971) Codes with simple automorphism groups. *Arch. Math.*, **22**, 459–66.

Sloane, N. J. A. (1973) Is there a (72, 36), $d = 16$ self-dual code? *IEEE Trans. Information Theory*, **19**, 251.

Sloane, N. J. A. (1975) *A Short Course on Error Correcting Codes*, Centre for Mechanical Sciences, Courses and Lectures No. 188, Springer-Verlag, Wien, New York.

Sloane, N. J. A. (1977) Error correcting codes and invariant theory: New applications of nineteenth century technique. 82–107.

Sloane, N. J. A. and Thompson, J. G. (1983) Cyclic self-dual codes. *IEEE Trans. Information Theory*, **29**, 364–6.

Solomon, G. and McEliece, R. (1966) Weights of cyclic codes. *J. Combinatorial Theory*, **1**, 459–75.

Spiegel, E. (1977) Codes of Z_m. *Information and Control*, **35**.

Spiegel, E. (1978) Codes over Z_m, revised. *Information and Control*, **37**.

Van Lint, J. H. (1971) *Coding Theory*, Lecture Notes in Mathematics, No. 201, Springer-Verlag, Berlin, Heidelberg, New York.

Van Lint, J. H. and MacWilliams, J. (1978) Generalised quadratic residue codes. *IEEE Trans. Information Theory*, **24**, 720–37.

Vermani, L. R. and Jindal, S. L. (1983) A note on maximum distance separable (MDS) codes. *IEEE Trans. Information Theory*, **29**, 136.

Vermani, L. R. and Yogesh Kumar, A note on quadratic residue codes (preprint).

Ward, H. N. (1974) Quadratic residue codes and symplectic groups. *J. Algebra*, **29**, 150–71.

Ward, H. N. (1983) Divisors of codeword weights. *IEEE Trans. Information Theory*, **29**, 337–42.

Zimmermann, K. H. (1992) On a complete decoding scheme for binary radical codes. *Arch. Math.*, **59**, 513–20.

Index

Algebraic extension 49
Automorphism group of
  a code 223
  a cyclic code 228
  extended quadratic residue code 237
  Hamming code 229
  linear code 226, 230

Basis 24
Berlekamp's algorithm 149
  a special case 157
Berman, S.D. 136
Binary representation 39
Block of a transitive group 235
Bound
  Gilbert–Varshamov 124–5
  Hamming 124
  for linear code 84
  Plotkin 127
  sphere packing 124

Check digit/symbol 42
Chinese remainder theorem
  for integers 155
  for polynomials 155
Code
  augmented 105
  augmented quadratic residue 189
  BCH 47, 65, 80, 85, 119
  cyclic 107–8, 118
  dual/orthogonal 21–2, 88, 91, 114
  equivalent 82, 83, 87
  expurgated 104–5
  expurgated quadratic residue 175, 189
  extended 102, 104
  extended quadratic residue 180, 186–8
  Golay 178, 188
  group 9–11, 28, 32, 42, 85
  Hadamard 248
  Hamming 41, 44, 45, 85, 98, 115, 116, 118, 119
  Hamming code as BCH 119
  Hamming code as linear 85
  Hamming code as perfect 123
  linear 81, 85
  matrix 9, 10, 15, 35
  maximum distance separable 208, 209, 221
  non-binary Hamming 120
  parity check 7, 9, 10, 13, 15
  perfect 123
  polynomial 24, 27, 28, 29, 30, 31, 91
  quadratic residue 173
  Reed–Solomon 220, 221
  self dual 93, 96, 97, 237, 239
  self dual cyclic 137, 141
  ternary 85
  triple repitition 8, 10, 13
  trivial MDS 212
Code invariance 134
Code word 2
Complementation 134
Coset leader 12, 13
Cyclotomic coset/class 143, 144, 145

Decoding by coset leader 12
Decoding failure 5
Decoding principle
  maximum likelihood 5
  nearest neighbourhood 5
  syndrome/parity check 17
Degree of an extension 49

Degree of polynomial 26
Difference ring/ring of quotients 47
Dimension 24
Distance 3
  minimum 10
Domain
  integral 26
  principal ideal 47
Double error 32
Dual of MDS code 209

Error
  corrected 6
  detected 5
  undetected 5, 11
Exponent 32
Extension field 49

Field 2
  finite/Galois 2, 51
  prime 49

Gaussian sum 182
Group
  Abelian 1
  cyclic 51
  imprimitive 235
  monomial 230
  primitive 235
  projective special linear 236
  transitive 235
Group algebra 135, 136

Hadamard transform 99
Hamming code as quadratic residue code 177
Hamming weight 23

Idempotent
  of cyclic code 129, 130, 194
  of quadratic residue codes 194, 197, 201
Intersection of words 87

Legendre symbol 182

MacWilliam's identity 98, 100, 102
Matrix
  binary Hadamard 248
  encoding/generating/generator 8, 9, 15, 18, 35, 108
  equivalent Hadamard 243
  Hadamard 242, 243
  monomial 229, 230
  normalized Hadamard 243
  parity check 15, 16, 17, 18, 19, 35, 45, 113
  permutation 83, 86, 87, 223
Multiplicative order mod *n* 143

Ordered *n*-tuple 2
Orthogonal vectors 88

Parity check equations 14
Parity check scheme 9, 10
Perron's theorem 185
Polynomial
  check 111
  encoding/generator 27, 28, 65, 220, 221
  irreducible 48, 53
  message 28
  minimal 50, 51
  monic 51
  primitive 58
  reducible 48
  symmetric 145
Prime subfield 49
Primitive element 52, 55, 140
Principal ideal 47

Quadratic non-residue 94, 172
Quadratic reciprocity law 197
Quadratic residue 94, 172

Rank 45
Redundency 42
Ring 1
  commutative 1

Sequence 2
Simple extension 59
Sloane, N.J.A. 234
Splitting field 59, 62
Subgroup 11
Sum of sequences 2
Sylow subgroup 236
Syndrome 17, 22

Thompson, J.G. 234

Vector space 24, 26

Weight 4
  of self dual code 93
  see also Hamming weight 23
Weight enumerator 98
  of dual code 101